GENETICS BASICS
EXPLORING DNA AND HEREDITY

GENETICS BASICS:
EXPLORING DNA AND HEREDITY

Edited by

Dr. Akansha Singh

Delve PUBLISHING

www.delvepublishing.com

Genetics Basics: Exploring DNA and Heredity

Dr. Akansha Singh

Delve Publishing
4164 Lakeshore Road
Burlington, ON L7L 1A4
Canada
www.delvepublishing.com
Email: orders@arclereducation.com

ISBN: 978-1-77469-933-1 (Hardcover)

Delve Publishing publishes wide variety of books and eBooks. For more information about Delve Publishing and its products, visit our website at www.delvepublishing.com.

ABOUT THE EDITOR

Dr. Akansha Singh is presently working as Project Scientist in Department of Genetics and Plant Breeding, Institute of Agricultural Sciences, Banaras Hindu University, India. She has also worked as Postdoctoral fellow, Mumbai University and as Associate Professor, Department of Genetics and Plant Breeding, College of Agriculture, Parul University, India She has obtained her Ph.D. (Ag) in Genetics and Plant breeding from Banaras Hindu University in the year 2012. She has been Awarded with ICAR, senior research fellowship in 2010 to pursue PhD. She has authored numerous national and international publications in journals of repute.

TABLE OF CONTENTS

LIST OF FIGURES

High-Speed and Label-Free Optical Diffraction Tomography. Cells. 2019; 8(11):1368. https://doi.org/10.3390/cells8111368, Creative Commons License)

Figure 8.3. Meiosis I & Meiosis II (Source: Sakuno T, Hiraoka Y. Rec8 Cohesin: A Structural Platform for Shaping the Meiotic Chromosomes. Genes. 2022; 13(2):200. https://doi.org/10.3390/genes13020200, Creative Commons License)

Figure 8.4. Phases of cell cycle (Source: Deng B, Song A, Zhang C. Cell-Cycle Dysregulation in the Pathogenesis of Diabetic Kidney Disease: An Update. International Journal of Molecular Sciences. 2023; 24(3):2133. https://doi.org/10.3390/ijms24032133, Creative Commons License)

Figure 8.5. The cell cycle involves many proteins (Source: Andrade-Tomaz M, de Souza I, Rocha CRR, Gomes LR. The Role of Chaperone-Mediated Autophagy in Cell Cycle Control and Its Implications in Cancer. Cells. 2020; 9(9):2140. https://doi.org/10.3390/cells9092140, Creative Commons License)

Figure 8.6. Phases of mitosis (Source: Lacroix B, Dumont J. Spatial and Temporal Scaling of Microtubules and Mitotic Spindles. Cells. 2022; 11(2):248. https://doi.org/10.3390/cells11020248, Creative Commons License)

Figure 8.7. Cell cycle checkpoint & DNA damage response defects as anticancer targets (Source: Piezzo M, Cocco S, Caputo R, Cianniello D, Gioia GD, Lauro VD, Fusco G, Martinelli C, Nuzzo F, Pensabene M, et al. Targeting Cell Cycle in Breast Cancer: CDK4/6 Inhibitors. International Journal of Molecular Sciences. 2020; 21(18):6479. https://doi.org/10.3390/ijms21186479, Creative Commons License)

LIST OF TABLES

PREFACE

"Genetics Basics: Exploring DNA & Heredity" offers an engaging and accessible journey into the fascinating world of genetics. This book is designed to be a comprehensive resource for students, enthusiasts, and curious minds seeking to unravel the mysteries of DNA and heredity.

In this opening chapter, readers will be introduced to the captivating field of genetics. The chapter sets the stage by exploring the historical milestones in genetics research and how the study of heredity has revolutionized our understanding of life itself. Readers will gain insights into the key figures who shaped the field and the foundational concepts that paved the way for modern genetic discoveries.

Chapter 2 delves into the heart of genetics: the structure and function of DNA. Readers will journey into the microscopic realm of nucleotides and learn about the iconic double-helix structure of DNA. The chapter illuminates the pivotal role of DNA as the blueprint of life, encoding the genetic information that shapes an organism's traits and characteristics.

Chapter 3 Heredity is the cornerstone of genetics, and in this chapter, readers will explore the principles that govern the passing of traits from one generation to the next. From Mendel's groundbreaking experiments with pea plants to the concept of dominant and recessive traits, this chapter elucidates the basic laws that underpin inheritance patterns.

Chapter 4 delves deeper into the core principles of genetics. Topics covered include codominance, incomplete dominance, and sex-linked inheritance, unraveling the complexity of genetic variations and how they contribute to the diversity of life.

Chapter 5 takes readers on a captivating journey into the world of the genetic code. It explores how DNA encodes genetic information, which is then translated into proteins through the fascinating process of transcription and translation. Readers will understand the pivotal role of proteins in the functioning of living organisms. Chapter 6 unravels the connection between genes and chromosomes.

Chapter 7 "Gene mutations" are essential components of genetic diversity but can also lead to genetic disorders. This chapter delves into the various types of gene mutations, their causes, and the intricate mechanisms of DNA repair that safeguard genetic stability.

The final chapter explores the connection between genetics and the cell cycle, shedding light on how genetic information is faithfully transmitted during cell division.

-Author

CHAPTER 1

INTRODUCTION TO GENETICS

Contents

1.1 INTRODUCTION

Genetics, encompassing the examination of heredity and inheritance, plays a central role in the fascinating trajectory of life. The complex realm of genetics encompasses a complex network of DNA, which serves as the fundamental code of life, and the intricate interplay of genes that governs the wide array of traits and attributes exhibited by organisms. In the pursuit of understanding genetics, we explore the underlying principles that govern the transmission of traits, delve into the enigmatic nature of DNA, and unravel the intricate mechanisms that influence the outcomes of biological organisms (Sturtevant & Beadle, 1939).

Genetics is an academic endeavor to comprehend the mechanisms by which characteristics and traits are inherited across successive generations. Throughout successive generations, the genetic information in an organism's DNA directs and regulates the development, functioning, and adaptation processes. The field of genetics has not only shed light on the fundamental principles of inheritance. Still, it has also yielded a deep comprehension of evolutionary processes, human health, and biological diversity (Waddington, 2016).

The field of genetics experienced a notable advancement through the groundbreaking contributions of Gregor Mendel, an Austrian monk of modest background. Mendel conducted innovative experiments involving pea plants during the mid-19th century. The groundbreaking research conducted by Mendel resulted in the identification of the basic concepts governing the transmission of traits from one generation to the next, which are currently recognized as Mendelian genetics. His investigations involving pea plants provided evidence of distinct hereditary units, subsequently recognized as genes. They revealed the principles that dictate their transfer from one generation to the succeeding one (Griffiths & Stotz, 2013).

The revelation of the DNA structure in the 1950s by James Watson and Francis Crick marked the onset of a transformative period in genetics. Deoxyribonucleic acid (DNA), the fundamental molecule responsible for transmitting hereditary information, exhibits a remarkable structure as a double-stranded helix. This intricate configuration enables DNA to effectively transmit the genetic instructions necessary for many biological processes that sustain life. The distinctive architecture of this biological molecule facilitates the accurate duplication and transfer of genetic material throughout cellular replication, thereby guaranteeing the perpetuation of life across successive cohorts (Halliburton & Halliburton, 2004).

The transmission of genetic information within cells adheres to the central dogma of genetics, as originally postulated by Francis Crick. This dogma delineates the chronological order of events: The transcription process involves the conversion of DNA into RNA, while the subsequent translation process entails the conversion of RNA into proteins. Proteins serve as essential components responsible for various cellular and organismal functions, encompassing structural support, functional activities, and regulatory processes (Crow, 2017).

The field of molecular genetics elucidates the complex interplay between genes and proteins that govern fundamental biological mechanisms. This statement explains the processes involved in gene expression, specifically highlighting the complex regulation of gene activation and deactivation, as well as the precise adjustment of protein synthesis to accommodate the ever-changing requirements of the cell (Dekkers & Hospital, 2002).

Genetic variation is the primary source of biodiversity, facilitating the vast assortment of life forms that populate our planet. Genetic diversity is generated by various mechanisms, including mutations and recombination, leading to accumulating a wide range of genetic variations. These variations serve as a substantial resource from which natural selection acts, influencing different species' survival and adaptive processes (Ewens, 2004).

Genetics, notably, assumes a pivotal role in shaping the essence of humanity. Human genetics encompasses investigating the transmission and manifestation of characteristics within the human population. The comprehension of human genetics has significant implications for our comprehension of evolution, health, and disease susceptibility. This field of genetics has elucidated the genetic underpinnings of hereditary disorders and the intricate interplay between genes and the environment in influencing human traits (Frankham et al., 2002).

Genetic disorders, which frequently result from mutations in particular genes, serve as poignant reminders of the intricate complexity inherent in genetics. These factors' impact extends to individuals and families, instigating a pursuit of comprehension, intervention, and mitigation.

Epigenetics, an intriguing domain within the field of genetics, investigates the influence of environmental factors on the expression of genes and the transmission of traits. This observation demonstrates the capacity of external factors to modulate gene activity, thereby influencing health and developmental outcomes across successive generations (Berger et al., 2009).

The progress made in genetic technologies has provided humanity with unprecedented opportunities to delve into the complex realm of genetics. The utilization of DNA sequencing enables the deciphering of the genetic code of organisms, thereby facilitating significant advancements in the field of genomics and personalized medicine. Gene editing technologies like CRISPR-

Cas9 possess significant potential for treating genetic disorders and advancing agriculture and biotechnology.

As society continues exploring the unexplored realms of genetics, acknowledging and addressing the emerging ethical implications is of utmost importance. Incorporating responsible practices in the utilization of genetic technologies, preserving individual privacy, and effectively managing the intricate nature of genetic information are fundamental aspects of the ongoing progression in the field of genetics (Nicholas, 2009).

The commencement of the study of genetics signifies the initiation of a remarkable journey as we go on a scientific expedition to unravel the enigmatic aspects of life. This intriguing expedition offers the opportunity to unveil the enigmas surrounding inheritance, heredity, and adaptation, thereby facilitating a more profound comprehension of life's vast array and complexity. As we explore the genetic code of organisms, we are positioned to uncover significant knowledge regarding the basic tenets that govern our existence and the intricate diversity of life on our planet (Nagylaki, 2013).

1.2 HISTORY

Despite its relatively recent emergence within the last century, the field of genetics has been utilized by humans for millennia (Lynch et al., 2009).

1.2.1 Pre-history

The initial indications of human comprehension and utilization of hereditary principles can be traced back to the domestication of animals and plants, which commenced around 10,000 to 12,000 years ago. The early nomadic societies relied on hunting and gathering as their primary means of sustenance. However, with the expansion of human populations, the accessibility of wild food resources gradually diminished. The decline mentioned above exerted significant pressure on societies to seek alternative sources of sustenance. Consequently, individuals initiated selectively modifying wild flora and fauna, leading to early agricultural practices and the establishment of permanent human settlements (Kendler & Klee, 2022). Individuals deliberately selected and cultivated indigenous flora and fauna in the early stages with favorable characteristics. Archaeological evidence about the velocity and trajectory of the domestication process substantiates the notion that individuals swiftly grasped a fundamental yet pivotal principle of heredity: the tendency for similar traits to be inherited within a given breed. By carefully choosing and mating specific plants or animals with favorable characteristics, it is possible to perpetuate these desired traits in subsequent generations (Barrett, 2019).

The emergence of agriculture is believed to have originated in the Middle East, specifically in present-day Turkey, Iraq, Iran, Syria, Jordan, and Israel. By approximately 10,000 years ago, domesticated plants and animals had become significant elements of the diet of numerous populations in this region. The initial instances of domesticated organisms encompassed various species such as wheat, peas, lentils, barley, dogs, goats, and sheep. The process of selective breeding has resulted in the development of goats and sheep with increased wool density and improved manageability, as well as the cultivation of cereal plant seeds that are larger and more convenient to harvest (Frampton, 2011). Sophisticated methods of genetics were being utilized in the Middle East as early as 4000 years ago. The Assyrians and Babylonians cultivated numerous cultivars of date palms, exhibiting variations in fruit dimensions, pigmentation, flavor, and maturation period. A bas-relief of Assyrian origin, dating back 2880 years, portrays using artificial fertilization techniques to manage the hybridization of date palms. During the same period, various cultures in Asia, Africa, and the Americas made significant advancements in cultivating additional crops and the domestication of various animal species (Dakubu, 1972).

1.2.2 Early Written Records

Examining ancient texts reveals that early human civilizations possessed knowledge regarding their hereditary characteristics. The ancient Hindu scriptures, which date back to 2000 years, ascribe various characteristics to the father and propose that dissimilarities among siblings can be attributed to influences from the mother. These writings above suggest that individuals should exercise caution when selecting potential life partners who possess undesirable traits that their offspring may inherit. The Talmud, a significant Jewish text comprising religious laws derived from ancient oral traditions, exhibits a remarkably precise comprehension of the hereditary transmission of hemophilia. According to the directive, if a woman gives birth to two sons who die due to bleeding following circumcision, any subsequent sons born to her should not undergo circumcision. Additionally, the directive suggests that the sons of her sisters should also not be circumcised, while the sons of her brothers should still undergo the procedure. This advice provides an accurate representation of the X-linked inheritance pattern observed in hemophilia cases (Mountain & Guelke, 2016).

The ancient Greeks demonstrated a meticulous examination of human reproduction and heredity. Alcmaeon, a Greek physician who lived around 520 B.C., performed anatomical dissections on animals and hypothesized that the brain served as the primary center for perception and also played a role in producing semen. The proposal mentioned above instigated an extensive philosophical discourse about the origin of semen and its significance in the transmission of genetic traits. The discourse reached its apex with the formulation of the notion of pangenesis, which posited that discrete particles, subsequently referred to as gemmules, transmit information from different bodily regions to the reproductive organs; after that, they are transmitted to the embryo during conception. Despite its inaccuracies, the notion of pangenesis exerted significant influence until the latter part of the 19th century (see Figure 1.1) (Mühlemann et al., 2020).

The concept of pangenesis prompted the ancient Greeks to propose the hypothesis of the inheritance of acquired features. According to this notion, traits acquired during an individual's lifetime are assimilated into their hereditary information and transmitted to their offspring. For instance, individuals who cultivated musical ability through dedicated study would give rise to progeny with an innate musical aptitude predisposition. Although once widely embraced, the concept of the inheritance of acquired characteristics has lost acceptance in contemporary academic discourse. Nonetheless, it continued to maintain popularity throughout the twentieth century.

Aristotle, the renowned Greek philosopher who lived from 384 to 322 B.C., demonstrated a profound fascination with heredity. The individual in question disapproved of the theories of pangenesis and the inheritance of acquired characteristics. He highlighted instances where individuals resemble their ancestors from previous generations more than their immediate parents. Furthermore, he emphasized that acquired characteristics, such as mutilated body parts, are not transmitted to subsequent generations. Aristotle posited that both males and females play integral roles in the reproductive process, each making distinct contributions and that a certain degree of conflict exists between these contributions (Pilbeam, 1996).

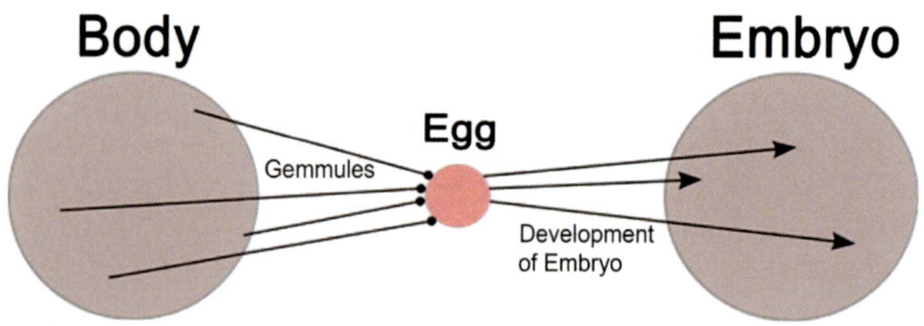

Figure 1.1. Pangenesis, an early concept of inheritance (Source:By Ian Alexander - Own work, CC BY-SA 4.0, https://commons.wikimedia.org/w/index.php?curid=62047504, Creative Commons License)

While the ancient Romans made limited contributions to the comprehension of human heredity, they effectively devised various animal and plant breeding methods. These techniques were primarily derived from empirical experimentation, lacking a comprehensive understanding of heredity. Limited advancements were made in the comprehension of genetics throughout the subsequent millennium. The concepts of pangenesis and the inheritance of acquired characteristics, which originated in ancient times, continued to be influential until the emergence of modern scientific thought during the seventeenth and eighteenth centuries. These ideas and methods employed in the selective breeding of plants and animals maintained their relevance throughout this transitional period in scientific understanding (Harper & Petermann, 2017).

1.2.3 The Rise of Modern Genetics

The emergence of rudimentary microscopes by Dutch spectacle makers during the late 16th century facilitated the groundbreaking discovery of cells by Robert Hooke (1653–1703) in 1665. Microscopes have bestowed upon naturalists novel and captivating perspectives on life. It is plausible that an excessive zeal for this nascent realm of minuscule entities contributed to the emergence of the concept of preformationism (Nye, 1993).

According to the preformationist theory, it was postulated that within the egg or sperm, there resided a minute replica of an adult organism known as a homunculus, which would undergo a process of enlargement throughout development.

The ovists posited that the homunculus was situated within the egg, while the spermists maintained that it resided within the sperm. The concept of preformationism posited that all hereditary characteristics would be exclusively derived from a single parent, either the father in the case of the presence of a homunculus within the sperm or the mother if it resided within the egg. Despite numerous observations indicating that offspring inherit a combination of traits from both parents, preformationism persisted as a widely accepted notion during the seventeenth and eighteenth centuries (Maisey et al., 2004).

An alternative early concept of heredity was the theory of blending inheritance, which posited that progeny are a composite or amalgamation of traits inherited from both parents.

This concept posited that the genetic material undergoes blending, analogous to the blending of blue and yellow pigments resulting in the creation of green paint. Once genetic differences are combined, they become indistinguishable in subsequent generations, similar to how the individual pigments of blue and yellow cannot be separated from green paint. Certain characteristics may exhibit blending inheritance, but it is now understood that individual genes do not blend (Caspari & Wolpoff, 2013).

According to Nehemiah Grew (1641 – 1712), plants engage in sexual reproduction by using pollen derived from the male gametes. Based on the provided information, a cohort of botanists initiated a series of experiments involving plant crossbreeding and the development of hybrid varieties. One of the prominent pioneers in early plant breeding was Joseph Gottleib Kölreuter (1733 – 1806), who conducted many crossbreeding experiments and microscopic examinations of pollen.

The researcher observed that a significant number of hybrid specimens displayed intermediate characteristics, falling between those exhibited by the parental varieties. Due to the crossing of plants exhibiting diverse traits, Kölreuter encountered difficulties identifying any overarching inheritance pattern. Despite this constraint, the contributions made by Kölreuter laid the groundwork for the contemporary investigation of genetics.

Following his initial research, several additional botanists embarked on conducting experiments involving hybridization. Among these scientists was Gregor Mendel (1822 – 1884), who contributed significantly to understanding fundamental principles of heredity. Mendel's findings, which remained undervalued for 45 years, established the fundamental basis for our contemporary comprehension of heredity. Consequently, he is widely acknowledged as the pioneer of the field of genetics (Mittler & Blumwald, 2010).

The advancements in the field of cytology during the 19th century significantly impacted the discipline of genetics. The cell nucleus was described by Robert Brown (1773 – 1858) in 1833. In 1839, Matthis Jacob Schleiden (1804 – 1881) and Theodor Schwann (1810 – 1882) advanced the cell theory, building upon the contributions of previous scholars. Based on this theoretical framework, it is posited that all forms of life are comprised of cellular structures, with the emergence of cells occurring solely from preexisting cells. Furthermore, the cell is regarded as the fundamental entity governing the organization and operation of biological organisms. Biologists initiated the examination of cells in order to investigate the mechanisms underlying the transmission of traits during the process of cell division (Moossavi & Bishehsari, 2019).

Charles Darwin, a prominent biologist of the 19th century, proposed the concept of evolution through natural selection and documented his theories in his seminal work, "On the Origin of Species," which was published in 1856. Darwin acknowledged the fundamental role of heredity in the process of evolution, and he conducted comprehensive genetic crosses involving pigeons and various other organisms. Nevertheless, his comprehension of inheritance remained inadequate, constituting a significant oversight within his theory of evolution (Muller, 1927). During the latter half of the nineteenth century, there was a surge in cytological research due to the emergence of the microtome, a device used to create thin tissue sections for microscopic analysis, and the advancement of histological stains. Several cytologists have provided evidence indicating that the nucleus plays a significant role in the fertilization process. In 1879, Walter Flemming (1843-1905) made notable observations regarding the process of chromosome division, subsequently providing a comprehensive account of mitosis in his published work. By 1885, there was widespread acknowledgment within the scientific community regarding the presence of genetic material within the nucleus (Wilson, 2005).

In the late nineteenth century, August Weismann (1834 – 1914) conclusively refuted the concept of the inheritance of acquired characteristics. The researcher conducted a series of experiments involving the removal of mice tails throughout 22 successive generations. The results of these experiments revealed that the length of the tails in subsequent generations exhibited a persistent tendency to remain long. As postulated by Weismann, the germ-plasm theory posits that the cells residing in the reproductive organs possess a comprehensive repertoire of genetic information, subsequently transmitted to the gametes (Allaby et al., 2008).

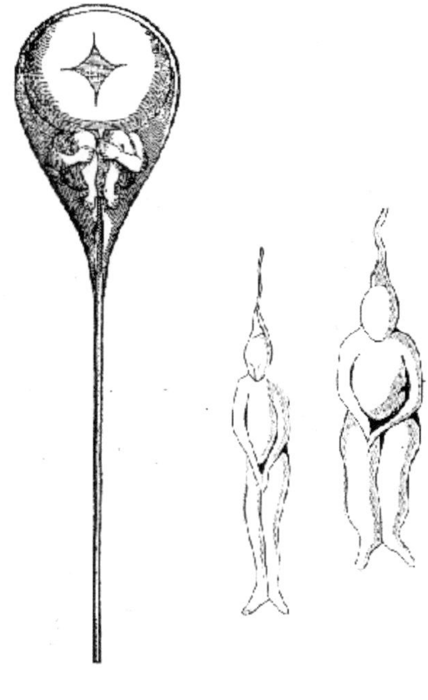

Figure 1.2 Perfomationism was a popular idea of inheritance in the 17th & 18th centuries; here is a drawing of a homunculus inside a sperm (Source: A tiny person ***inside a sperm, as drawn by*** Nicolaas Hartsoeker ***in 1695, Creative Commons License).***

Fun fact:

Humans share 98% of their DNA with the chimpanzees, revealing our close evolutionary relationship with these primates.

1.2.4 20th Century Genetics

The year 1900 marked a significant turning point in the field of genetics. The seminal 1866 publication by Gregor Mendel, which presented his experiments on pea plants and elucidated the fundamental principles of heredity, underwent a process of "rediscovery". The recognition of the importance of his findings prompted other biologists to promptly initiate comparable genetic investigations on various organisms such as mice, chickens, and others. The findings of these investigations demonstrated that numerous traits adhere to Mendel's principles (Christensen, 1999).

In 1902, Walter Sutton (1877 – 1916) proposed that genes are situated on chromosomes. In 1910, Thomas Hunt Morgan (1866 – 1945) made the groundbreaking discovery of the initial genetic mutant within the fruit fly species. Subsequently, Morgan employed fruit flies as a model organism to elucidate numerous intricacies in the field of transmission genetics. The establishment of the field of population genetics in the 1930s can be attributed to the contributions of Ronald A. Fisher (1890 – 1962), John B. S. Haldane (1892 – 1964), and Sewall Wright (1889 – 1988) (Skopek, 2011).

The utilization of bacteria and viruses by geneticists commenced during the 1940s. The expeditious reproduction and uncomplicated genetic systems exhibited by these organisms facilitated meticulous examination of gene organization and structure. During approximately the same period, a growing body of evidence emerged indicating that DNA was the storage site for genetic information. In 1953, James Watson and Francis Crick made a significant contribution to the field of molecular genetics by elucidating the three-dimensional configuration of DNA. This breakthrough marked the beginning of a new era in scientific understanding.

The elucidation of the chemical structure of DNA and the mechanism through which it governs the arrangement of amino acids in proteins had been accomplished by 1966. The advent of molecular genetics facilitated the initiation of the inaugural recombinant DNA experiments in 1973, thereby instigating a subsequent paradigm shift in the field of genetic research. The pioneering work of Walter Gilbert, born in 1932, and Frederick Sanger, born in 1918, led to the establishment of DNA sequencing techniques in 1977. The polymerase chain reaction (PCR), a method utilized to amplify minuscule quantities of DNA rapidly, was pioneered by Kary Mullis (b. 1944) and a team of researchers in 1986. In the year 1990, the United States witnessed the inaugural application of gene therapy for treating human genetic diseases, coinciding with the commencement of the Human Genome Project. In 1995, the scientific community achieved a significant milestone by successfully determining the DNA sequence of a free-living organism, *Haemophilus influenzae*. Subsequently, a year later, the first complete sequence of a eukaryotic organism, specifically yeast, was reported. In the early years of the twenty-first century, the sequencing of the human genome was successfully accomplished, marking the commencement of a transformative period in the field of genetics (Figure 1.3) (Kerminen et al., 2021).

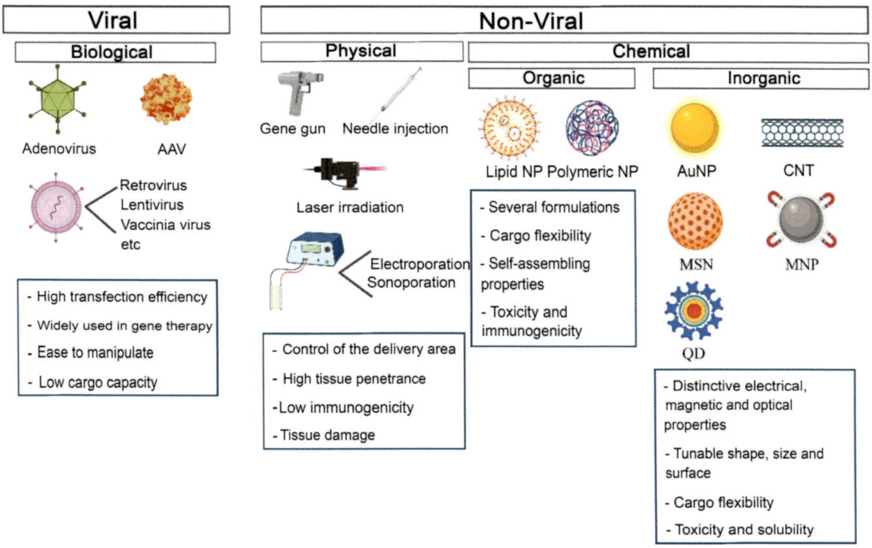

Figure 1.3 Gene therapy applies genetic engineering to the treatment of human disease (Source: J. Mirón-Barroso S, Domènech EB, Trigueros S. Nanotechnology-Based Strategies to Overcome Current Barriers in Gene Delivery. International Journal of Molecular Sciences. 2021; 22(16):8537. https://doi.org/10.3390/ijms22168537, Creative Commons License)

1.2.5 Future Of Genetics

The rate at which genetic information is expanding has been observed to double approximately every few years. Each year, numerous genome sequences of various organisms are incorporated into DNA databases, contributing to the continuous expansion of our understanding of heredity by providing new insights into gene structure and function. The information above enhances our comprehension of various biological processes and evolutionary associations. The rapid influx of genetic data necessitates the ongoing advancement of complex computer algorithms to store, retrieve, compare, and analyze genetic information. Consequently, this has led to the emergence of the interdisciplinary field known as bioinformatics, which combines principles from molecular biology and computer science (Schork et al., 2000).

In forthcoming times, the emphasis of DNA-sequencing endeavors will transition from examining genomes across various species to exploring intraspecific variations. It is a plausible conjecture that every individual may have access to their complete genome sequence in the future. The utilization of novel genetic microchips enables the concurrent analysis of numerous RNA molecules, thereby facilitating the acquisition of comprehensive insights into the behavior of numerous genes within a specific cell (Bernard et al., 2018). This technology enables the generation of a detailed depiction of cellular responses to external stimuli, environmental pressures, and pathological conditions. The utilization of genetics in the agricultural, chemical, and healthcare sectors is anticipated to undergo further growth and development. Certain experts have suggested that biotechnology may hold a comparable significance to the electronics industry in the twentieth century in shaping the advancements of the twenty-first century. The expanding breadth of genetics research will give rise to considerable ethical, social, and economic concerns (Guttmacher et al., 2010).

The following concise exposition on the history of genetics is not meant to be exhaustive; instead, its purpose is to impart an understanding of the increasingly rapid progression of developments in genetics (Gjøen & Bentsen, 1997).

Did You Know?

Identical twins share the same DNA, but did you know that even they can have subtle genetic differences? These small variations, known as somatic mutations, can occur during their development and contribute to their individuality.

1.3 IMPORTANCE OF GENETICS

The field of genetics, which encompasses the examination of genetics and the transfer of characteristics from one generation to another, possesses considerable importance in comprehending the underlying mechanisms that influence life on our planet. Genetics plays a pivotal role in unraveling the enigmas of inheritance, evolution, and the intricate interconnectedness of life, encompassing a vast spectrum ranging from minuscule microorganisms to intricate multicellular organisms such as humans. Genetics, with its extensive ramifications across multiple disciplines, assumes a central and influential position in shaping our comprehension of health, agriculture, biodiversity, and the human race's evolution (Ursin et al., 2009).

Genetics offers a framework for understanding the transmission of traits from one generation to the next. The observation above elucidates the intricate dynamics of genetic interactions in determining an individual's phenotypic traits, encompassing a wide range of attributes such as eye color and predisposition to specific ailments. Comprehending the fundamental principles of heredity holds significant importance not only in acquiring foundational biological knowledge but also in facilitating practical applications for predicting and managing genetic traits and diseases (Notter, 1999).

Genetics has profoundly impacted the medical field, leading to the development of precision healthcare. By analyzing an individual's genetic composition, healthcare providers can customize treatments to accommodate their unique requirements and genetic predispositions. Implementing a customized approach has resulted in notable progress in diagnostics, disease prevention, and targeted therapies, offering the potential for enhanced efficacy and efficiency in medical interventions (Govindaraj et al., 2015).

Identifying genetic disorders is considered one of the most crucial aspects of the field of genetics. Genetic testing enables the timely identification of hereditary conditions, thereby equipping individuals and families with the knowledge necessary to make well-informed choices about their health and reproductive strategies. The implementation of early intervention and counseling services has been shown to have a substantial positive impact on the overall well-being and quality of life of individuals affected by genetic disorders and their families (Aartsma-Rus et al., 2016).

Genetics offers a robust methodology for investigating the evolutionary trajectory of organisms on our planet. Through examining genetic variations and the comparative analysis of genomes among different species, researchers can reconstruct the evolutionary connections between organisms. This knowledge holds significant value in comprehending the origins of species, the mechanisms of adaptation, and the patterns of biodiversity that have influenced the world's ecosystems.

Genetics significantly influences agricultural progress by enhancing crop productivity, bolstering resistance against diseases, and enhancing nutritional composition. Scientists have employed genetic engineering and selective breeding techniques to cultivate crops that exhibit resilience in challenging environmental circumstances, possess resistance against pests, and provide augmented nutritional advantages. The advancements above play a crucial role in addressing the global issue of food security, given the ongoing growth of the world's population (Hoedemaekers et al., 2010). The field of genetics plays a substantial role in the preservation and maintenance of biodiversity. By examining genetic diversity within populations and species, conservationists can discern populations at risk, effectively manage genetic variability, and formulate strategies to safeguard species that are on the brink of extinction. This knowledge facilitates the development of efficient conservation programs, thereby preserving the intricate equilibrium of ecosystems and protecting the diverse array of life forms on our planet (Palstra & Ruzzante, 2008).

Genetics also plays a significant role in the pharmaceutical industry. Various pharmaceuticals and food additives are produced through the genetic modification of fungi and bacteria, resulting in enhanced efficiency in their synthesis. The biotechnology sector utilizes molecular genetic methodologies to facilitate commercial development and large-scale production of molecules. Commercial production of growth hormone, insulin, and clotting factors is currently facilitated through the utilization of genetically modified bacteria (Figure 1.4). Molecular genetics methodologies have effectively developed bacterial strains capable of extracting minerals from ore, degrading hazardous chemicals, and impeding detrimental frost formation on agricultural crops (Lyhne et al., 2001).

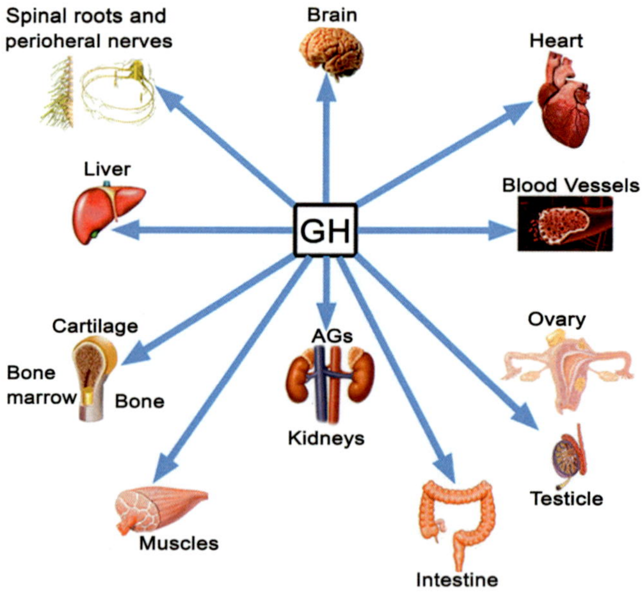

Figure 1.4. Growth hormone is produced by genetically engineered bacteria (Source: Caicedo D, Díaz O, Devesa P, Devesa J. Growth Hormone (GH) and Cardiovascular System. International Journal of Molecular Sciences. 2018; 19(1):290. https://doi.org/10.3390/ijms19010290, Creative Commons License)

The field of genetics has provided valuable insights into the historical aspects of humanity and the patterns of migration. By examining genetic markers, researchers can discern human lineages and patterns of migration, thereby illuminating the historical trajectory of our predecessors across various continents and over extensive periods of time. The comprehension above of human genetic diversity cultivates a perception of interconnection and common lineage among diverse populations across the globe (Lande, 1988).

The discipline of biotechnology is greatly indebted to the field of genetics. Recombinant DNA technology, a prominent tool within the biotechnology field, enables scientists to manipulate organisms by introducing desired traits, thereby facilitating significant advancements in medicine, agriculture, and industrial applications. The utilization of genetic engineering and gene editing techniques, such as CRISPR-Cas9, presents the possibility of precise alterations, therapeutic interventions, and revolutionary progress in diverse sectors.

The field of medicine is significantly influenced by genetics. Medical professionals acknowledge that numerous diseases and disorders exhibit a hereditary element, encompassing widely recognized genetic disorders like sickle-cell anemia and Huntington's disease, as well as prevalent ailments such as asthma, diabetes, and hypertension. Significant progress in the field of molecular genetics has facilitated valuable revelations regarding the fundamental characteristics of cancer, thereby enabling the creation of numerous diagnostic examinations. Gene therapy, which involves genetically modifying genes to treat human diseases, has now become a tangible and practical approach (Kalyna et al., 2012).

As genetics progresses, the significance of ethical considerations becomes more pronounced. Ensuring the responsible utilization of genetic technologies, safeguarding genetic privacy, and effectively addressing ethical quandaries are of utmost importance in order to harness the potential of genetics for the collective benefit of humanity while simultaneously upholding individual rights and values.

Genetics is of utmost importance as it serves as the fundamental basis for comprehending various aspects of life, including inheritance and evolutionary processes. Genetics plays a pivotal role in influencing medical treatments, agricultural advancements, understanding human ancestry, and exploring biodiversity. Its contributions have been instrumental in shaping our current knowledge and guiding our future endeavors. The responsible and ethical utilization of genetics plays a crucial role in effectively harnessing its potential for advancing society and conserving the diverse array of life forms that encompass our environment. As the investigation into the genetic code of life progresses, we embark upon an expedition of exploration, wherein the boundless potential of genetics lies in anticipation (Lacy, 1997).

1.4 ROLE OF GENETICS IN BIOLOGY

An appreciation of genetics is significant for individuals across various disciplines, but it is important in biology education. The field of genetics offers a fundamental principle that unifies the study of biology: all living organisms utilize nucleic acids as their genetic material and employ a shared mechanism for encoding their genetic information. The field of genetics serves as a fundamental basis for the investigation and understanding of numerous other branches within the realm of biology. As an illustrative instance, evolution denotes the process of genetic modification occurring over successive periods; hence, a comprehensive comprehension of fundamental genetics is imperative for investigating evolution. The field of developmental biology heavily relies on the principles of genetics, as the formation of tissues and organs is intricately regulated by the expression of genes (Figure 1.5). Genetic methods are becoming more prevalent in various disciplines, including taxonomy, ecology, and animal behavior. A comprehensive comprehension of genes and genetic methodologies is indispensable in the study of nearly every domain within the realms of biology and medicine (O'Brien, 1994).

The significance of genetics in the field of biology is fundamental and extensive, as it intricately intersects with all facets of the life sciences. The field of genetics plays a fundamental role in elucidating the enigmas surrounding heredity, variation, and evolution, thereby shaping our comprehension of the realm of living organisms (Porteous et al., 2006).

The fundamental focus of biology centers around examining genes, which serve as the fundamental elements of heredity, and the subsequent transfer of these genes from one generation to the next. The field of genetics explores the underlying mechanisms of heredity, elucidating the processes by which traits are transmitted via DNA and how genetic variations contribute to the diversity observed among individuals and populations. This knowledge constitutes the fundamental basis for comprehending various species' origins, evolution, and interdependence (Wang et al., 2018).

Molecular genetics serves as a potent instrument for the analysis of cellular mechanisms, elucidating the processes by which genes are expressed, regulated, and engage in mutual interactions. The complex orchestration of gene expression regulates the synthesis of proteins, which serve as the fundamental components of biological machinery. This intricate process is important in comprehending cellular mechanisms, developmental processes, and overall well-being (Rêdowicz, 2002).

The field of human genetics is of utmost importance in the realm of biology, as it delves into the genetic underpinnings of various diseases, identifies genetic factors that contribute to risk, and offers valuable insights into the origins and variations within the human population. The advancement of personalized medicine has created opportunities for tailoring treatments to an individual's distinct genetic composition, resulting in enhanced healthcare outcomes (Finnerty et al., 2010).

The field of genetics in the realm of biology encompasses not only human beings but also encompasses all forms of living organisms. Genetics plays a pivotal role in the agricultural sector by facilitating the development of crops that possess augmented characteristics, including but not limited to increased yield, enhanced resistance against diseases, and improved nutritional composition. The phenomenon above plays a significant role in enhancing comprehension of plant and animal breeding, thereby influencing agricultural methodologies and bolstering the imperative aspect of food security.

The field of conservation biology utilizes genetic principles and techniques in order to safeguard and maintain biodiversity. Through the examination of genetic diversity within populations and species, scientists can discern populations that are at risk, monitor species that are in danger of extinction, and formulate conservation strategies that are efficient and successful. The field of genetics provides us with the ability to understand the interconnections within ecosystems and the significance of safeguarding genetic diversity, which serves as the foundation for ecological resilience (Nguyen & Xu, 2008).

In evolutionary biology, genetics plays a crucial role in elucidating the phylogenetic tree, delineating the genetic affinities among different species, and elucidating the genetic variations that underlie adaptations to environmental alterations. Understanding genetic diversity and natural selection provides insight into the mechanisms that propel evolutionary change over extended periods.

The influence of genetics on the field of biology extends beyond disciplinary confines, establishing connections between biology and various domains such as medicine, agriculture, ecology, and forensic science. Genetics is a unifying factor across a wide range of organisms, from minute microorganisms to intricate multicellular entities. It plays a pivotal role in deciphering the intricate genetic blueprint governing the remarkable life forms (Nordestgaard & Langsted, 2016).

The expanding body of knowledge in genetics is increasingly influencing the field of biology. Genetics enables the investigation of the genetic foundations of intricate diseases and the examination of the genetic origins of behavioral characteristics in animals, thereby facilitating novel avenues of research and enriching our comprehension of the captivating realm of living organisms. The tool in question continues to be of utmost importance, driving advancements in biology and holding the potential to unravel further enigmas surrounding the intricate blueprint of life (Maher et al., 2001).

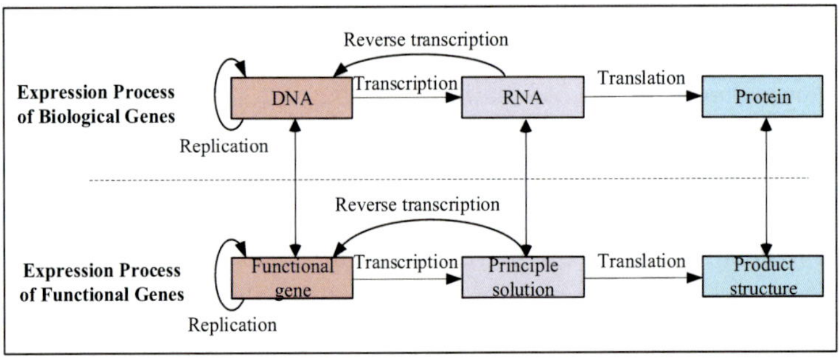

Figure 1.5 Biological gene expression process (Source: Wang H, Zhang P, Zhang Z, Zhang Y, Wang Y. Product Innovation Design Process Model Based on Functional Genes Extraction and Construction. *Applied Sciences*. 2022; 12(24):12990. https://doi.org/10.3390/app122412990, Creative Commons License)

1.5 DIVISION OF GENETICS

Historically, the field of genetics has been categorized into three primary subdisciplines, namely transmission genetics, molecular genetics, and population genetics (see Figure 1.6).

Transmission genetics, also referred to as classical genetics, is a field that encompasses the fundamental principles of genetics and elucidates the mechanisms by which traits are inherited from one generation to the subsequent. This section pertains to the correlation between chromosomes and heredity, the organization of genes on chromosomes, and the process of gene mapping. The primary emphasis lies on the individual organism, encompassing the mechanisms by which it acquires its genetic composition and transmits its genetic material to subsequent generations (Burow et al., 2001).

Molecular genetics is focused on the chemical composition of genes, encompassing the processes of encoding, replicating, and expressing genetic information. The cellular processes of replication, transcription, and translation transfer genetic information from one molecule to another. Additionally, gene regulation encompasses the processes that govern the expression of genetic information. The primary emphasis in the field of molecular genetics pertains to the gene, encompassing its structural composition, arrangement, and operational characteristics (Beauchamp et al., 2011).

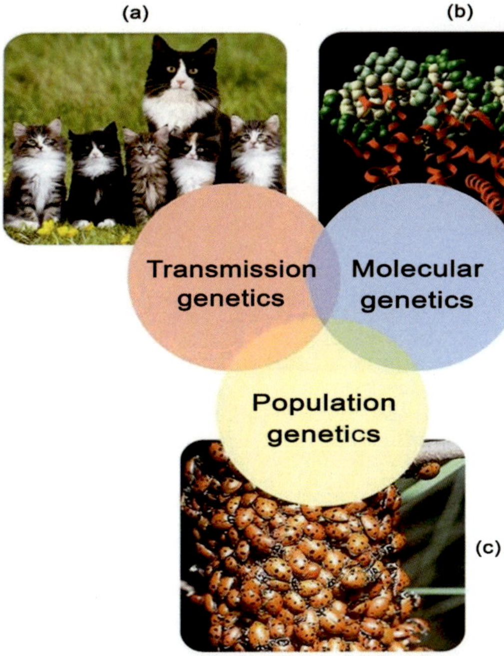

Figure 1.6. Genetics can be subdivided into three inter-related fields (Source: Alan Carey, Creative Commons License)

Population genetics investigates the genetic makeup of populations, which are groups of individuals belonging to the same species. It examines how this genetic composition undergoes temporal and spatial changes. Population genetics is a discipline that primarily focuses on the study of evolution due to its inherent connection to genetic change. Population genetics primarily concerns itself with the collective genetic makeup of a given population.

The categorization of genetics into three distinct groups is a commonly accepted and practical approach. However, it is important to acknowledge that these divisions are not rigid, and there is considerable overlap between them. Furthermore, each of these primary subdivisions can be further subdivided into numerous specialized fields, including but not limited to chromosomal genetics, biochemical genetics, and quantitative genetics. The field of genetics can be categorized based on the specific organism under study, such as fruit fly, corn, or bacterial genetics. Within each of these organisms, research can be conducted at various levels, including transmission, molecular, and population genetics. The field of modern genetics is characterized by its extensive breadth, as it encompasses a multitude of interconnected subdisciplines and specialized areas of study (Dobzhansky, 1955).

Did You Know?

Modern DNA sequencing technology allows for the rapid sequencing of the entire human genome, which consists of billions of base pairs, within a relatively short timeframe of a few hours. The aforementioned statement serves as evidence of the remarkable progress made in the fields of genetics and biotechnology.

1.6 BASIC CONCEPTS IN GENETICS

In this discussion, we will examine several fundamental genetic concepts.

There are two fundamental categories of cells: eukaryotic and prokaryotic. From a structural perspective, cells can be categorized into two fundamental types. However, it is important to note that the evolutionary history of cells presents a more intricate narrative. Prokaryotic cells are characterized by the absence of a nuclear membrane and the absence of membrane-bound organelles. In contrast, eukaryotic cells are more intricate in structure, as they possess a nucleus and membrane-bound organelles, including chloroplasts and mitochondria (Katscher, 2004).

A gene is considered the fundamental unit of heredity. At its most fundamental level, a gene can be conceptualized as a discrete entity responsible for encoding a hereditary trait.

Genes exist in various allelic forms, known as alleles. Multiple alleles can exist for a gene, each representing a distinct form that determines a specific characteristic. An illustration of this concept can be observed in the genetic makeup of cats, where a specific gene responsible for determining coat color can be found in alleles that encode either black or orange fur (Choi et al., 1999).

Genetic information is responsible for encoding observable traits or phenotypes. The differentiation between traits and genes is a fundamental concept within the field of genetics. Inheritance of traits does not occur directly. Genes are inherited and, in conjunction with environmental factors, play a crucial role in determining the manifestation of traits. The genotype of an individual organism refers to its genetic information, while its phenotype refers to the observable traits. An illustration of this concept can be seen in the A blood type, which represents a phenotype. The genotype is the underlying genetic information responsible for encoding the blood type A antigen.

The transmission of genetic information occurs through the molecules of DNA and RNA. The molecular composition of nucleic acids, namely deoxyribonucleic acid (DNA) and ribonucleic acid (RNA), carries genetic information. Nucleic acids are macromolecules composed of nucleotide monomers, wherein each nucleotide is comprised of a sugar moiety, a phosphate group, and a nitrogenous base. DNA is composed of four distinct nitrogenous bases, namely adenine (A), cytosine (C), guanine (G), and thymine (T). The arrangement or sequence of these bases within the DNA molecule serves as the fundamental code for storing and transmitting genetic information. Most organisms store their genetic information in the form of DNA, while a minority of viruses utilize RNA as their genetic material. The abbreviations A, C, G, and U are used to represent the four nitrogenous bases found in RNA.

Chromosomes serve as the carriers of genetic information within cells, as they are composed of DNA molecules and associated proteins. Each species is characterized by a specific number of chromosomes within its cells. For instance, bacterial cells typically contain a single chromosome, while human cells possess 46 chromosomes, and pigeon cells possess 80 chromosomes. Every chromosome contains a substantial quantity of genes (Azofeifa et al., 1995).

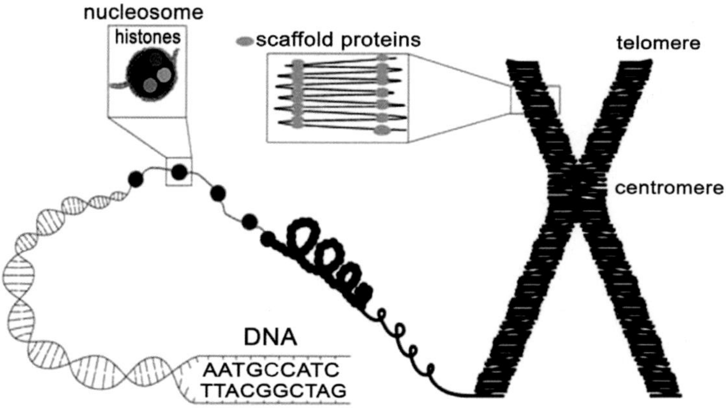

Figure 1.7. Genes are carried on chromosomes (Source: Deakin JE, Potter S, O'Neill R, Ruiz-Herrera A, Cioffi MB, Eldridge MDB, Fukui K, Marshall Graves JA, Griffin D, Grutzner F, et al. Chromosomics: Bridging the Gap between Genomes and Chromosomes. Genes. 2019; 10(8):627. https://doi.org/10.3390/genes100806, Creative Commons License)

Chromosomes separate through the processes of mitosis and meiosis- The processes of mitosis and meiosis ensure the proper distribution of an organism's chromosomes to each daughter cell, guaranteeing the preservation of a complete set of genetic material. Mitosis refers to the process by which replicated chromosomes are segregated during the division of somatic cells, excluding those involved in sexual reproduction. Meiosis refers to the process by which replicated chromosomes in sex cells undergo pairing and subsequent separation, ultimately producing gametes, which are specialized reproductive cells (Hall et al., 2003).

Genetic information is transferred from DNA to RNA to protein. Numerous genes are responsible for encoding traits through the determination of protein structure. The process of genetic information transfer involves the initial transcription of DNA into RNA, followed by the subsequent translation of RNA into the amino acid sequence that constitutes a protein.

Mutations are enduring alterations in genetic information that can be passed down through generations. Gene mutations exclusively impact the genetic information of a singular gene, while chromosome mutations modify the number or structure of chromosomes, consequently influencing multiple genes.

Certain traits are affected by multiple factors. For instance, the determination of human height is influenced by numerous genetic factors, encompassing hundreds of genes, in conjunction with environmental variables like nutritional intake.

Evolution can be conceptualized as a dual-stage mechanism, wherein the first stage involves the emergence of genetic variation, followed by the second stage, wherein certain genetic variants experience an increase in frequency. In contrast, others undergo a decrease in frequency (Longdon et al., 2014).

Fun fact:

There are over 20,000 genes in the human genome, responsible for encoding the instructions that make us who we are. Each gene plays a unique role in determining our traits, from eye color to personality traits.

SUMMARY

The introductory chapter of "Genetics: Exploring DNA and Heredity" establishes the foundation for an enlightening exploration into genetics, a captivating field dedicated to the study of heredity and the transmission of traits. This text explores the significant historical events that have influenced the development of the field of genetics, beginning with early observations of heredity and culminating in Gregor Mendel's pioneering experiments involving pea plants. These experiments served as a fundamental basis for the establishment of contemporary genetics.

This chapter underscores the significant role of genetics in our comprehension of life. The field of genetics elucidates the enigmas surrounding the process of inheritance, facilitating our understanding of the mechanisms by which traits are transmitted from one generation to the succeeding one. The study of evolution, health, and the remarkable variety of life forms on Earth allows us to gain knowledge and understanding in these areas. Genetics holds significant importance in various domains such as medicine, agriculture, and conservation, exerting a profound influence on our current state and laying the groundwork for future advancements.

The introductory chapter of "Genetics: Exploring DNA and Heredity" provides an overview of genetics, establishing a foundation for a profound exploration into the realm of heredity, traits, and the remarkable mechanisms of genetic inheritance. Beginning with the rudimentary observations of heredity made by ancient civilizations, this narrative traces the historical trajectory that culminated in the groundbreaking contributions of Gregor Mendel, widely regarded as the progenitor of modern genetics. Through this exploration, readers are invited to embark on a historical odyssey that has significantly influenced our comprehension of the fundamental blueprint of life.

Across the annals of time, the relentless pursuit to unravel the enigmatic intricacies of genetics has served as a captivating narrative of human curiosity and unwavering resolve. The foundation of genetics emerged from the ancient cultures' initial fascination with selective breeding to the systematic experiments conducted by Mendel in the 19th century involving pea plants. Mendel's seminal findings elucidated the principles governing inheritance, thereby introducing the fundamental notions of dominant and recessive traits and laying the foundation for the development of contemporary genetics.

Genetics' role in elucidating life's intricacies is of utmost importance. The comprehension of genetics facilitates the acquisition of knowledge regarding the transmission of traits and characteristics across successive generations, thereby establishing the fundamental principles of heredity. This tool offers a robust means of investigating the complexities of evolution, elucidating the interconnections that bind together all organisms, and demonstrating the common lineage shared by a wide array of species.

The utilization of genetics transcends the boundaries of the scientific domain. Genetics assumes a pivotal role in the realm of medicine, particularly in the context of personalized healthcare. Through the utilization of genetic testing, healthcare professionals are empowered to devise treatment plans that are customized to the unique needs of individual patients while also enabling the timely identification of genetic disorders. Genetics has the potential to bring about significant advancements in healthcare outcomes, ranging from predictive medicine to gene therapies, thereby offering promising opportunities for revolutionary transformations.

Genetics has also brought about a revolution in the field of agriculture. The comprehension of plant genetics has facilitated the advancement of genetically modified crops possessing augmented characteristics, such as heightened resistance to pests and increased tolerance to environmental stressors. The advancements above are pivotal in effectively tackling global challenges such as food security and sustainable agriculture.

Both conservationists and biologists rely on genetic methods to safeguard biodiversity and mitigate the risk of extinction for endangered species. By examining genetic diversity within populations, scholars acquire valuable knowledge regarding the well-being and adaptability of ecosystems, thereby enhancing their ability to devise conservation strategies and uphold the intricate equilibrium of the natural world.

Genetics plays a fundamental role that permeates across diverse disciplines within the field of biology. The field of molecular genetics has contributed significantly to our comprehension of gene expression, while human genetics has provided insights into the genetic underpinnings of various diseases. Together, these areas of study have enhanced our knowledge of the intricate nature of life. The statement reveals the intricate interdependence of life on Earth, wherein each organism is a result of the complex genetic interactions that determine their distinct traits and attributes. The DNA molecule, characterized by its complex double helix configuration, contains the genetic instructions necessary for developing and operating living organisms, orchestrating a cohesive interplay of genetic data.

Upon engaging with this compelling prologue to genetics, individuals are encouraged to embark on an intellectual expedition that holds the potential to unravel the enigmatic intricacies of heredity and the vast array of biological variations. Genetics plays a pivotal role in unraveling the enigmatic intricacies of life's overarching blueprint, encompassing both the idiosyncrasies that render individuals distinct and the common lineage that unites all organisms. Genetics, with its extensive historical foundations, wide-ranging significance, and limitless possibilities, presents an attractive invitation to delve into its captivating realm. In this realm, the intricate blueprint of life is gradually revealed through the analysis of individual genetic sequences.

Genetics occupies a central position within the field of biology, serving as a unifying force that links diverse disciplines and provides insights into the intricate nature of life. This statement explains the underlying mechanisms contributing to genetic variation, the processes involved in evolution, and the intricate relationship between genes and the environment. The field of genetics encompasses a diverse range of specialized branches, including classical, molecular, and human genetics, among others. These branches investigate distinct facets of heredity and inheritance, spanning from classical genetics to molecular genetics and extending further into human genetics.

The chapter provides an overview of fundamental principles in genetics, including the importance of DNA as hereditary material, the transmission of genetic information through transcription and translation processes, and the cell cycle's involvement in DNA replication and cellular division. These fundamental concepts serve as the foundation of genetics, facilitating our comprehension of how genes encode the vital information necessary for the diverse array of processes in life.

The book's introductory chapter catalyzes readers' curiosity, enticing them to embark on a journey exploring the intricacies of genetics. The statement underscores the profound and indispensable nature of genetics, its historical origins, and its extensive ramifications. The present chapter functions as an introductory passage that grants access to a realm of intellectual curiosity, pledging to disclose the enigmatic aspects of heredity and the multitude of variations in life forms as individuals engage in a more profound exploration of the enthralling field of genetics.

REFERENCES

1. Aartsma-Rus, A., Ginjaar, I. B., & Bushby, K. (2016). The importance of genetic diagnosis for Duchenne muscular dystrophy. *Journal of medical genetics*, *53*(3), 145-151.
2. Allaby, R. G., Fuller, D. Q., & Brown, T. A. (2008). The genetic expectations of a protracted model for the origins of domesticated crops. *Proceedings of the National Academy of Sciences*, *105*(37), 13982-13986.
3. Azofeifa, J., Voit, T., Hübner, C., & Cremer, M. (1995). X-chromosome methylation in manifesting and healthy carriers of dystrophinopathies: concordance of activation ratios among first degree female relatives and skewed inactivation as cause of the affected phenotypes. *Human Genetics*, *96*, 167-176.
4. Barrett, J. H. (2019). An environmental (pre) history of European fishing: past and future archaeological contributions to sustainable fisheries. *Journal of Fish Biology*, *94*(6), 1033-1044.
5. Beauchamp, J. P., Cesarini, D., Johannesson, M., van der Loos, M. J. M., Koellinger, P. D., Groenen, P. J. F., ... & Christakis, N. A. (2011). Molecular genetics and economics. *Journal of Economic Perspectives*, *25*(4), 57-82.
6. Berger, S. L., Kouzarides, T., Shiekhattar, R., & Shilatifard, A. (2009). An operational definition of epigenetics. *Genes & development*, *23*(7), 781-783.
7. Bernard, A., Lheureux, F., & Dirlewanger, E. (2018). Walnut: past and future of genetic improvement. *Tree genetics & genomes*, *14*(1), 1.
8. Burow, M. D., Simpson, C. E., Starr, J. L., & Paterson, A. H. (2001). Transmission genetics of chromatin from a synthetic amphidiploid to cultivated peanut (Arachis hypogaea L.): broadening the gene pool of a monophyletic polyploid species. *Genetics*, *159*(2), 823-837.
9. Caspari, R., & Wolpoff, M. H. (2013). The process of modern human origins: the evolutionary and demographic changes giving rise to modern humans. *The origins of modern humans: biology reconsidered*, 5, 355-391.
10. Choi, D. W., Zhu, B., & Close, T. J. (1999). The barley (Hordeum vulgare L.) dehydrin multigene family: sequences, allele types, chromosome assignments, and expression characteristics of 11 Dhn genes of cv Dicktoo. *Theoretical and Applied Genetics*, *98*, 1234-1247.
11. Christensen, K. (1999). The 20th century Danish facial cleft population—epidemiological and genetic-epidemiological studies. *The Cleft palate-craniofacial journal*, *36*(2), 96-104.

12. Crow, J. F. (2017). *An introduction to population genetics theory*. Scientific Publishers. 84(546), 9871-9877.
13. Dakubu, M. K. (1972). Linguistic pre-history and historical reconstruction: the Ga-Adangme migrations. *Transactions of the Historical Society of Ghana*, *13*(1), 87-111.
14. Dekkers, J. C., & Hospital, F. (2002). The use of molecular genetics in the improvement of agricultural populations. *Nature Reviews Genetics*, *3*(1), 22-32.
15. Dobzhansky, T. (1955, January). A review of some fundamental concepts and problems of population genetics. In *Cold Spring Harbor symposia on quantitative biology.* Cold Spring Harbor Laboratory Press. (Vol. 20, pp. 1-15).
16. Ewens, W. J. (2004). *Mathematical population genetics: theoretical introduction*. New York: Springer. (Vol. 2, pp. 582-585).
17. Finnerty, J. R., Wang, W. X., Hébert, S. S., Wilfred, B. R., Mao, G., & Nelson, P. T. (2010). The miR-15/107 group of microRNA genes: evolutionary biology, cellular functions, and roles in human diseases. *Journal of Molecular Biology*, *402*(3), 491-509.
18. Frampton, S. (2011). Patents, priority disputes and the value of credit: towards a history (and pre-history) of intellectual property in medicine. *Medical history*, *55*(3), 319-324.
19. Frankham, R., Briscoe, D. A., & Ballou, J. D. (2002). *Introduction to conservation genetics*. Cambridge university press. 67, 230-235.
20. Gjøen, H. M., & Bentsen, H. B. (1997). Past, present, and future of genetic improvement in salmon aquaculture. *ICES Journal of Marine Science*, *54*(6), 1009-1014.
21. Govindaraj, M., Vetriventhan, M., & Srinivasan, M. (2015). Importance of genetic diversity assessment in crop plants and its recent advances: an overview of its analytical perspectives. *Genetics research international*, *20, 152-155.*
22. Griffiths, P., & Stotz, K. (2013). *Genetics and philosophy: an introduction*. Cambridge University Press. 2(34), 2310-2312.
23. Guttmacher, A. E., McGuire, A. L., Ponder, B., & Stefánsson, K. (2010). Personalized genomic information: preparing for the future of genetic medicine. *Nature Reviews Genetics*, *11*(2), 161-165.
24. Hall, I. M., Noma, K. I., & Grewal, S. I. (2003). RNA interference machinery regulates chromosome dynamics during mitosis and meiosis in fission yeast. *Proceedings of the National Academy of Sciences*, *100*(1), 193-198.

25. Halliburton, R., & Halliburton, R. (2004). *Introduction to population genetics*. Upper Saddle River: Pearson/Prentice Hall. 8, 987422.
26. Harper, P. S., & Petermann, H. I. (2017). The International Workshops on Genetics, Medicine and History: An Overview, 2003–2015. *History of Human Genetics: Aspects of Its Development and Global Perspectives*, 8(7), 7-38.
27. Hoedemaekers, Y. M., Caliskan, K., Michels, M., Frohn-Mulder, I., van der Smagt, J. J., Phefferkorn, J. E., ... & Majoor-Krakauer, D. F. (2010). The importance of genetic counseling, DNA diagnostics, and cardiologic family screening in left ventricular noncompaction cardiomyopathy. *Circulation: Cardiovascular Genetics*, *3*(3), 232-239.
28. Kalyna, M., Simpson, C. G., Syed, N. H., Lewandowska, D., Marquez, Y., Kusenda, B., ... & Brown, J. W. (2012). Alternative splicing and nonsense-mediated decay modulate expression of important regulatory genes in Arabidopsis. *Nucleic acids research*, *40*(6), 2454-2469.
29. Katscher, F. (2004). The history of the terms prokaryotes and eukaryotes. *Protist*, *155*(2), 257.
30. Kempthorne, O. (1957). An introduction to genetic statistics. 9, 875-877.
31. Kendler, K. S., & Klee, A. (2022). Introduction to "The beginnings of systematic studies of the genetics of schizophrenia: 1916–1933". *Schizophrenia Bulletin*, *48*(1), S4-S7.
32. Kerminen, S., Cerioli, N., Pacauskas, D., Havulinna, A. S., Perola, M., Jousilahti, P., ... & Pirinen, M. (2021). Changes in the fine-scale genetic structure of Finland through the 20th century. *PLoS Genetics*, *17*(3), e1009347.
33. Lacy, R. C. (1997). Importance of genetic variation to the viability of mammalian populations. *Journal of mammalogy*, *78*(2), 320-335.
34. Lande, R. (1988). Genetics and demography in biological conservation. *Science*, *241*(4872), 1455-1460.
35. Longdon, B., Brockhurst, M. A., Russell, C. A., Welch, J. J., & Jiggins, F. M. (2014). The evolution and genetics of virus host shifts. *PLoS Pathogens*, *10*(11), e1004395.
36. Lyhne, N., Sjølie, A. K., Kyvik, K. O., & Green, A. (2001). The importance of genes and environment for ocular refraction and its determiners: a population based study among 20–45 year old twins. *British Journal of Ophthalmology*, *85*(12), 1470-1476.
37. Lynch, H. T., Lynch, P. M., Lanspa, S. J., Snyder, C. L., Lynch, J. F., & Boland, C. R. (2009). Review of the Lynch syndrome: history,

molecular genetics, screening, differential diagnosis, and medicolegal ramifications. *Clinical genetics*, *76*(1), 1-18.

38. Maher, E. A., Furnari, F. B., Bachoo, R. M., Rowitch, D. H., Louis, D. N., Cavenee, W. K., & DePinho, R. A. (2001). Malignant glioma: genetics and biology of a grave matter. *Genes & development*, *15*(11), 1311-1333.
39. Maisey, J. G., Naylor, G. J., & Ward, D. J. (2004). Mesozoic elasmobranchs, neoselachian phylogeny and the rise of modern elasmobranch diversity. *Mesozoic fishes*, *3*, 17-56.
40. Mittler, R., & Blumwald, E. (2010). Genetic engineering for modern agriculture: challenges and perspectives. *Annual review of plant biology*, *61*, 443-462.
41. Moossavi, S., & Bishehsari, F. (2019). Microbes: possible link between modern lifestyle transition and the rise of metabolic syndrome. *Obesity Reviews*, *20*(3), 407-419.
42. Mountain, D. C., & Guelke, J. K. (2016). Genetics, genealogy, and geography. *Geography and Genealogy*, 7, 153-174.
43. Mühlemann, B., Vinner, L., Margaryan, A., Wilhelmson, H., de la Fuente Castro, C., Allentoft, M. E., ... & Sikora, M. (2020). Diverse variola virus (smallpox) strains were widespread in northern Europe in the Viking Age. *Science*, *369*(6502), 8977.
44. Muller, H. J. (1927). Artificial transmutation of the gene. *Science*, *66*(1699), 84-87.
45. Nagylaki, T. (2013). *Introduction to theoretical population genetics*. Springer Science & Business Media. (Vol. 21, pp. 246-248).
46. Nguyen, D., & Xu, T. (2008). The expanding role of mouse genetics for understanding human biology and disease. *Disease models & mechanisms*, *1*(1), 56-66.
47. Nicholas, F. W. (2009). *Introduction to veterinary genetics*. John Wiley & Sons. (Vol. 1, pp. 123-125).
48. Nordestgaard, B. G., & Langsted, A. (2016). Lipoprotein (a) as a cause of cardiovascular disease: insights from epidemiology, genetics, and biology. *Journal of lipid research*, *57*(11), 1953-1975.
49. Notter, D. R. (1999). The importance of genetic diversity in livestock populations of the future. *Journal of animal science*, *77*(1), 61-69.
50. Nye, R. A. (1993). The rise and fall of the eugenics empire: recent perspectives on the impact of biomedical thought in modern society. *The Historical Journal*, *36*(3), 687-700.
51. O'Brien, S. J. (1994). A role for molecular genetics in biological

conservation. *Proceedings of the National Academy of Sciences*, *91*(13), 5748-5755.

52. Palstra, F. P., & Ruzzante, D. E. (2008). Genetic estimates of contemporary effective population size: what can they tell us about the importance of genetic stochasticity for wild population persistence?. *Molecular Ecology*, *17*(15), 3428-3447.
53. Pilbeam, D. (1996). Genetic and morphological records of the Hominoidea and hominid origins: a synthesis. *Molecular phylogenetics and evolution*, *5*(1), 155-168.
54. Porteous, D. J., Thomson, P., Brandon, N. J., & Millar, J. K. (2006). The genetics and biology of DISC1—an emerging role in psychosis and cognition. *Biological psychiatry*, *60*(2), 123-131.
55. Rêdowicz, M. J. (2002). Myosins and pathology: genetics and biology. *Acta Biochimica Polonica*, *49*(4), 789-804.
56. Schork, N. J., Fallin, D., & Lanchbury, J. S. (2000). Single nucleotide polymorphisms and the future of genetic epidemiology. *Clinical genetics*, *58*(4), 250-264.
57. Skopek, J. M. (2011). Principles, exemplars, and uses of history in early 20th century genetics. *Studies in History and Philosophy of Science Part C: Studies in History and Philosophy of Biological and Biomedical Sciences*, *42*(2), 210-225.
58. Sturtevant, A. H., & Beadle, G. W. (1939). An introduction to genetics. *An introduction to genetics.41, 7400-7405.*
59. Ursin, G., Lillie, E. O., Lee, E., Cockburn, M., Schork, N. J., Cozen, W., ... & Mack, T. (2009). The relative importance of genetics and environment on mammographic density. *Cancer Epidemiology Biomarkers & Prevention*, *18*(1), 102-112.
60. Waddington, C. H. (2016). *An introduction to modern genetics*. Routledge. Vol. 3, pp. 654-656.
61. Wang, G., Zhao, D., Spring, D. J., & DePinho, R. A. (2018). Genetics and biology of prostate cancer. *Genes & development*, *32*(17-18), 1105-1140.
62. Wilson, E. O. (2005). Kin selection as the key to altruism: its rise and fall. *Social research*, 2(1), 159-166.

CHAPTER 2

DNA STRUCTURE & FUNCTION

Contents

2.1 INTRODUCTION

The complex and sophisticated double helical structure of DNA, characterized by its deceptively straightforward composition but far-reaching ramifications, serves as the fundamental genetic code underlying all living organisms. DNA functions as the universal medium of heredity, governing a wide range of biological processes that characterize living organisms, spanning from tiny microbes to the most intricate life forms on our planet. The present introductory chapter initiates an engaging exploration into the realm of DNA, delving into its structural and functional aspects and the profound insights it offers into the fundamental nature of life (Travers & Muskhelishvili, 2015).

The narrative surrounding the discovery of DNA is characterized by remarkable scientific ingenuity and unwavering perseverance. The origins of this phenomenon can be traced back to the initial observations made by researchers such as Friedrich Miescher during the latter part of the 19th century. Miescher's investigations led to the identification of a perplexing substance located within the cellular nuclei. In 1953, James Watson, along with Francis Crick, made a significant scientific discovery by elucidating the double helix framework of DNA. This milestone achievement was recognized with the prestigious Nobel Prize. This groundbreaking revelation revolutionized our comprehension of genetics and left an indelible impact on the field (Herbert & Rich, 1999).

The fundamental basis of DNA's aesthetic appeal resides in its structural composition. DNA is comprised of four nucleotide bases, namely adenine (A), guanine (G), cytosine (C), and thymine (T). These bases form complementary base pairs, specifically A-T and C-G, which are connected by hydrogen bonds. The intertwining of these nucleotide base pairs forms the well-known double helix structure, wherein the sugar-phosphate backbone functions as a robust support. The distinctive architecture of this structure not just guarantees the integrity of genetic data but also facilitates precise duplication during cellular division and the inheritance of characteristics across successive generations (Wang et al., 1983).

The basic importance of DNA lies in its role as a repository of genetic data. Every deoxyribonucleic acid (DNA) sequence contains the essential instructions required for the construction and sustenance of an organism. Genes, which are distinct segments of DNA, contain the encoded instructions necessary for the synthesis of proteins. These proteins serve as vital molecular entities responsible for carrying out essential cellular functions. The genetic code, which consists of a three-letter code composed of nucleotide bases, governs the arrangement of amino acids that constitute proteins, thereby directing the various functions that support biological existence (Watt & Hickson, 1994).

The role of DNA extends beyond its primary function of genetic information storage, encompassing the crucial responsibility of facilitating its accurate replication. The procedure of replication of DNA is characterized by its exceptional accuracy and precision, as it involves the unwinding of the double helix of DNA and the utilization of each strand to be a template for the synthesis of a strand that is complementary to it. The replication occurs within the cell cycle, a crucial mechanism involved in cell division. This process is vital as it ensures the precise transmission of genetic material to daughter cells, thereby maintaining a high level of fidelity (Spiegel et al., 2020).

The intricate orchestration of cellular activities involved in gene expression, whereby genetic information facilitates protein synthesis, represents a multifaceted process. The initial stage of the process, known as transcription, encompasses the transformation of deoxyribonucleic acid (DNA) into ribonucleic acid (RNA), a molecule consisting of a single strand that serves as a carrier for genetic information. Subsequently, the RNA molecule proceeds to engage in the translation process, wherein it is deciphered by ribosomes to generate the specific arrangement of amino acids, which ultimately constitute proteins. The complex interaction of molecules regulates the extensive range of biological processes necessary for sustaining life (Kool et al., 2000).

The comprehension of the structure and function of DNA has brought about a paradigm shift in medicine. The investigation of gene mutations and their correlation with diseases has unveiled novel avenues in diagnostics, therapies, and genetic counseling. The utilization of genetic testing while sequencing has become essential in the identification of hereditary disorders, the prediction of disease susceptibility, and the customization of individualized treatments, thereby enhancing medical results for numerous individuals (Cheng, 1995).

In addition to its utilization in medicine, DNA testing has also been applied in forensic science. The distinctive and exceptionally personalized characteristics of DNA sequences render it a formidable instrument for individual identification, crime resolution, and establishing biological connections. The utilization of DNA profiling has emerged as a conventional technique in the realm of criminal investigations, facilitating the administration of justice with unparalleled precision (Reece & Maxwell, 1991).

The revelation of DNA's structural composition and its functional properties has concurrently incited a paradigm shift in the field of biotechnology. The application of recombinant DNA technology enables scientists to manipulate DNA by integrating genes from a single organism into another, resulting in the development of genetically modified organisms (GMOs) possessing advantageous characteristics. The field of biotechnology has had a profound impact on various sectors such as agriculture, medicine, along with industry, thereby presenting novel opportunities for tackling worldwide challenges and enhancing the quality of human existence (Waring, 1968).

The exploration of DNA structure and function is a captivating endeavor that delves into the fundamental framework of life. From the time of its initial discovery to its significant contributions in the fields of genetics, medicine, and other related disciplines, DNA has consistently served as a source of inspiration for both scientists and individuals without specialized knowledge. The comprehension of the organization and operation of DNA provides access to the concealed mechanisms of heredity in living organisms, establishing a unifying link among all life forms by means of the universal language of genetics. As we explore the complexities of DNA, we gain access to a realm of scientific advancements, revelations, and possibilities that extend well beyond the boundaries of human imagination. This trajectory propels us towards a future in which the enigmatic properties of DNA persistently unfold, offering new insights and opportunities (Champoux, 2001).

The length of DNA, if fully extended, would span a distance of approximately 10 billion miles, equivalent to multiple round trips between Earth and Pluto.

2.2 HISTORY

The investigation into the structure of DNA has a historical trajectory spanning a century, despite the relatively recent emergence of our comprehension regarding how DNA functions as a carrier of genetic information (Strimmer & Pybus, 2001).

2.2.1 Early Studies of DNA

Johann Friedrich Miescher completed his medical education in Switzerland in the year 1868. Motivated by the influence of his uncle, who held the belief that comprehending diseases necessitated an understanding of tissue chemistry, Miescher embarked on a journey to Tubingen, Germany, to pursue academic studies under the guidance of Ernst Felix Hoppe-Seyler, a prominent figure in the nascent discipline of biochemistry. Under the guidance of Hoppe-Seyler, Miescher directed his focus toward the chemical analysis of pus, an element that holds significant relevance in medicine (Razin & Riggs, 1980). Pus comprises leukocytes characterized by their prominent nuclei, which are isolated through a method devised by Miescher. The limited quantity of nuclear material acquired by the individual was inadequate to conduct a comprehensive chemical analysis. However, the individual did ascertain the presence of a unique substance within the material, characterized by a mildly acidic nature and a notable concentration of phosphorus. The substance composed of protein and DNA, referred to as nuclei, was identified by Miescher. One of his pupils subsequently renamed the substance nucleic acid (Reid, 1987).

By the year 1887, scholars had reached the determination that the nucleus serves as the physical foundation of heredity. The composition of chromatin has been demonstrated to comprise nucleic acid and proteins. However, it remains necessary to ascertain which components are genetic information. During the latter part of the 19th century, Albrecht Kossel conducted additional research on the chemical composition of DNA. His investigations led to the discovery that DNA is comprised of four nitrogenous bases, namely guanine, adenine, cytosine, & thymine, which are commonly referred to as A, C, G, and T, respectively (Pirrung, 2002).

During the first half of the 20th century, Rockefeller University, located in New York City, emerged as a prominent hub for scientific investigations about nucleic acids. Phoebus Aaron Levene became a member of the Institute in the year 1905, where he dedicated the subsequent four decades to the investigation of the chemical properties of DNA. The researcher observed that DNA is composed of numerous interconnected, recurring entities, wherein each entity encompasses a sugar, a mineral called phosphate, and a base, thereby constituting a nucleotide (Caruthers, 1991).

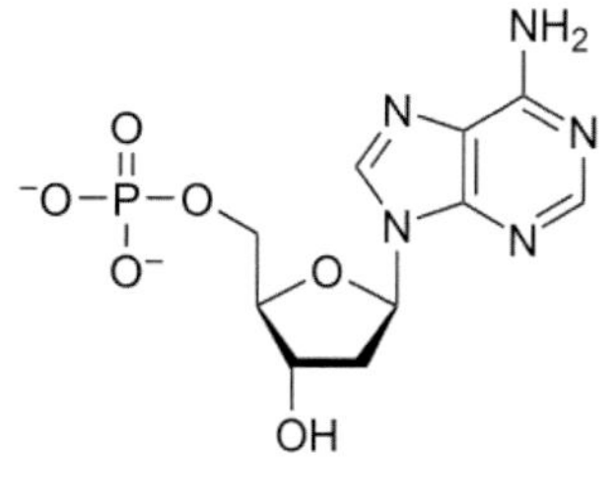

Figure 2.1. Nucleotide (Source: By Hbf878 - Own work, CC0, https://commons.wikimedia.org/w/index.php?curid=74023664, Creative Commons License)

The user's statement posits an erroneous proposition regarding the composition of DNA. Specifically, it suggests that DNA is comprised of a linear arrangement of four-nucleotide units, encompassing each of the four bases—cytosine, adenine, guanine, and thymine—in a particular order. The tetranucleotide theories posited that the regularity of DNA's structure rendered it unsuitable as DNA. The tetranucleotide concept played a significant role in shaping the notion that protein serves as genetic material. This theory posited that the genetic material consists of four nucleotides, and it was observed that proteins, which possess a repertoire of 20 distinct amino acids, exhibit a remarkable degree of structural diversity (Szyf & Bick, 2013).

As further investigations into the chemical properties of DNA were conducted during the years 1940 to 1950, the prevailing perception of DNA as an ordinary and unchanging molecule gradually evolved. Erwin Chargaff and his colleagues conducted meticulous measurements of the quantities of all four bases present in DNA derived from diverse organisms. Their findings revealed significant variations in the base composition of DNA across different organisms. The discovery presented herein has refuted the tetranucleotide theory. It has been found that a consistent pattern exists within each species regarding the ratios of bases. Specifically, the quantity of adenine always matches that of thiamine (A = T), while the quantity of guanine always matches that of cytosine (G = C). The discoveries above came to be recognized as Chargaff's principles (Sanger, 1981).

2.3 DNA AS THE SOURCE OF GENETIC INFORMATION

During the period when chemists were elucidating the structure of DNA, biologists were engaged in efforts to ascertain the origin of genetic information. Two separate sets of experiments were conducted, one involving bacteria and one involving viruses, which yielded crucial evidence supporting that DNA, instead of protein, served as the genetic material (Leu et al., 2011).

2.3.1 The discovery of the transforming principle

The initial indication that DNA served as the medium for transmitting hereditary information emerged through the substantiation of its involvement in a biological occurrence known as transformation. The observation of the phenomenon can be attributed to Fred Griffith, an English physician with a particular focus on Streptococcus pneumonia, the bacterium responsible for pneumonia, who first documented it in 1928. Griffith successfully isolated various strains of *S. pneumonia*, including type I, II, III, and others. In the pathogenic variants of a particular strain, individual bacteria are enveloped by a polysaccharide layer, resulting in a visually uniform appearance of the bacterial colony when cultivated on an agar plate. These variants are commonly denoted as S, representing their smooth characteristic. Griffith's research revealed that certain virulent strains underwent sporadic mutations, resulting in the emergence of nonvirulent variants characterized by the absence of a polymers coat and the formation of rough-textured colonies on agar plates. These nonvirulent strains are commonly denoted as R strains (Watson et al., 1993).

Griffith's research focused on elucidating the origins of various varieties of S. respiratory infections and investigating the factors contributing to the virulence of certain types while others remained non-virulent. The researcher observed that introducing a small quantity of live type IIIS bacteria into mice resulted in pneumonia and subsequent mortality. Upon conducting autopsies, the researcher discovered significant quantities of type IIIS microbes within the bloodstream of the mice (Fig. 2.2a). Upon injecting type IIR microbes into mice, Griffith observed that the mice exhibited survival, and subsequent analysis revealed the absence of bacteria in the blood of the animals (Figure 2.2b). Griffith possessed the knowledge that the process of boiling effectively eradicated all bacteria and rendered them non-virulent. Griffith observed that the mice survived the injection by administering substantial quantities of heat-killed form IIIS microbes to mice. The subsequent analysis revealed the absence of type IIIS microbes in their bloodstream (Figure 2.2c) (Wyatt, 1975).

The outcomes of these studies were frequently observed. Nevertheless, Griffith was taken aback by the results when he administered a minute quantity of viable type IIR bacteria, in conjunction with a substantial quantity of heat-inactivated type IIIS bacteria, to the mice. Due to the nonvirulent nature of both the kind IIR bacteria and the heat-killed kind IIIS bacteria, it was anticipated that the mice would survive. Remarkably, within 5 days after administering injections, the mice exhibited susceptibility to pneumonia infection, ultimately leading to their demise (Figure 2.2d). Upon analysis of the blood samples extracted from the coronary arteries of the mice, as mentioned above, Griffith observed viable type IIIS bacteria. Moreover, it was observed that these bacteria maintained their type IIIS traits across multiple generations, indicating that the infectivity trait could be passed down through heredity (Lederberg, 1994).

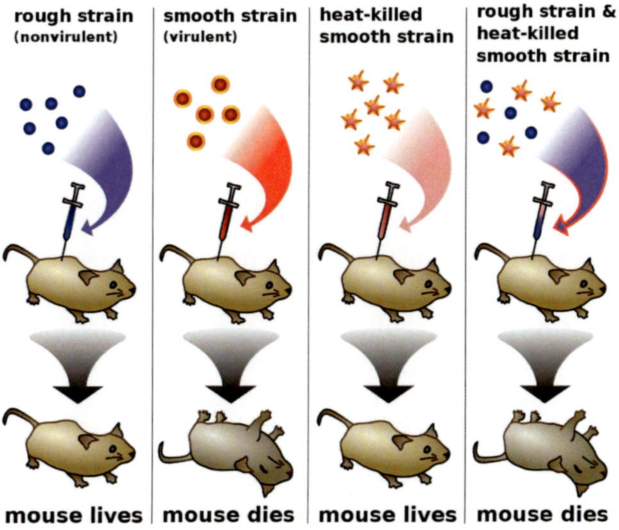

Figure 2.2. Griffith's experiments demonstrated transformation in bacteria (Source: By No machine-readable author provided. Madprime assumed (based on copy-

right claims). - No machine-readable source provided. Own work assumed (based on copyright claims)., CC BY-SA 3.0, https://commons.wikimedia.org/w/index.php?curid=1819839, Creative Commons License).

Griffith's findings encompassed multiple potential theories, all of which were duly taken into account by the researcher. Initially, it is plausible that he may have failed to adequately sterilize the type IIIS bacteria, resulting in a small number of viable bacteria within the culture. If live bacteria were introduced into the mice, they would have undergone replication and subsequently induced pneumonia. Griffith was aware of the unlikelihood of this possibility, as he had exclusively employed heat-killed type IIIS microbes within the control experiment, which had never elicited pneumonia in the mice (Eichmann & Krause, 2013).

An alternative perspective posited that the living type IIR bacteria underwent a mutation, resulting in the acquisition of the deadly S form. The mutation above would result in the manifestation of pneumonia in the mice; however, it would generate type IIS bacteria, as opposed to the type IIIS bacteria discovered by Griffith in the deceased mice. The transition of type II bacteria to type III bacteria would necessitate numerous mutations, and the probability of these mutations occurring concurrently was exceedingly remote (Dahm, 2005).

Griffith ultimately concluded that the type IIR bacteria transformed, acquiring the genetic makeup and virulence of the deceased type IIIS bacteria. The alteration, as mentioned earlier, resulted in a lasting, hereditary modification in the bacteria. While Griffith lacked comprehension regarding the mechanism of this transformation, he postulated that a certain component within the polysaccharide layer of the deceased bacteria could potentially be accountable. The substance in question was referred to as the transforming principle by the individual (Olby, 1985).

2.3.2 Identification of the transforming principle

During the period when Griffith's report was published, Oswald Avery held the position of a microbiologist within the Rockefeller Institute. Initially, Avery harbored skepticism; however, after the successful replication of Griffith's experiments by other microbiologists utilizing alternative bacteria, thereby confirming the occurrence of transformation, Avery embarked upon elucidating the characteristics of the transformative entity (Atchley, 1982).

Following an extensive decade-long research endeavor, Avery, Colin MacLeod, and Maclyn McCarty achieved the isolation and purification of the transformative substance. The researchers demonstrated that the substance possessed a chemical composition that closely resembled that of DNA while exhibiting notable distinctions from the chemical composition of proteins.

Enzymes, including trypsin and chymotrypsin, recognized for their ability to hydrolyze proteins, exhibited no discernible impact on the transforming material. The enzyme ribonuclease, which can degrade RNA molecules, exhibited no discernible impact. Enzymes with the ability to degrade DNA, however, effectively nullify the biological functionality of the substance responsible for transformation (Figure 2.3) (Herriott, 1971).

The researchers Avery, MacLeod, and McCarty demonstrated that the purified transforming substance exhibited a precipitation rate similar to that of purified DNA and absorbed ultraviolet light on the same wavelengths as DNA. The findings, published in 1944, presented persuasive evidence supporting the notion that DNA is the transforming principle and thus harbors genetic information. Despite the persistence of certain biologists, there remains a prevailing preference for the hypothesis positing protein as the genetic material (Anami et al., 2013).

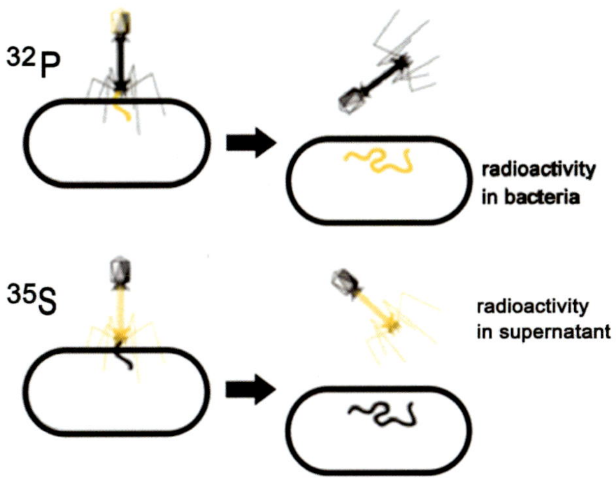

Figure 2.3. Avery, Macleod & McCarty's experiment revealed the nature of transforming principle (Source: https://bio.libretexts.org/Courses/City_College_of_San_Francisco/Introduction_to_Genetics/01%3A_Overview_DNA_and_Genes/1.01%3A_DNA_is_the_Genetic_Material, Creative Commons License)

2.3.3 The Hershey-Chase experiment

Another piece of evidence supporting the notion that DNA is the genetic material emerged from a research investigation on the T2 virus by Alfred Hershey and Martha Chase. T2 is a bacteriophage, also known as a phage, which is capable of infecting the bacteria *Escherichia coli* (Fig. 2.4a). The replication process of a phage involves the attachment of the phage to the external membrane of a bacterial cell, followed by the injection of its genetic material into the cell. Within the cell, the phage DNA undergoes replication and subsequently instructs the cell to generate the phage protein. The DNA of the phage becomes enclosed by proteins, resulting in the generation of offspring phages that cause lysis, or the rupture, of the cell and subsequent release (as depicted in Figure 2.5b) (Dawson, 2003). During the period in which the Hershey-Chase study was conducted, which culminated in the publication of their paper in 1952, there existed a lack of comprehensive understanding among biologists regarding the precise mechanisms underlying phage replication. The available knowledge indicated that the genome of the T2 phage is composed of roughly equal proportions of protein and nucleic acid. Additionally, it was understood that the initial step of phage infection involves attachment to the cell's wall and that subsequent production of progeny phages occurs within the cell. The transmission of genetic information from the affecting phage to the progeny is necessary for the progeny to exhibit the same attributes as the infecting phage. However, the mechanism by which this transmission occurs remains unidentified (Van Valen et al., 2012).

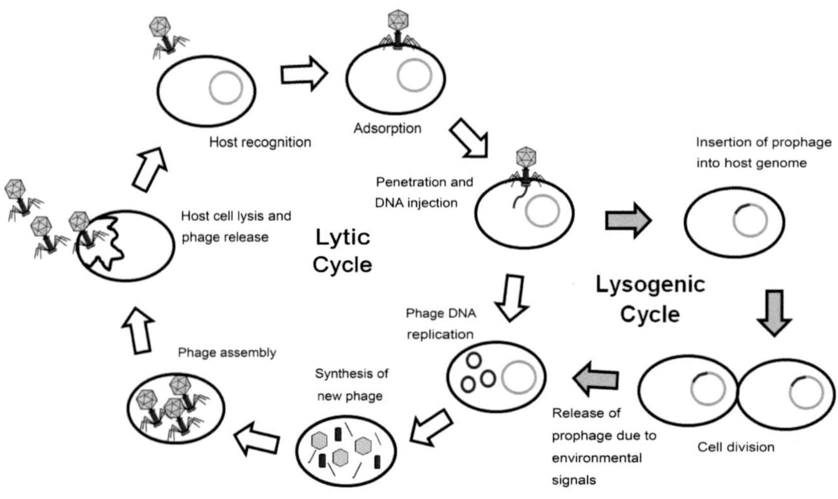

Figure 2.4. T2 is a bacteriophage that infects E.coli. Its life cycle (Source: Doss J, Culbertson K, Hahn D, Camacho J, Barekzi N. A Review of Phage Therapy against Bacterial Pathogens of Aquatic and Terrestrial Organisms. *Viruses*. 2017; 9(3):50. https://doi.org/10.3390/v9030050, Creative Commons License)

Hershey and Chase conducted a series of experiments to ascertain the mode of transmission, either the phage amino acids or the phage DNA, during phage reproduction. To track the trajectory of protein along with DNA, researchers employed radioactive isotopes of phosphorus and sulfur. Using a radioactive isotope as a tracer enables the identification of the precise location of a particular molecule, as the presence of the isotope within any molecule renders it radioactive, facilitating its straightforward detection. DNA is composed of phosphorus but does not contain sulfur. Therefore, Hershey and Chase employed 32P to track the replication of phage DNA. Proteins are known to contain sulfur but not phosphorus. Therefore, researchers utilized the isotope 35S to track the movement and behavior of proteins (Wu, 2010).

Initially, Hershey and Chase cultivated *Escherichia coli* in a culture medium supplemented with 32P. Subsequently, they proceeded to infect the bacterial cells with T2 bacteriophages, ensuring that the resulting phages possessed DNA that was distinctly labeled with 32P (refer to Figure 2.5). Subsequent cultivation of *Escherichia coli* was conducted in a medium supplemented with 35S, followed by introducing T2 bacteriophages to infect these bacterial cells. This ensured that all newly produced phages possessed protein components specifically labeled with 35S. Subsequently, Hershey and Chase proceeded to introduce distinct sets of unmarked *Escherichia coli* samples to the 35S- along with 32P-labeled bacteriophages. Following a suitable incubation period for phage infection, the cells of *E. coli* were subjected to mechanical disruption using a blender, resulting in the removal of the protein coats (referred to as "ghosts") previously occupied by the phages while leaving the cell walls intact. The protein coats were isolated, and the contaminating bacterial cells were cultured. Ultimately, the cellular membranes ruptured, leading to the emergence of novel phage particles (Hartman, 1979).

Upon infection of the bacteria by phages labeled with 35S, a significant portion of the radioactivity was observed to be associated with the protein ghosts. At the same time, only a small fraction was retained within the cells. Moreover, it was observed that upon the emergence of the cell, the newly formed phages exhibited a negligible amount of radioactivity, as depicted in Figure 2.5. The findings of this study suggest that while the protein part of a phage is essential for the process of infection, it does not penetrate the cell and is not inherited by subsequent generations of phages (Hodgkin, 2001).

In contrast, in the experiment conducted by Hershey and Chase, bacteria were infected with phages labeled with 32P. Subsequently, the protein remnants were eliminated, yet the bacteria retained their radioactivity. One notable observation was that following the lysis of cells and the subsequent emergence of new progeny phages, many of these phages exhibited radioactivity originating from 32P. This finding served as evidence that the DNA derived from the affecting phages had been transmitted to the offspring phages (refer to Figure 2.5). The findings of this study provide confirmation that the genetic makeup of phages is DNA rather than protein (Stahl, 1998).

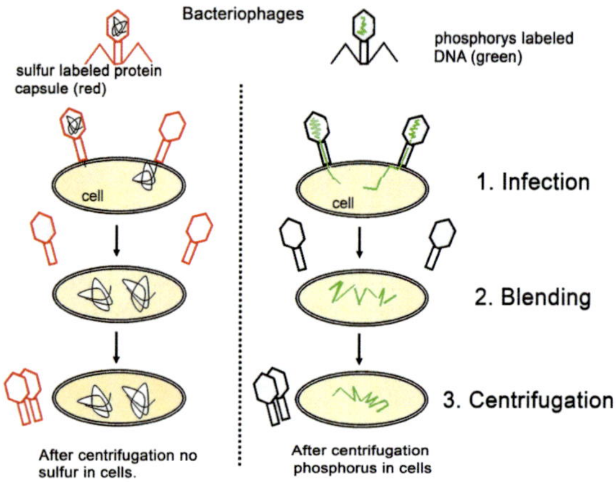

Figure 2.5. Hershey & Chase demonstrated that DNA carries the genetic information in bacteriophages (Source: By Thomasione - Modified from German Wikipedia [1], GFDL, https://commons.wikimedia.org/w/index.php?curid=15034881, Creative Commons License)

Did You Know?

A mere 1.5% of the human DNA is responsible for encoding proteins, while the remaining portion assumes crucial functions in the regulation of genes.

2.4 WATSON & CRICK'S DISCOVERY OF THE THREE-DIMENSIONAL STRUCTURE OF DNA

The investigations about the characteristics of genetic material laid the foundation for a significant breakthrough in biology, namely the elucidation of the three-dimensional configuration of DNA through James Watson, along with Francis Crick in 1953 (Stevens, 2003).

Watson's doctoral research focused on studying bacteriophage, a virus that infects bacteria. As a result of his academic pursuits, he possessed knowledge of Avery's scientific contributions, which enabled him to comprehend the significant role of DNA in the field of genetics. After completing his doctoral degree, Watson joined the esteemed Cavendish Laboratory at Cambridge University in England. This institution housed a cohort of esteemed researchers who were actively engaged in investigating the intricate three-dimensional configuration of macromolecules. One of the researchers involved in this endeavor was Francis Crick, and she was concurrently pursuing his doctoral degree. Watson and Crick swiftly established a personal and professional relationship (Poltev et al., 2020).

The foundational principles of DNA chemistry were previously elucidated by Miescher, Kossel, Levene, and other researchers. Their work established that DNA is composed of nucleotides, each of which comprises a sugar, base, and phosphate group. Nevertheless, the precise manner in which the nucleotides interlock within the three-dimensional configuration of the molecules remained considerably ambiguous (Gendron et al., 2001).

William Ashbury initiated a study in 1947 to investigate the three-dimensional arrangement of DNA through the utilization of X-ray diffraction, as depicted in Figure 2.6. However, his diffraction images must possess a higher level of resolution to unveil the underlying structure. The investigation conducted by a research team at King's College in London, under the guidance of Maurice Wilkins and Rosalind Franklin, involved the examination of DNA's structure through X-ray diffraction techniques. This approach yielded notably enhanced visual representations of the molecule. However, the collaborative efforts of Wilkins and Franklin were ultimately hindered by interpersonal conflicts, preventing them from achieving a comprehensive understanding of the molecular structure (Pan et al., 2014).

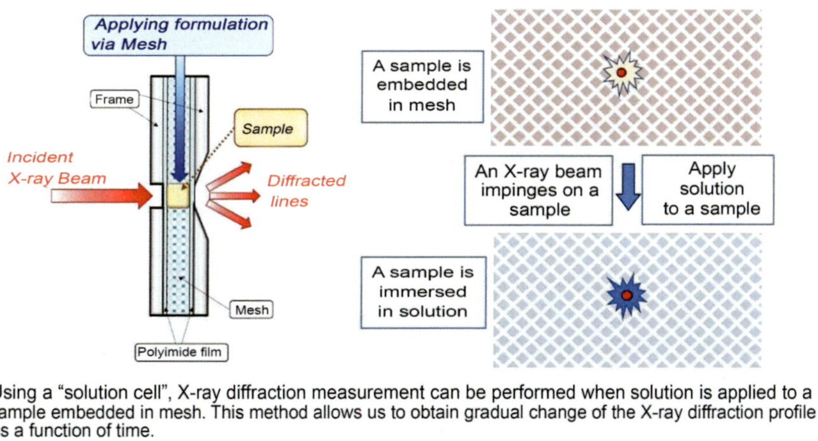

Figure 2.6. X-ray diffraction provides information about the structure of molecules (Source: Hatta I. Stratum Corneum Structure and Function Studied by X-ray Diffraction. Dermato. 2022; 2(3):79-108. https://doi.org/10.3390/dermato2030009, Creative Commons License)

Watson and Crick investigated the makeup of DNA, employing a methodology that did not involve the collection of novel data. Instead, they utilized existing information about the chemical properties of DNA to construct model molecules (see Figure 2.7). Through the application of principles derived from structural chemistry, researchers successfully constrained the potential configurations that DNA could potentially adopt. Watson and Crick conducted empirical investigations by constructing physical models composed of wire and metal plates. By utilizing their models, the researchers could assess the compatibility of a given structure with both chemical principles and X-ray images (Sabri et al., 2020).

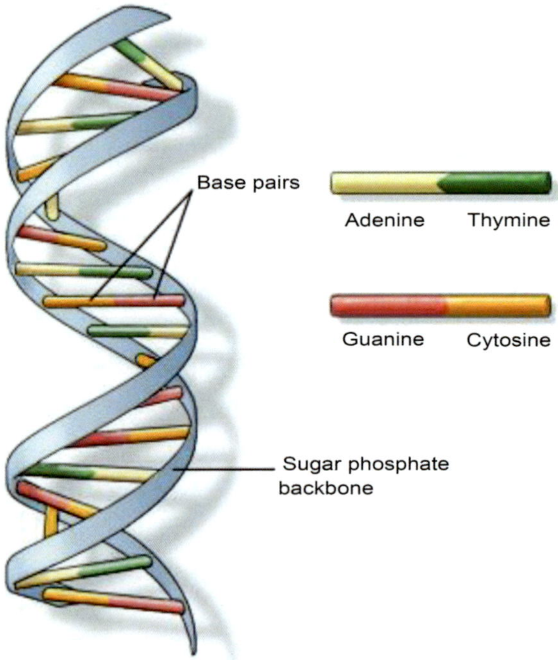

Figure 2.7. Watson & Crick Model of DNA (Source: Badawi HF, Laamarti F, El Saddik A. Devising Digital Twins DNA Paradigm for Modeling ISO-Based City Services. Sensors. 2021; 21(4):1047. https://doi.org/10.3390/s21041047, Creative Commons License)

The breakthrough in deciphering the structure occurred when Watson discerned the potential for adenine to form a bond with thymine and guanine to form a bond with cytosine. This observation provided an explanation of the base ratios previously identified by Chargaff. The model proposed by Watson and Crick elucidated the structural composition of DNA, revealing that it comprises two intertwined strands of nucleotides arranged in a right-handed helical structure. The outer portion of the helix is composed of sugars and phosphates, while the bases occupy the interior. In 1953, a captivating depiction of their model was published in the esteemed scientific journal Nature. Simultaneously,

Wilkins and Franklin conducted the publication of their X-ray diffraction info, thereby providing experimental evidence in support of the theory proposing the helical nature of DNA (Ke, 2014).

The resolution of DNA's structure has been widely regarded as the most significant biological breakthrough of the twentieth century. The Nobel Prize in 1962 was awarded to Watson, Crick, and Wilkins in recognition of their significant contribution to the field of science (Klug, 2004). Rosalind Franklin's eligibility for the discussed prize was precluded due to her unfortunate demise from cancer in 1957.

Did You Know?

Deoxyribonucleic acid (DNA) exhibits a remarkable capacity to store an impressive amount of information, estimated at approximately 215 million gigabytes per gram. This characteristic renders DNA a highly promising medium for data storage purposes.

2.5 RNA AS GENETIC MATERIAL

DNA typically carries genetic information in the majority of organisms. Nevertheless, a small number of viruses employ RNA instead of DNA as their genetic code. In 1956, Heinz Fraenkel-Conrat, along with Bea Singer, conducted an experiment to validate this particular observation. Their study focused on tobacco mosaic virus (TMV), a pathogenic virus that targets tobacco plants, leading to infection and subsequent disease manifestation. The virus that causes tobacco mosaics is characterized by a single RNA molecule enveloped by a cylindrical structure composed of protein molecules arranged in a helical manner. The experimental findings of Fraenkel-Conrat demonstrated that by isolating the RNA alongside protein components of TMV, subsequent recombination of these constituents resulted in the production of fully functional viral particles capable of causing infection (Alejska et al., 2001). Singer and Fraenkel-Conrat proceeded to generate hybrid viruses by combining RNA and protein components derived from distinct varieties of TMV (Fig. 2.8). When these hybrid viruses infected tobacco leaves, it resulted in the production of new viral particles. The newly generated viral offspring displayed an identical genetic makeup to the original strain that produced the RNA extracted and did not manifest the traits of the pressure that contributed to the protein. The findings of this study suggest RNA carries the genetic data in TMV (Hill & Hillova, 1974).

In 1956, Alfred Gierer and Gerhard Schramm conducted an experiment wherein they established that RNA derived from TMV possesses the capability to infect tobacco plants and facilitate the synthesis of novel TMV particles. This finding served as conclusive evidence supporting the notion that RNA serves as a carrier of genetic instructions (Nelson et al., 2000).

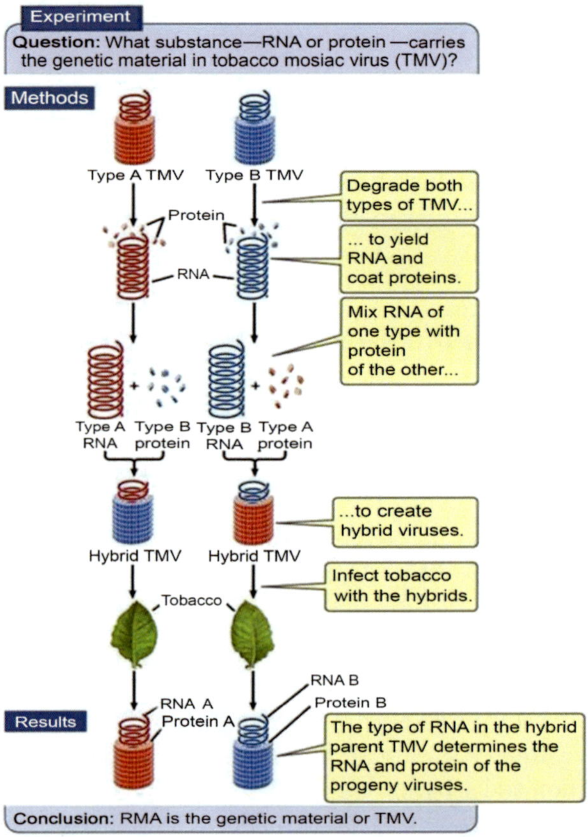

Figure 2.8. Fraenkal-Conrat and Singer's experiment demonstrated that, in the tobacco mosaic virus, RNA carries the genetic information (Source: A. Barrington, Creative Commons License)

Fun fact:

It is noteworthy that humans exhibit a 50% genetic similarity with bananas, thereby underscoring the existence of evolutionary links between the two species.

2.6 THE STRUCTURE OF DNA

The structure of DNA, despite its relative simplicity, possesses a remarkable elegance and aesthetic appeal that surpasses that of other macromolecules. The consideration of DNA structure encompasses three distinct levels that increase complexity, namely the primary, secondary, along with tertiary components of DNA. The main structure of DNA pertains to its nucleotide composition and the manner in which the nucleotides are interconnected. The term "secondary structure" pertains to the stable three-dimensional arrangement of DNA, specifically the helical structure elucidated by Watson and Crick (Johnson et al., 2008).

2.6.1 The Primary Structure Of DNA

The fundamental composition of DNA is comprised of a sequence of nucleotides that are interconnected through phosphodiester linkages. DNA nucleotides are commonly found in the form of a macromolecule, characterized by their considerable length. As an illustration, it is worth noting that each human chromosome contains a single DNA molecule, which, when fully extended, would measure several centimeters in magnitude. Despite its considerable magnitude, DNA possesses a comparatively uncomplicated configuration: it functions as a polymer, a sequence comprised of numerous recurring components interconnected. As previously stated, the recurring components of DNA consist of nucleotides, which are composed of three fundamental constituents: (1) a type of sugar, (2) an amino acid group, and (3) a nitrogenous base (Lindahl, 1993).

The pentose sugars found in nucleic acids are characterized by having five carbon atoms, which are designated as 1', 2', 3', and so on (Figure 2.9). The pentagonal ring links four carbon atoms with one oxygen atom, while the fifth carbon atom (designated as 5') extends vertically from the ring structure. Each carbon atom is attached to hydrogen or hydroxyl group (OH) atoms. The structural composition of the sugars in DNA and RNA exhibits slight variations. RNA possesses a hydroxyl group connected to the 2-carbon atom of its ribose sugar. In contrast, DNA's sugar, deoxyribose, lacks this hydroxyl group and instead has the hydrogen atom at the same position. Consequently, DNA contains one less oxygen atom in its overall structure compared to RNA. The distinction between these two molecules leads to the nomenclature of ribonucleic acid (RNA) and deoxyribonucleic acid (DNA). The cellular enzymes communicating with DNA or RNA can distinguish between these nucleic acids due to a subtle chemical distinction, resulting in distinct functions for each. Moreover, the presence of an additional oxygen atom within the RNA nucleotide renders it more susceptible to chemical reactivity and less thermodynamically stable compared to DNA. Due to this rationale, DNA is more aptly suited to function as the stable repository of genetic data (Li, 1997).

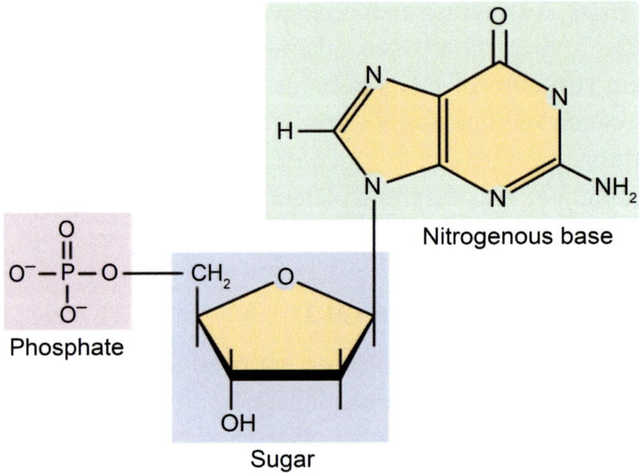

Figure 2.9 A nucleotide contains either a ribose sugar (in RNA) or a deoxyribose sugar (in DNA) (Source: Https://bio.libretexts.org/Learning_Objects/Laboratory_Experiments/General_Biology_Labs/Biology_I_Laboratory_Manual/Module_11%3A_DNA_Structure_and_Function, Creative Commons License)

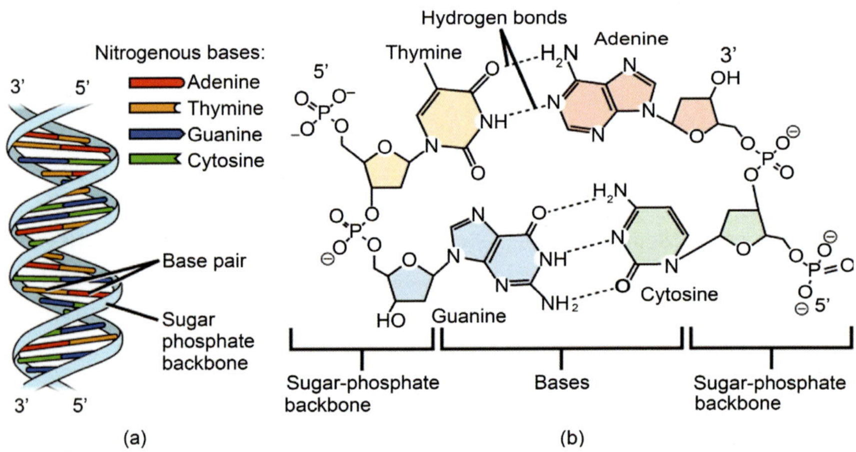

Figure 2.10. A nucleotide contains either a purine or a pyrimidine base (Source: Https://bio.libretexts.org/Learning_Objects/Laboratory_Experiments/General_Biology_Labs/Biology_I_Laboratory_Manual/Module_11%3A_DNA_Structure_and_Function, Creative Commons License)

The nitrogenous base is the second constituent of a nucleotide, and it can exist in two forms: purine or pyrimidine (Figure 2.10). Purines are composed of a hexagonal ring fused with a pentagonal ring, while pyrimidines consist solely of a hexagonal ring. Both DNA and RNA consist of two purines, namely adenine (A) and guanine (G). These purines differ in the arrangement between their dual bonds and the bonds that are attached to the six-membered ring. Nucleic acids encompass a trio of pyrimidines, namely thymine (T), cytosine (C), and uracil (U).

Cytosine is observed in both deoxyribonucleic acid (DNA) and ribonucleic acid (RNA); nevertheless, thymine is exclusively confined to DNA, while uracil is exclusively encountered in RNA. The three pyrimidines exhibit variations in the composition of groups or atoms that are bonded with the carbon atoms within the ring structure and in the number of double bonds present in the ring. The covalent bond between the nitrogenous base and the 1'-carbon atom within the sugar is a consistent feature in nucleotides (see Figure 2.9). The combination of a substance (or ribose), sugar, and a base is commonly known as a nucleoside (Grosberg et al., 1993).

A phosphate group, which comprises an atom of phosphorus bonded to the four oxygen atoms (Fig. 2.11), is the third constituent of a nucleotide. Phosphate groups are present in all nucleotides and commonly possess a negative charge, thereby contributing to the acidic nature of DNA. In a nucleotide, the phosphate group is consistently attached to the carbon atom located at the 5' position of the sugar molecule (Joyce et al., 1982).

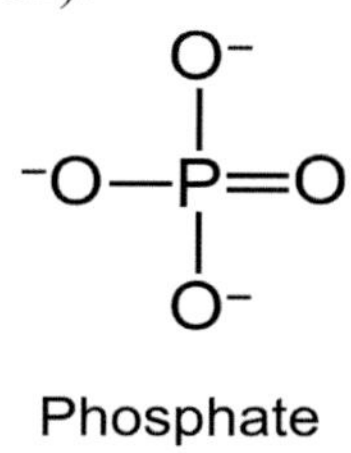

Figure 2.11. A nucleotide contains a phosphate group (Source: McCarty, Creative Commons License)

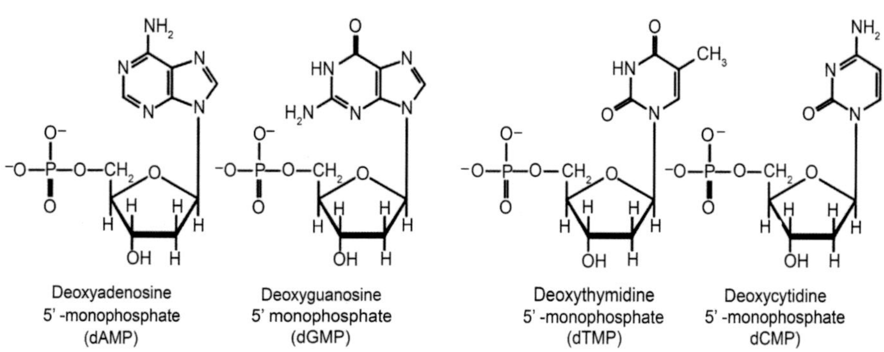

Figure 2.12. Four types of DNA nucleotides (Source: Roberto, Creative Commons License)

The DNA nucleotides are commonly referred to as deoxyribonucleotides or deoxyribonucleoside 5'-monophosphateseither in scientific literature. The presence of four distinct types of bases in DNA results in the existence of four distinct DNA nucleotides (Fig. 2.12). The RNA nucleotides that possess the same function as their DNA counterparts are referred to as RNA molecules or ribonucleoside 5'-monophosphates. RNA molecules occasionally possess uncommon bases, and they have modified variants of the four prevalent bases (Udvardy & Schedl, 1991).

Polynucleotide strands refer to chains of nucleotides composed of repeating units of sugar-phosphate backbones and nitrogen. Deoxyribonucleic acid (DNA) consists of a series of nucleotides linked together by covalent bonds. These bonds connect the 5'-phosphate group of a single nucleotide to the 3'-carbon atom of the subsequent nucleotide (see Figure 2.13). The covalent bonds known as phosphodiester linkages are robust in nature. When a sequence of nucleotides is connected through these bonds, it forms a polynucleotide strand.

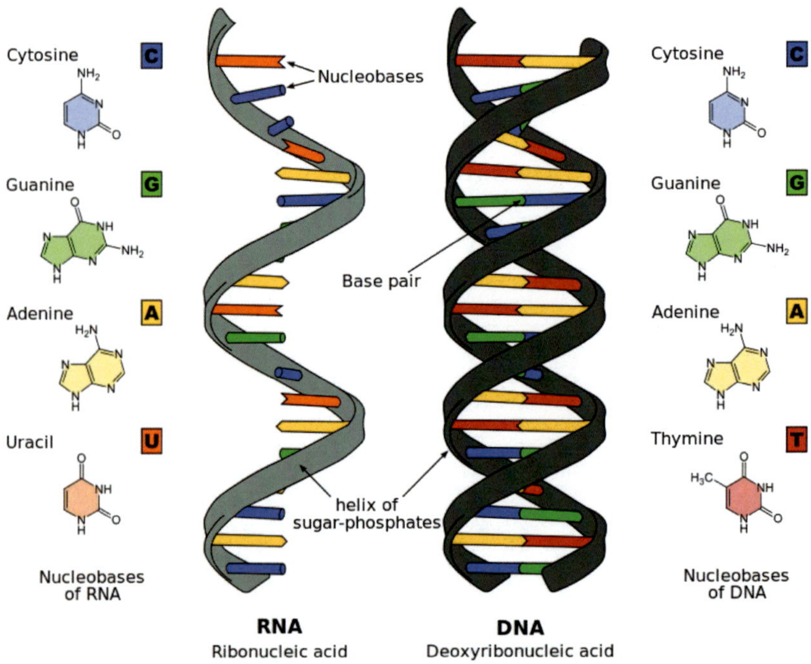

Figure 2.13. DNA consists of two polynucleotide chains that are antiparallel and complementary, and RNA consists of a single nucleotide chain (Source: By File:Difference DNA RNA-DE.svg: Sponk / *translation: Sponk - Chemical structures of nucleobases by Roland1952, CC BY-SA 3.0, https://commons.wikimedia.org/w/index.php?curid=9810855, Creative Commons License).

The nucleotide strand comprises a repetitive arrangement of sugars and phosphates, forming its backbone. The bases, on the other hand, extend outward from the elongated axis of the strand. The neutralization of the negative charges on phosphate groups is commonly achieved through interacting with positively charged entities such as proteins, metals, and other molecules (Nakano et al., 2004).

The directionality, or polarity, is a significant characteristic of the polynucleotide. At the terminal extremity of the nucleotide strand, a group of phosphates is exclusively bonded to the carbon atom located at the 5' position of the sugar moiety. Hence, the terminal portion of the strand is commonly denoted as the 5' end. The opposite terminus of the nucleic acid chain, commonly known as the 3' end, is characterized by the presence of a hydroxyl (OH) group that is covalently bonded to the carbon atom located at the 3' position of the sugar moiety (Richmond & Davey, 2003).

The polynucleotide strands are formed by the connection of RNA nucleotides through phosphodiester linkages, as depicted in Figure 2.13 (Tao et al., 1989).

2.6.2 Secondary Structure Of DNA

The second structure of genetic material pertains to its three-dimensional conformation, specifically its fundamental helical arrangement. The secondary structure of DNA can exhibit diverse configurations, contingent upon its base pattern and the environmental conditions under which it is situated (Bram, 1971).

The concept of the double helix One essential attribute of the secondary structure of DNA is its composition of two polynucleotide strings that are intertwined, forming a double helix. The sugar-phosphate linkages are positioned on the exterior of the helix, while the bases are arranged in a stacked manner within the molecule (refer to Figure 2.13). The two nucleotide strands exhibit an antiparallel orientation, running in opposite directions. This antiparallel nature is characterized by the 5' end of one strand being positioned opposite to the 3' end of the second strand. Two distinct types of molecular forces facilitate the cohesion between the strands. Hydrogen bonds join the bases on opposite strands (Figure 2.13). The strength of these bonds is comparatively lower when compared to the covalent bonds of phosphodiester that link both phosphate and sugar groups of adjacent nucleotides. As demonstrated, separating the two nucleotide strands of DNA is essential for various significant functions. This division can be easily achieved due to the relatively effortless disruption and reformation of hydrogen bonds (McMurray, 1999).

The characteristics of the bond formed by hydrogen introduce a constraint on the potential bases capable of forming pairs. In the context of DNA base pairing, it is typically observed that adenine forms exclusive hydrogen bonds with thymine, facilitated by two hydrogen bonds. Similarly, cytosine pairs exclusively with guanine, forming three hydrogen bonds (Figure 2.13). The strength of the C-G pairing is greater than that of the A-T pairing due to the formation of three hydrogen bonds between C and G, compared to the formation of only two bonds of hydrogen between A and T. The principle of base pairing dictates that the presence of an adenine (A) on one DNA strand necessitates the presence of a thymine (T) at the appropriate place on the complementary strand. Similarly, the presence of a guanine (G) on one strand mandates the presence of a cytosine (C) on the other strand. The two polynucleotide strands comprising the molecules of DNA are inherently non-identical, yet they exhibit a complementary relationship (Yonezawa et al., 1992).

The second mechanism responsible for the bonding of both DNA strands involves the interaction occurring among the stacked pair of bases. The stacking interactions observed in DNA molecules play a significant role in their stability, as they do not necessitate a specific base sequence. Hence, the nucleotide sequence for the DNA molecule can exhibit variability, thereby enabling DNA to serve as a carrier of genetic information. Consequently, the two strands of the DNA molecules are not identical but exhibit a complementary nature (Bochman et al., 2012). Various secondary structures as observed, DNA typically comprises two polynucleotide strands which are antiparallel and complementary, with the exception of single stranded DNA molecules found in certain viruses. The molecular conformation of DNA can exhibit variability in its three-dimensional structure, contingent upon the environmental conditions in which it is situated and, in certain instances, the specific arrangement of its base sequence (Ronaghi et al., 1999).

The structure of DNA described by Watson and Crick, which exists in three dimensions, is commonly referred to as the B-DNA structure (Figure 2.14). This configuration is observed when an ample amount of water surrounds the molecule and there isn't an atypical base sequence within the DNA. These conditions are anticipated to be prevalent in cellular environments. Under physiological conditions, the B-DNA structure is widely regarded as the most stable setting for an arbitrary arrangement of nucleotides. The prevailing evidence strongly supports its predominance within the cellular environment (Hud et al., 1994).

B-DNA exhibits an alpha-helical structure characterized by a right-handed, or counterclockwise, spiral orientation. The DNA molecule exhibits a characteristic of having an estimated 10 base pairings (bp) per 360-degree rotation of its double helix. Consequently, each base pair undergoes a torsional twist of approximately 36 degrees in relation to the neighboring bases, as depicted in Figure 2.14. The distance between base pairs is measured to be

0.34 nanometers (nm), resulting in a total span of 3.4 nm for each full rotation of the molecule. The helix of the DNA molecules has a diameter of 2 nm, with the bases positioned perpendicular to its long axis (Andera et al., 1994). The utilization of a space-filling model effectively demonstrates that B-DNA possesses a comparatively slender and elongated configuration, as depicted in Figure 2.14. The coiling of the nucleic acid strands within the DNA helix results in the formation of distinct grooves, known as major and minor grooves. These structural features play a crucial role in facilitating the interaction between DNA-binding proteins and the DNA molecule, thereby influencing the regulation of genetic information expression. Table 2.1 presents a comprehensive overview of the distinctive attributes associated with the B-DNA structure, as well as the characteristics exhibited by alternative secondary structures that manifest under specific conditions or in the presence of atypical base sequences (Baase & Johnson, 1979).

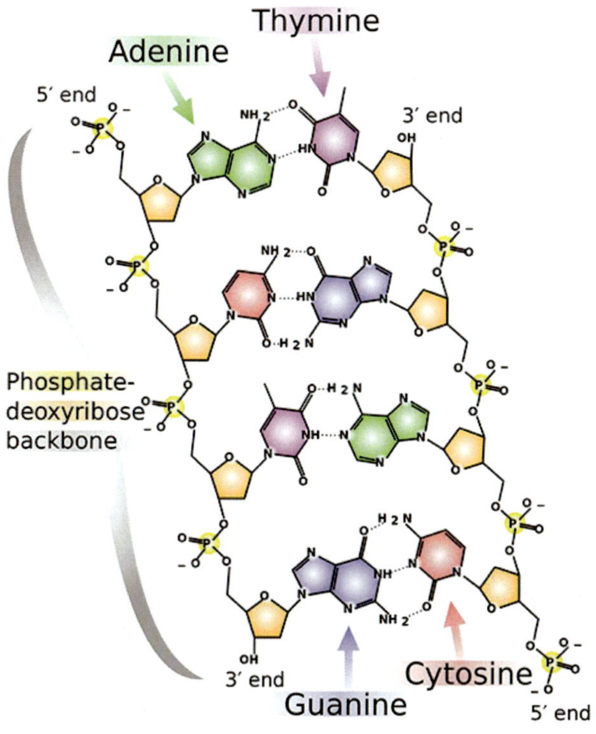

Figure 2.14. Watson & Wricks model of double helical DNA (Source: Corona-Ruiz M, Hernandez-Cabrera F, Cantú-González JR, González-Amezcua O and Javier Almaguer F (2019) A Stochastic Phylogenetic Algorithm for Mitochondrial DNA Analysis. Front. Genet. 10:66. doi: 10.3389/fgene.2019.00066, Creative Commons License)

The A-DNA structure is an additional secondary conformation that DNA can adopt, particularly in conditions of reduced hydration. Similar to B-DNA, A-DNA exhibits an alpha (right-handed) helical structure (Figure 2.15a). However, A-DNA is characterized by its shorter and wider conformation compared to B-DNA (Figure 2.15b), with its bases inclined away from the central axis of the molecule. Limited evidence suggests that the existence of A-DNA under physiological conditions is scarce (Higgs, 2000).

Table 2.1. Characteristics Of DNA secondary structure (Source: Handt, Creative Commons License)

Characteristics	**A-DNA**	**B-DNA**	**Z-DNA**
Conditions required to produce	75% H2O	92% H2O	Alternating purine structure
Helix Direction	Right-handed	Right-Handed	Left-Handed
Average base pairs per turn	11	10	12
Rotation per base pair	32.7º	36º–30º	
Distance between adjacent bases	0.26nm	0.34nm	0.37nm
Diameter	2.3nm	1.9nm	1.8nm
Overall Shape	Short & Wide	Long & Narrow	Elongated & Narrow

Z-DNA, depicted in Figure 2.15c, is characterized by a distinct secondary structure that exhibits a left-handed helical conformation. The sugar-phosphate backbones in this configuration exhibit a zigzag pattern, thereby leading to the nomenclature Z-DNA (denoting zigzag). Z-DNA structures can be formed under physiological conditions in the presence of specific base sequences, such as consecutive sequences of C and G. Certain segments of active genes have the ability to adopt a Z-DNA conformation, thereby implying that Z-DNA might have a regulatory function in the process of gene transcription (Gao et al., 2006).

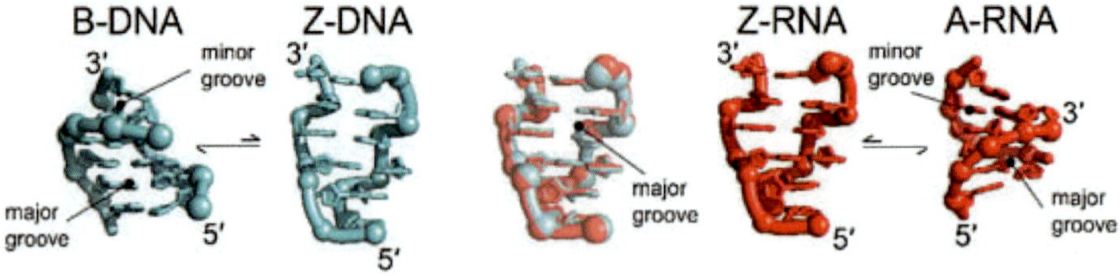

Figure 2.15 DNA can assume several different secondary structures. These structures depends on the base sequence of the DNA & the conditions under which it's

placed (Source: Krall JB, Nichols PJ, Henen MA, Vicens Q, Vögeli B. Structure and Formation of Z-DNA and Z-RNA. Molecules. 2023; 28(2):843. https://doi.org/10.3390/molecules28020843, Creative Commons License)

Additional secondary structures may potentially arise in specific circumstances or with unique base sequences, and pertinent attributes of these structures are provided in Table 2.1. Rarely, if ever, do structures apart from B-DNA exist within cells (Baikar & Malpathak, 2010).

Did You Know?

The horse fossil contains the most ancient DNA sample discovered to date, with an estimated age of 700,000 years. This finding provides valuable insights into the characteristics and genetic makeup of ancient horse species.

2.6.3 Local variation in secondary structures

The secondary structure of DNA is often portrayed as a stable and unchanging entity. The numerical values presented in Figure 2.14 pertaining to the parameters of B-DNA are representative averages, with slight variations observed across different regions of the molecule. The angular displacement between base pairs throughout an individual molecule of B-DNA can range from 27 degrees to a maximum of 42 degrees. The observed diversity in DNA structure within a given locality can be attributed to variations in the surrounding environmental conditions, including the presence of amino acids, metals, and ions that have the potential to interact with DNA. The local structure of DNA is also influenced by the base sequence (Kumar & Bansal, 2015).

2.7 SPECIAL STRUCTURES IN DNA & RNA

The stability and helical supplementary structure of double-stranded DNA are a result of the pairing of bases upon opposite nucleotide strands. Single-stranded DNA and RNA, particularly RNA, which is predominantly single-stranded, do not possess the stabilizing effect of paired nucleotide strands. Consequently, they do not exhibit any shared secondary structure. Sequences present within a singular strand of nucleotides have the potential to exhibit complementarity, allowing them to form pairs through the establishment of hydrogen bonds. This interaction results in the formation of a double-stranded region, as depicted in Figure 2.16. Internal base pairing confers a secondary structure upon a single-stranded molecule. Indeed, internal base pairing across individual strands of amino acids can lead to a wide range of secondary structures (Noy et al., 2004).

A frequently observed secondary structure in single-stranded nucleotides is known as a hairpin. This structure arises when segments of nucleotides upon the same strand exhibit inverted complementarity. The sequences 5' TGCGAT 3' and 5' ATCGCA 3' serve as instances of inverted complements. When these sequences are present on identical nucleotide strands, they can align and create a hairpin structure. A hairpin structure is comprised of a segment of nucleotide bases that are paired together (known as the stem), and in certain cases, may also contain unpaired bases located between the paired bases (known as the loop). When the contiguous sequences of the complementary strands are nearby, the resulting hairpin structure exhibits a stem region but lacks a loop. Hairpins play a significant role in regulating various aspects of information transmission. RNA molecules can possess multiple hairpins, which enables them to undergo intricate folding processes resulting in complex structures (Masai & Tanaka, 2020).

The term "inverted repeats" refers to sequences within double-stranded DNA that exhibit an inverted replication pattern, resulting in the formation of identical but reversed sequences. The subsequent sequence of nucleotides, consisting of two strands, serves as an illustration of inverted repeats (Creacy et al., 2008):

$$5' - AAAG \ldots CTTT - 3'$$

$$3' - TTTC \ldots GAAA - 5'$$

It is worth noting that the sequences upon the two strands exhibit similarity when read in the 5' to 3' direction. However, due to the opposing polarities of the two strands, their sequences have been reversed when observed coming from left to right. A self-complementary inverted repeat, characterized by its ability to form a complementary sequence with itself, is an intriguing genetic feature (Clore & Gronenborn, 1985):

$$5' - ATCGAT - 3'$$

$$3' - TAGCTA - 5'$$

Additionally, the term "palindrome" refers to a phrase or word that maintains its original form when reading both forwards and backward, exemplified by the word "rotator." Inverted repeats exhibit palindromic characteristics as the nucleotide sequences on both strands are identical, albeit in a reverse orientation. When a reversed repeat exhibits perfect palindromic characteristics, the sequence of the double-stranded DNA is identical when read in both the forward and backward directions (Šponer et al., 2012).

An alternative secondary structure, known as a cruciform, may be generated through the formation of an inverted repeat, wherein a hairpin structure is formed within each of the single-stranded scenes.

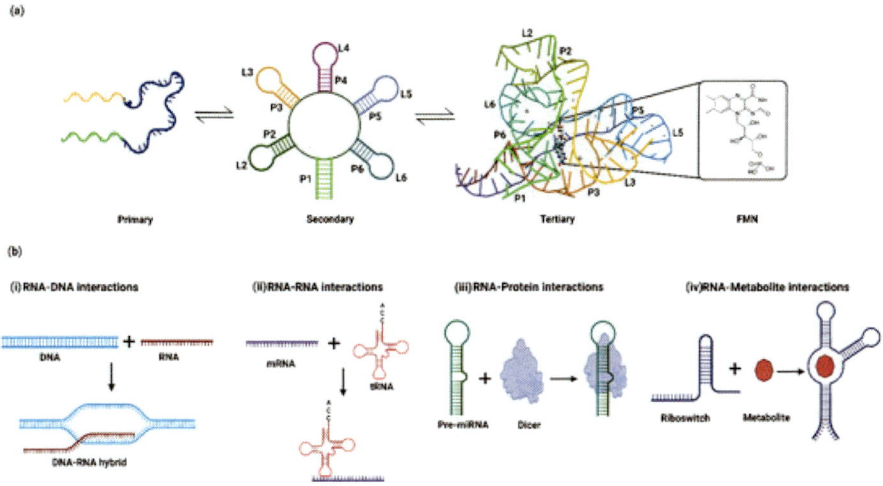

Figure 2.16. Both DNA & RNA can form special secondary structures. (Source: Gupta P, Khadake RM, Panja S, Shinde K, Rode AB. Alternative RNA Conformations: Companion or Combatant. Genes. 2022; 13(11):1930. https://doi.org/10.3390/genes13111930, Creative Commons License)

2.8 DNA METHYLATION

Several potential modifications can be made to the fundamental structure of DNA. As mentioned above, the alterations play a crucial role in the manifestation of the gene material. One particular alteration that occurs is DNA methylation, a process that occurs when groups of methyl groups (–CH3) have been enzymatically added to specific places on the nucleotide bases (Singal & Ginder, 1999).

Methylation of adenine and cytosine is a prevalent occurrence in bacterial genomes, whereas in eukaryotic genomes, cytosine methylation is the most frequently observed form of DNA methylation. Bacterial DNA is commonly subjected to methylation to discerning it from exogenous, non-methylated DNA that viruses may introduce. Bacteria employ specialized proteins known as restriction enzymes to cleave any non-methylated viral DNA (Robertson, 2005).

Cytosine bases in eukaryotic DNA are frequently subjected to methylation, resulting in the formation of 5-methylcytosine (Figure 2.17). The degree of cytosine methylation exhibits variation, with approximately 5% of cytosine bases being methylated in most animal cells, while certain plant species display methylation in over 50% of their cytosine bases. In contrast, the absence of cytosine methylation has been observed in yeast cells, while Drosophila exhibits only minimal methylation levels (approximately 1 methylated cytosine base per 12,500 nucleotides). The underlying reasons for the significant variation in methylation levels among eukaryotic organisms remain unclear. The process of methylation is observed with the highest frequency on cytosine DNA strands that are adjacent to guanine DNA strands on the same DNA strand (Moore et al., 2013):

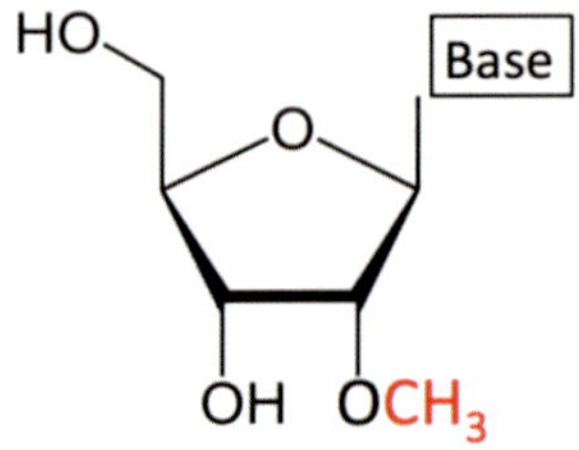

Figure 2.17. In eukaryotic DNA, cytosine bases are often methylated to form 5-methylcytosine (Source: Dimitrova DG, Teysset L, Carré C. RNA 2′-O-Methylation (Nm) Modification in Human Diseases. *Genes*. 2019; 10(2):117. https://doi.org/10.3390/genes10020117, Creative Commons License)

…GC…   *…CG…*

The methylation process in eukaryotic cells is frequently associated with regulating gene expression. Sequences that undergo methylation commonly exhibit reduced transcription levels, whereas sequences that lack methylation are actively transcribed. Methylation can potentially impact the structural conformation of the DNA molecule in three-dimensional space (Jeltsch & Jurkowska, 2014).

2.8.1 Bends In DNA

Certain specific base sequences, such as a consecutive arrangement of four or more adenine-thymine base pairs, induce curvature in the DNA double helix. DNA bending plays a significant role in the binding of specific proteins to DNA molecules, potentially exerting regulatory control over the transcriptional activity of certain genes (Kim et al., 1989).

Proteins can induce curvature in the DNA helix by binding to particular DNA sequences (Fig. 2.18). The SRY protein, derived from a gene on the Y chromosome, is crucial in determining sex in mammals. It can bind to specific DNA sequences, primarily within the minor groove, activating adjacent genes responsible for developing male characteristics. When known as SRY protein interacts with the DNA, it induces a significant bending of the molecule, resulting in an approximate angle of 80 degrees. The observed distortion for the DNA helix appears to enhance the affinity for additional proteins, thereby promoting the transcriptional activation of genes responsible for expressing male traits (Ganguly et al., 1993).

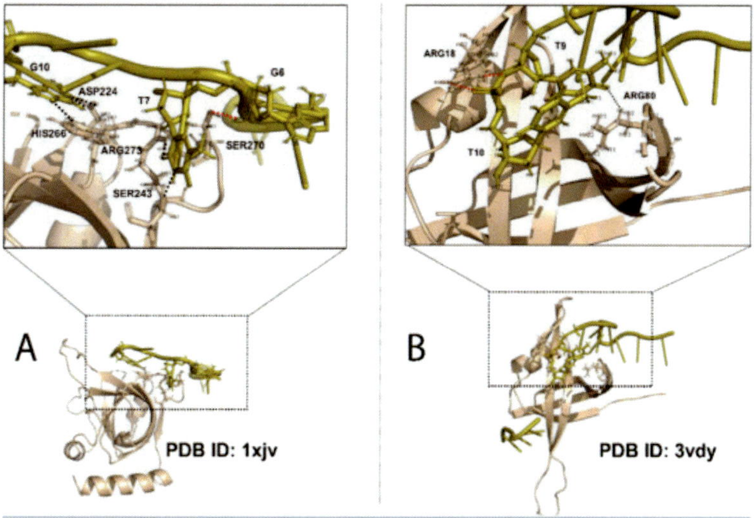

Figure 2.18. The DNA helix can be bent by the binding of proteins to the DNA molecule (Source: Guo J-T, Malik F. Single-Stranded DNA Binding Proteins and Their Identification Using Machine Learning-Based Approaches. *Biomolecules*. 2022; 12(9):1187. https://doi.org/10.3390/biom12091187, Creative Commons License).

2.9 FUNCTION OF DNA

Deoxyribonucleic acid (DNA) is widely recognized as the component of DNA that functions as the foundational blueprint of life. DNA, which is present in virtually all living cells, contains the essential instructions required for the construction, maintenance, and regulation of organisms. The central role of genetic information storage is crucial in comprehending the intricacies associated with the diverse forms and functionalities observed in living organisms. This section focuses on the functional aspects of DNA, examining its various roles encompassing the encoding of genetic information and the coordination of essential life-sustaining processes (Gelfand, 1995).

2.9.1 Encoding Genetic Information

The fundamental essence of DNA functions resides in its capacity to serve as the primary storage site for genetic information. Every deoxyribonucleic acid (DNA) molecule is composed of a distinct arrangement of nucleotide bases, namely adenine (A), thiamine (T), the cytosine (C), and guanine, or G, which are organized in a specific sequence along the backbone of sugar-phosphate. The genetic code is a sequence of three-letter codes composed of nucleotide bases. This code governs the arrangement of amino acids, which are responsible for forming proteins. Proteins, in turn, serve as molecular entities that carry out vital functions within the cell.

Genes, which are distinct sections of DNA, serve as the fundamental components of heredity by containing the essential instructions required for the construction and sustenance of an organism. DNA molecules serve as the fundamental templates for protein synthesis, providing instructions that govern the wide array of biological processes necessary for the maintenance and continuation of living organisms. Every gene contains the genetic code necessary for the synthesis of a specific protein or group of proteins, which ultimately govern an organism's attributes, encompassing both its physical characteristics and biochemical functions (Matsuura et al., 1997).

2.9.2 DNA Replication: Ensuring Genetic Fidelity

The faithful replication of DNA is a crucial component of its functionality. DNA replication represents a remarkable demonstration of meticulousness and precision, taking place within the cell cycle as cells undergo division. Before cell division, the double helix of DNA undergoes unwinding, wherein each strand assumes the role of a template for synthesizing a novel complementary strand. Enzymes, including DNA polymerase, facilitate the process of nucleotide base addition, thereby ensuring the fidelity of DNA replication by maintaining the complementarity between the newly synthesized DNA strand and the template strand. The rigorous nature of this meticulous process ensures the precise transmission of genetic information from a single generation of living cells to the subsequent generation with unparalleled accuracy.

DNA replication serves the dual purpose of facilitating the transfer of genetic data during cell division and fulfilling a vital function in various processes, such as DNA repair, thereby ensuring the accurate correction and preservation of the genetic code. DNA replication plays a crucial role in maintaining the existence of life by ensuring the integrity of genetic material (Hoekstran et al., 2017).

2.9.3 Gene Expression: From DNA to Proteins

DNA is responsible for encoding genetic information, while gene expression refers to how this information is utilized to synthesize proteins. The process of gene expression encompasses two primary stages, namely transcription and translation.

Transcription represents the initial stage of gene expression, wherein a specific segment of DNA is transcribed into an RNA molecule that is complementary in nature. The RNA polymerase enzyme exhibits binding affinity towards a distinct genomic region known as the promoter, thereby commencing the process of RNA synthesis by utilizing the DNA sequence as its template. The recently synthesized RNA molecule, messenger RNA (mRNA), transports the genetic information from the DNA into the cellular apparatus accountable for protein synthesis.

The subsequent phase, known as translation, occurs within the cell's cytoplasm. During this process, ribosomes decipher the genetic instructions conveyed by mRNA and synthesize a sequence of amino acids accordingly. The correspondence between an amino acid in particular and each set of 3 nucleotide bases, known as a codon, in the messenger RNA (mRNA) is established. During translation, ribosomes traverse the mRNA molecule, sequentially decoding each codon. This sequential decoding prompts the attraction of the equivalent amino acids, which are joined together to establish a polypeptide chain. This chain serves as a precursor to the eventual formation of an entirely operational protein.

The arrangement of amino acids within a polypeptide sequence plays a crucial role in determining the structural and functional properties of proteins, thereby facilitating their ability to perform specific cellular or organismal functions. The process of gene expression is highly regulated, enabling cells to exert control over the activation of specific genes and the subsequent production of proteins. This regulation ensures the precise synthesis of proteins, both in terms of timing and quantity, thereby facilitating the proper functioning of cellular processes (Hoyos et al., 1989).

2.9.4 Proteins: The Molecular Machines of Life

Proteins play a crucial role in cellular processes, exhibiting a diverse range of functions that are indispensable for sustaining life. Proteins play a crucial role in various cellular and organismal functions, encompassing various activities, including catalyzing biochemical reactions and providing structural support and form to cells and tissues.

The diversity of proteins is derived from the multitude of possible sequences of amino acids they can possess. Although the number of amino acids is limited to 20, proteins exhibit a vast range of functions due to their ability to incorporate many amino acids in highly specific configurations. This intricate arrangement allows for an almost limitless diversity in protein functionality (Kinbara & Aida, 2005).

2.9.5 Regulatory Proteins: Orchestrating Cellular Processes

DNA additionally encodes regulating proteins, which possess a pivotal function in governing gene expression and orchestrating cellular processes. Regulatory proteins can function as transcription factors, whereby they bind to specific DNA regions in proximity to genes and subsequently modulate their expression by either augmenting or suppressing it.

The regulatory proteins play a crucial role in facilitating gene expression according to specific messages or developmental cues, enabling cells to effectively adapt to dynamic environments or carry out intricate processes such as embryonic development and differentiation.

2.9.6 DNA Repair: Safeguarding Genetic Integrity

Ensuring the preservation of the genome is of utmost importance for the stability and well-being of an organism. Nevertheless, DNA is vulnerable to harm from many sources, encompassing radiation, chemicals, and errors during replication.

Cells possess intricate repair processes for DNA that diligently monitor the quality of DNA and promptly rectify any errors or destruction that may arise, thereby safeguarding against potential harm. The repair mechanisms in question serve to promptly rectify any mutations or alterations in the genetic code, thereby mitigating the potential for deleterious outcomes such as cancer or genetic disorders (Sørensen & Syljuåsen, 2012).

2.9.7 Inheritance and Heredity

The role of DNA in inheritance, along with heredity, is considered highly significant. Organisms transmit their genes to their progeny during reproduction, thereby ensuring the continuation of the species. During sexual reproduction, offspring acquire a composite genetic material from both progenitors. The genetic composition of an offspring is derived from both the mother and the father, with each parent contributing approximately 50% of the DNA. This process results in a distinct genetic makeup crucial in fostering the diversity observed in living organisms.

Meiosis, a specialized type of cellular division, ensures that gametes, namely sperm and egg cells, possess only half of the typical number of chromosomes. This crucial process facilitates the amalgamation of genetic material via two parents during the fertilization process. This biological phenomenon leads to the emergence of genetic diversity, thereby guaranteeing that every subsequent generation possesses a distinct amalgamation of characteristics inherited from their predecessors.

2.9.8 Genetic Mutations

Genetic mutations encompass alterations in the sequence of DNA, which can manifest spontaneously or as a result of external stimuli. While certain mutations may have detrimental effects, resulting in genetic conditions or diseases, others can have a neutral or advantageous impact, thereby contributing to genetic diversity and evolution.

The process of evolution is dependent on mutations in genes that give rise to beneficial characteristics, which can subsequently be inherited by succeeding generations, thereby augmenting an organism's ability to survive and reproduce. The phenomenon of natural selection, influenced by the interaction between genetic diversity and environmental factors, plays a crucial role in shaping the multitude of life forms and facilitating their ability to adjust to evolving circumstances throughout history (Blackford et al., 2009).

2.9.9 Applications of DNA

In addition to its fundamental significance within cellular processes, DNA has brought about transformative advancements in various disciplines, including healthcare, the field of biotechnology, criminal justice, along with anthropology. DNA sequencing is a revolutionary technological advancement that allows for the precise identification of the sequence of bases that are found within a DNA molecule. This technique has emerged as a fundamental tool in various fields, such as medical study, medical diagnosis, and personalized medicine.

Genetic testing enables the detection of hereditary disorders, the anticipation of disease susceptibility, and the customization of treatment strategies according to an individual's genetic composition. DNA profiling has emerged as a highly potent instrument within forensic science, facilitating the precise identification of individuals, resolution of criminal cases, and establishment of biological connections with unparalleled accuracy.

In summary, DNA as the primary storage of genetic information plays a fundamental role in the vast array of diversity and intricate nature observed in living organisms. DNA plays a central role in the complexities of life, encompassing tasks such as encoding genetic instructions and coordinating the production of proteins that govern cellular functions. The capacity of organisms to faithfully replicate, regulate gene expression, and repair any damage plays a crucial role in safeguarding their stability and overall health (Wagenknecht, 2006).

The comprehension of DNA's function has not only enhanced our understanding of genetics yet has also brought about significant transformations in disciplines such as medicine, the field of biotechnology, and forensic science. The investigation of DNA has revealed the concealed aspects of genetic transmission and hereditary traits, enabling us to trace the evolutionary roots and acknowledge the interdependence among all organisms.

As the exploration of DNA progresses, novel understandings regarding the intricacies of life are acquired, thereby facilitating the emergence of pioneering breakthroughs and advancements. The role of DNA serves as an example of the remarkable complexity of life's genetic information, motivating researchers and scientists to investigate the boundless potential and opportunities embedded within the fundamental structure of life (Anderson & Record, 1982).

Did You Know?

DNA profiling aids forensic science, identifying individuals with unique DNA sequences.

SUMMARY

The chapter titled "DNA Structure alongside Function" delves into the intricate nature of deoxyribonucleic acid (DNA) and explores its significant contributions to the fundamental blueprint of life. The chapter provides the historical backdrop, emphasizing the seminal revelation of the DNA double-helix structure through James Watson and Francis Crick. This discovery fundamentally revolutionized our comprehension of genetics.

The fundamental aspect of DNA's functionality resides in its capacity as the storage site for genetic information. The structure of a DNA molecule is characterized by a distinct arrangement of nucleotide bases, namely thymine (T), adenine (A), cytosine (C), and the base guanine (G), which are organized in a specific sequence along the backbone composed of sugar-phosphate. The genetic code is a sequence consisting of three-letter codes composed of nucleotide bases, which govern the arrangement of amino acids responsible for protein synthesis. Genes, which are distinct segments of DNA, serve as the fundamental elements of heredity responsible for encoding the requisite information essential for the construction and sustenance of an organism. These genes are pivotal in determining an organism's attributes, encompassing its physical characteristics and biochemical mechanisms.

The intricate process further demonstrates the role of DNA as the fundamental genetic material for DNA replication. Throughout the cell cycle, the DNA molecule undergoes unwinding, wherein each strand assumes the role of a template for synthesizing a novel complementary strand. The rigorous procedure guarantees the accurate transmission of the genome from one cellular generation to the subsequent, thereby upholding the uninterrupted progression of life.

The chapter additionally delves into the concept of gene expression, which refers to the mechanism through which the genetic information encoded in DNA is employed to produce proteins. The initial stage in the process, known as transcription, entails the production of RNA by utilizing the structure of DNA as a blueprint. The resultant mRNA molecule then conveys the genetic information from the DNA into the cellular apparatus accountable for protein synthesis. The process of translation, which is the second stage, takes place within the cytoplasm of the cell. During this stage, ribosomes interpret the genetic information encoded in mRNA and synthesize a chain of amino acids accordingly, resulting in the formation of proteins.

Proteins, which are integral components of cellular machinery, exhibit a remarkable diversity of functions that are indispensable for sustaining life. Proteins play a crucial role in a wide range of cell and organismal functions, encompassing biochemical reactions facilitated by enzymes as well as the provision of nourishment and form to cells and tissues through structural proteins.

In addition to its role in encoding genetic data and controlling gene expression, DNA functions as a reservoir of genetic diversity through the occurrence of mutations. While certain mutations may have detrimental effects, others can exhibit a neutral or advantageous nature, thereby augmenting genetic diversity and facilitating the progression of evolution.

The section concludes by emphasizing the diverse range of applications that DNA has in multiple disciplines, including medicine, the field of biotechnology, criminal justice, and anthropology. The utilization of DNA sequencing has emerged as a fundamental aspect in the realm of biomedical studies and clinical diagnostics. This technological advancement has facilitated the detection of hereditary disorders, anticipation of disease susceptibility, and the development of tailored treatment strategies. DNA profiling has brought about a significant transformation in forensic science, facilitating the precise identification of individuals, resolution of criminal cases, and establishment of biological relationships with unprecedented accuracy.

In essence, the publication "DNA Structure or Function" provides a comprehensive exploration of the intricate nature and profound importance of DNA as a fundamental blueprint of biological existence. The fundamental significance of DNA in comprehending the diversity and intricacy of living organisms is underscored by its function as the storage site for genetic information, its role in inheritance along with heredity, and its function as the blueprint for protein synthesis. As the scientific community delves deeper into the complexities of DNA, valuable knowledge is being acquired regarding the fundamental aspects of life. This ongoing exploration not only provides us with a better understanding of biological processes but also paves the way for significant advancements and transformative breakthroughs that will have far-reaching implications in biology and beyond.

REFERENCES

1. Alejska, M., Kurzyńska-Kokorniak, A., Broda, M., Kierzek, R., & Figlerowicz, M. (2001). How RNA viruses exchange their genetic material. *Acta Biochimica Polonica*, *48*(2), 391-407.
2. Anami, S., Njuguna, E., Coussens, G., Aesaert, S., & Van Lijsebettens, M. (2013). Higher plant transformation: principles and molecular tools. *International Journal of Developmental Biology*, *57*(6-7-8), 483-494.
3. Andera, L., Spangler, C. J., Galeone, A., Mayol, L., & Geiduschek, E. P. (1994). Interrelations of secondary structure stability and DNA-binding affinity in the bacteriophage SPO1-encoded type II DNA-binding protein TF1. *Journal of Molecular Biology*, *236*(1), 139-150.
4. Anderson, C. F., & Record Jr, M. T. (1982). Polyelectrolyte theories and their applications to DNA. *Annual Review of Physical Chemistry*, *33*(1), 191-222.
5. Atchley, W. A. (1982). Transforming principle of the pneumococcus: Rosetta Stone to the new biology. *Transactions of the American Clinical and Climatological Association*, *93*, 22.
6. Baase, W. A., & Johnson Jr, W. C. (1979). Circular dichroism and DNA secondary structure. *Nucleic Acids Research*, *6*(2), 797-814.
7. Baikar, S., & Malpathak, N. (2010). Secondary metabolites as DNA topoisomerase inhibitors: a new era towards designing of anticancer drugs. *Pharmacognosy reviews*, *4*(7), 12.
8. Blackford, A., Parmigiani, G., Kensler, T. W., Wolfgang, C., Jones, S., Zhang, X., ... & Hruban, R. H. (2009). Genetic mutations associated with cigarette smoking in pancreatic cancer. *Cancer research*, *69*(8), 3681-3688.
9. Bochman, M. L., Paeschke, K., & Zakian, V. A. (2012). DNA secondary structures: stability and function of G-quadruplex structures. *Nature Reviews Genetics*, *13*(11), 770-780.
10. Bram, S. (1971). The secondary structure of DNA in solution and in nucleohistone. *Journal of Molecular Biology*, *58*(1), 277-288.
11. Caruthers, M. H. (1991). Chemical synthesis of DNA and DNA analogs. *Accounts of chemical research*, *24*(9), 278-284.
12. Champoux, J. J. (2001). DNA topoisomerases: structure, function, and mechanism. *Annual review of biochemistry*, *70*(1), 369-413.
13. Cheng, X. (1995). Structure and function of DNA methyltransferases. *Annual review of biophysics and biomolecular structure*, *24*(1), 293-318.

14. Clore, G. M., & Gronenborn, A. M. (1985). Probing the three dimensional structures of DNA and RNA oligonucleotides in solution by nuclear Overhauser enhancement measurements. *FEBS letters*, *179*(2), 187-198.
15. Creacy, S. D., Routh, E. D., Iwamoto, F., Nagamine, Y., Akman, S. A., & Vaughn, J. P. (2008). G4 resolvase 1 binds both DNA and RNA tetramolecular quadruplex with high affinity and is the major source of tetramolecular quadruplex G4-DNA and G4-RNA resolving activity in HeLa cell lysates. *Journal of Biological Chemistry*, *283*(50), 34626-34634.
16. Dahm, R. (2005). Friedrich Miescher and the discovery of DNA. *Developmental biology*, *278*(2), 274-288.
17. Dawson, M. (2003). Martha Chase dies. *Genome Biology*, *4*(1), 1-4.
18. Eichmann, K., & Krause, R. M. (2013). Fred Neufeld and pneumococcal serotypes: foundations for the discovery of the transforming principle. *Cellular and Molecular Life Sciences*, *70*, 2225-2236.
19. Ganguly, A., Rock, M. J., & Prockop, D. J. (1993). Conformation-sensitive gel electrophoresis for rapid detection of single-base differences in double-stranded PCR products and DNA fragments: evidence for solvent-induced bends in DNA heteroduplexes. *Proceedings of the National Academy of Sciences*, *90*(21), 10325-10329.
20. Gao, Y., Wolf, L. K., & Georgiadis, R. M. (2006). Secondary structure effects on DNA hybridization kinetics: a solution versus surface comparison. *Nucleic acids research*, *34*(11), 3370-3377.
21. Gelfand, M. S. (1995). Prediction of function in DNA sequence analysis. *Journal of computational biology*, *2*(1), 87-115.
22. Gendron, P., Lemieux, S., & Major, F. (2001). Quantitative analysis of nucleic acid three-dimensional structures. *Journal of Molecular Biology*, *308*(5), 919-936.
23. Grosberg, A., Rabin, Y., Havlin, S., & Neer, A. (1993). Crumpled globule model of the three-dimensional structure of DNA. *Europhysics letters*, *23*(5), 373.
24. Hartman, P. S. (1979). *Synergistic inactivation of Escherichia coli and phage T7 by near-ultraviolet radiation and hydrogen peroxide*. Missouri Univ.74, 745-750.
25. Herbert, A., & Rich, A. (1999). Left-handed Z-DNA: structure and function. *Genetica*, *106*, 37-47.
26. Herriott, R. M. (1971). Effects on DNA: transforming principle. In *Chemical Mutagens: Principles and Methods for Their Detection Volume.* Boston, MA: Springer US. (Vol. 2, pp. 175-217).

27. Higgs, P. G. (2000). RNA secondary structure: physical and computational aspects. *Quarterly reviews of biophysics*, *33*(3), 199-253.
28. Hill, M., & Hillova, J. (1974). RNA and DNA forms of the genetic material of C-type viruses and the integrated state of the DNA form in the cellular chromosome. *Biochimica et Biophysica Acta (BBA)-Reviews on Cancer*, *355*(1), 7-48.
29. Hodgkin, J. (2001). Hershey and his heaven. *Nature Cell Biology*, *3*(3), E77-E77.
30. Hoekstra, T. P., Depken, M., Lin, S. N., Cabanas-Danés, J., Gross, P., Dame, R. T., ... & Wuite, G. J. (2017). Switching between exonucleolysis and replication by T7 DNA polymerase ensures high fidelity. *Biophysical journal*, *112*(4), 575-583.
31. Hoyos, B., Ballard, D. W., Böhnlein, E., Siekevitz, M., & Greene, W. C. (1989). Kappa B-specific DNA binding proteins: role in the regulation of human interleukin-2 gene expression. *Science*, *244*(4903), 457-460.
32. Hud, N. V., Milanovich, F. P., & Balhorn, R. (1994). Evidence of novel secondary structure in DNA-bound protamine is revealed by Raman spectroscopy. *Biochemistry*, *33*(24), 7528-7535.
33. Jeltsch, A., & Jurkowska, R. Z. (2014). New concepts in DNA methylation. *Trends in biochemical sciences*, *39*(7), 310-318.
34. Johnson, R. R., Johnson, A. C., & Klein, M. L. (2008). Probing the structure of DNA− carbon nanotube hybrids with molecular dynamics. *Nano letters*, *8*(1), 69-75.
35. Joyce, C. M., Kelley, W. S., & Grindley, N. D. (1982). Nucleotide sequence of the *Escherichia coli* polA gene and primary structure of DNA polymerase I. *Journal of Biological Chemistry*, *257*(4), 1958-1964.
36. Ke, Y. (2014). Designer three-dimensional DNA architectures. *Current Opinion in Structural Biology*, *27*, 122-128.
37. Kim, J., Zwieb, C., Wu, C., & Adhya, S. (1989). Bending of DNA by gene-regulatory proteins: construction and use of a DNA bending vector. *Gene*, *85*(1), 15-23.
38. Kinbara, K., & Aida, T. (2005). Toward intelligent molecular machines: directed motions of biological and artificial molecules and assemblies. *Chemical reviews*, *105*(4), 1377-1400.
39. Klug, A. (2004). The discovery of the DNA double helix. *Journal of Molecular Biology*, *335*(1), 3-26.
40. Kool, E. T., Morales, J. C., & Guckian, K. M. (2000). Mimicking the structure and function of DNA: insights into DNA stability and replication. *Angewandte Chemie International Edition*, *39*(6), 990-1009.

41. Kumar, P., & Bansal, M. (2015). Identification of local variations within secondary structures of proteins. *Acta Crystallographica Section D: Biological Crystallography*, *71*(5), 1077-1086.
42. Lederberg, J. (1994). The transformation of genetics by DNA: an anniversary celebration of Avery, MacLeod and McCarty (1944). *Genetics*, *136*(2), 423.
43. Leu, K., Obermayer, B., Rajamani, S., Gerland, U., & Chen, I. A. (2011). The prebiotic evolutionary advantage of transferring genetic information from RNA to DNA. *Nucleic acids research*, *39*(18), 8135-8147.
44. Li, W. (1997). The study of correlation structures of DNA sequences: a critical review. *Computers & chemistry*, *21*(4), 257-271.
45. Lindahl, T. (1993). Instability and decay of the primary structure of DNA. *nature*, *362*(6422), 709-715.
46. Masai, H., & Tanaka, T. (2020). G-quadruplex DNA and RNA: Their roles in regulation of DNA replication and other biological functions. *Biochemical and Biophysical Research Communications*, *531*(1), 25-38.
47. Matsuura, M., Saldanha, R., Ma, H., Wank, H., Yang, J., Mohr, G., ... & Lambowitz, A. M. (1997). A bacterial group II intron encoding reverse transcriptase, maturase, and DNA endonuclease activities: biochemical demonstration of maturase activity and insertion of new genetic information within the intron. *Genes & development*, *11*(21), 2910-2924.
48. McMurray, C. T. (1999). DNA secondary structure: a common and causative factor for expansion in human disease. *Proceedings of the National Academy of Sciences*, *96*(5), 1823-1825.
49. Moore, L. D., Le, T., & Fan, G. (2013). DNA methylation and its basic function. *Neuropsychopharmacology*, *38*(1), 23-38.
50. Nakano, S. I., Karimata, H., Ohmichi, T., Kawakami, J., & Sugimoto, N. (2004). The effect of molecular crowding with nucleotide length and cosolute structure on DNA duplex stability. *Journal of the American Chemical Society*, *126*(44), 14330-14331.
51. Nelson, K. E., Levy, M., & Miller, S. L. (2000). Peptide nucleic acids rather than RNA may have been the first genetic molecule. *Proceedings of the National Academy of Sciences*, *97*(8), 3868-3871.
52. Noy, A., Perez, A., Lankas, F., Luque, F. J., & Orozco, M. (2004). Relative flexibility of DNA and RNA: a molecular dynamics study. *Journal of Molecular Biology*, *343*(3), 627-638.
53. Olby, R. (1985). The Stuff of Heredity: Discoveries Recounted: The Transforming Principle. Discovering That Genes Are Made of DNA. Maclyn McCarty. Norton, New York, 1985. 252 pp., illus. $14.95. The

Commonwealth Fund Book Program. *Science*, *229*(4716), 851-851.

54. Pan, K., Kim, D. N., Zhang, F., Adendorff, M. R., Yan, H., & Bathe, M. (2014). Lattice-free prediction of three-dimensional structure of programmed DNA assemblies. *Nature Communications*, *5*(1), 5578.
55. Pirrung, M. C. (2002). How to make a DNA chip. *Angewandte Chemie International Edition*, *41*(8), 1276-1289.
56. Poltev, V., Anisimov, V. M., Dominguez, V., Deriabina, A., Gonzalez, E., Garcia, D., ... & Rivas, F. (2020). Current Problems in Computer Simulation of Variability of Three-Dimensional Structure of DNA. In *Advances in Quantum Systems in Chemistry, Physics, and Biology: Selected Proceedings of QSCP-XXIII (Kruger Park, South Africa, September 2018) 23*. Springer International Publishing. (Vol. 3, pp. 233-253).
57. Razin, A., & Riggs, A. D. (1980). DNA methylation and gene function. *Science*, *210*(4470), 604-610.
58. Reece, R. J., & Maxwell, A. (1991). DNA gyrase: structure and function. *Critical reviews in biochemistry and molecular biology*, *26*(3-4), 335-375.
59. Reid, B. R. (1987). Sequence-specific assignments and their use in NMR studies of DNA structure. *Quarterly reviews of biophysics*, *20*(1-2), 1-34.
60. Richmond, T. J., & Davey, C. A. (2003). The structure of DNA in the nucleosome core. *Nature*, *423*(6936), 145-150.
61. Robertson, K. D. (2005). DNA methylation and human disease. *Nature Reviews Genetics*, *6*(8), 597-610.
62. Ronaghi, M., Nygren, M., Lundeberg, J., & Nyrén, P. (1999). Analyses of secondary structures in DNA by pyrosequencing. *Analytical biochemistry*, *267*(1), 65-71.
63. Sabri, M. Z., Hamid, A. A. A., Hitam, S. M. S., & Rahim, M. Z. A. (2020). The assessment of three dimensional modelling design for single strand DNA aptamers for computational chemistry application. *Biophysical chemistry*, *267*, 106492.
64. Sanger, F. (1981). Determination of nucleotide sequences in DNA. *Science*, *214*(4526), 1205-1210.
65. Singal, R., & Ginder, G. D. (1999). DNA methylation. *Blood, The Journal of the American Society of Hematology*, *93*(12), 4059-4070.
66. Sørensen, C. S., & Syljuåsen, R. G. (2012). Safeguarding genome integrity: the checkpoint kinases ATR, CHK1 and WEE1 restrain CDK activity during normal DNA replication. *Nucleic acids research*, *40*(2), 477-486.

67. Spiegel, J., Adhikari, S., & Balasubramanian, S. (2020). The structure and function of DNA G-quadruplexes. *Trends in Chemistry*, *2*(2), 123-136.
68. Šponer, J., Mládek, A., Šponer, J. E., Svozil, D., Zgarbová, M., Banáš, P., ... & Otyepka, M. (2012). The DNA and RNA sugar–phosphate backbone emerges as the key player. An overview of quantum-chemical, structural biology and simulation studies. *Physical Chemistry Chemical Physics*, *14*(44), 15257-15277.
69. Squires, G. L. (2003). The discovery of the structure of DNA. *Contemporary physics*, *44*(4), 289-305.
70. Stahl, F. W. (1998). Hershey. *Genetics*, *149*(1), 1-6.
71. Stevens, R. C. (2003). The cost and value of three-dimensional protein structure. *Drug Discovery World*, *4*(3), 35-48.
72. Strimmer, K., & Pybus, O. G. (2001). Exploring the demographic history of DNA sequences using the generalized skyline plot. *Molecular Biology and Evolution*, *18*(12), 2298-2305.
73. Szyf, M., & Bick, J. (2013). DNA methylation: a mechanism for embedding early life experiences in the genome. *Child development*, *84*(1), 49-57.
74. Tao, N. J., Lindsay, S. M., & Rupprecht, A. (1989). Structure of DNA hydration shells studied by Raman spectroscopy. *Biopolymers: Original Research on Biomolecules*, *28*(5), 1019-1030.
75. Travers, A., & Muskhelishvili, G. (2015). DNA structure and function. *The FEBS journal*, *282*(12), 2279-2295.
76. Udvardy, A., & Schedl, P. (1991). Chromatin structure, not DNA sequence specificity, is the primary determinant of topoisomerase II sites of action in vivo. *Molecular and cellular biology*, *11*(10), 4973-4984.
77. Van Valen, D., Wu, D., Chen, Y. J., Tuson, H., Wiggins, P., & Phillips, R. (2012). A single-molecule Hershey-Chase experiment. *Current Biology*, *22*(14), 1339-1343.
78. Wagenknecht, H. A. (2006). Electron transfer processes in DNA: mechanisms, biological relevance and applications in DNA analytics. *Natural Product Reports*, *23*(6), 973-1006.
79. Wang, J. C., Peck, L. J., & Becherer, K. (1983, January). DNA supercoiling and its effects on DNA structure and function. In *Cold Spring Harbor symposia on quantitative biology.* Cold Spring Harbor Laboratory Press. (Vol. 47, pp. 85-91).
80. Waring, M. J. (1968). Drugs which affect the structure and function of DNA. *Nature*, *219*, 1320-25.

81. Watson, D. A., Musher, D. M., Jacobson, J. W., & Verhoef, J. (1993). A brief history of the pneumococcus in biomedical research: a panoply of scientific discovery. *Clinical infectious diseases*, *17*(5), 913-924.

82. Watt, P. M., & Hickson, I. D. (1994). Structure and function of type II DNA topoisomerases. *Biochemical Journal*, *303*(3), 681.

83. Wu, D. (2010, March). How Bacteriophage Genomes Get Inside Cells. In *APS March Meeting Abstracts* (Vol. 2010, pp. A7-005).

84. Wyatt, H. V. (1975). Knowledge and prematurity: the journey from transformation to DNA. *Perspectives in Biology and Medicine*, *18*(2), 149-156.

85. Yonezawa, A., Kuwahara, J., Fujii, N., & Sugiura, Y. (1992). Binding of tachyplesin I to DNA revealed by footprinting analysis: significant contribution of secondary structure to DNA binding and implication for biological action. *Biochemistry*, *31*(11), 2998-3004.

CHAPTER 3

PRINCIPLES OF HEREDITY

Contents

3.1 INTRODUCTION

Within the domain of biological sciences, a limited number of individuals have exerted a significant and enduring influence, and among them is Gregor Mendel, widely recognized as the seminal figure in the field of genetics. The individual's groundbreaking research on pea plants during the 19th century established the fundamental principles underpinning our contemporary comprehension of heredity. This chapter explores Mendel's pioneering findings, which reveal the fundamental principles that dictate the inheritance of characteristics from one generation to the succeeding one (Riggs et al., 2013).

In order to fully understand the complexities of heredity, it is imperative to possess a comprehensive understanding of genetic terminology. This chapter will examine the linguistic aspects of genetics, introducing fundamental concepts including alleles, genotype, phenotype, and dominance (Bateson, 1903).

Monohybrid crosses play a fundamental role in comprehending the patterns of inheritance for a single trait. Through the analysis of progeny derived from the union of individuals possessing distinct alleles for a single gene, we elucidate patterns of heredity that establish the fundamental basis for intricate genetic situations (Horin, 1998).

The ability to anticipate the result of genetic crosses is an essential proficiency within the field of genetics. By comprehending the fundamental concepts of dominance, segregation, and independent assortment, one can make predictions regarding the phenotypic traits that will be expressed in the progeny resulting from a particular mating event. This knowledge significantly contributes to our understanding of the patterns of inheritance associated with specific traits (Froggatt & Nevin, 1971).

As we progress further into the chapter, we are introduced to several key concepts: multiplication, addition, binomial expansion, and probability. These concepts collectively establish the mathematical foundation for determining the probability of specific outcomes in genetic crosses. These tools facilitate the generation of informed predictions regarding the distribution of traits within offspring populations.

The testcross is a significant tool utilized in genetics research to ascertain the genotype of an individual exhibiting a dominant phenotype. By performing a cross between the individual of interest and a known homozygous recessive individual, it is possible to infer the unknown genotype by examining the observable traits exhibited by the resulting offspring (Marsh, 1911).

During our investigation of heredity, we will also encounter instances of incomplete dominance, wherein neither allele exhibits complete dominance over the other. The observed phenotypes demonstrate a fascinating combination of characteristics inherited from both parents, contributing to a more intricate comprehension of genetic heredity.

The concept of multiple loci crosses provides insight into the intricate nature of inheritance, particularly when multiple genes are implicated in the determination of a singular trait. The interplay between various genes introduces additional levels of complexity to the predictability of offspring characteristics (Berry, 1967).

In brief, the chapter titled "Principles of Heredity" elucidates the genetic foundations of the inheritance process. Our exploration begins with the remarkable contributions of Gregor Mendel, whose research established the fundamental principles of genetics and served as the bedrock for developing genetics as an academic field. This study delves into the fundamental principles that govern heredity, encompassing the mastery of genetic terminology and the ability to predict offspring outcomes. By examining monohybrid crosses, applying the multiplication and addition rules, and exploring the realm of probability, we acquire significant knowledge regarding the transmission of traits. Furthermore, the utilization of test crosses, incomplete dominance, and crosses involving multiple loci serve to enhance our comprehension of the intricate nature of genetic inheritance. Throughout the course of this chapter, we will delve into the complexities of heredity and explore the mechanisms that confer life with its remarkable array of variations (Mange, 1959).

3.2 HISTORY

The historical development of the principles of heredity encompasses a significant duration of scientific investigation and exploration, ultimately leading to our contemporary comprehension of genetics. The narrative commences by delving into historical records of ancestral observations about the transmission of traits in flora and fauna, thereby culminating in preliminary concepts surrounding the phenomenon of heredity. Nevertheless, it was only during the 19th century that Gregor Mendel's pioneering investigations of pea plants established contemporary genetics' fundamental principles. The elucidation of the structure of DNA in the 20th century, along with subsequent discoveries, significantly contributed to the advancement of our knowledge regarding heredity (Cobb, 2006).

The concept of heredity has been a topic of scholarly inquiry since antiquity. The ancient civilizations, including the Egyptians and Greeks, made observations regarding the tendency of offspring to exhibit specific traits commonly shared with their parents. A theory of inheritance known as "blending inheritance" was postulated by Aristotle, wherein he posited that traits from parents undergo a process of amalgamation in offspring, akin to the blending of colors in paint. Although this concept maintained dominance for numerous centuries, it failed to explain the observed consistency of characteristics over successive generations (Schmale et al., 1994). During the 17th and 18th centuries, two divergent theories arose in the field of heredity, namely preformationism and epigenesis. The preformationists subscribed to the notion that diminutive, fully-developed entities known as homunculi were present within either the egg or sperm and subsequently underwent growth throughout the development process. On the contrary, the theory of epigenesis posited that organisms underwent a gradual developmental process originating from a rudimentary state, wherein environmental factors molded their traits. Although these theories exerted significant influence, they lacked the empirical evidence required for substantiation (Lynch et al., 1993). During the mid-19th century, Gregor Mendel, a member of the Augustinian monastic order residing in Brno (currently in the Czech Republic), conducted pioneering experiments that would significantly transform our comprehension of inheritance principles. From 1856 to 1863, Gregor Mendel conducted a series of carefully controlled experiments involving the crossbreeding of pea plants to study and document the patterns of inheritance exhibited by their various traits (Delhommeau et al., 2000).

Mendel's seminal contributions encompassed the elucidation of fundamental principles such as dominance, segregation, and independent assortment. The researcher discovered that certain traits exhibited dominance over others during the inheritance process while others remained latent. Additionally, it was observed by the researcher that every trait was governed by two "factors" (currently referred to as alleles). These factors were found to segregate and assort independently during the formation of gametes, thereby resulting in the inheritance of distinct traits (Müller-Wille & Rheinberger, 2007).

Regrettably, Mendel's scientific endeavors remained largely unacknowledged throughout his lifetime, with his significant contributions to genetics only being rediscovered and duly acknowledged at the onset of the 20th century.

During the early 20th century, significant progress in the fields of microscopy and cytology yielded valuable knowledge regarding the underlying physical mechanisms of heredity. Biologists, exemplified by Thomas Hunt Morgan and his team of researchers in the "fly room," conducted investigations on the fruit fly species *Drosophila melanogaster*, leading to the identification of a correlation between distinct characteristics and their respective positions on chromosomes (Hirsch, 1997). The field of genetics experienced a significant paradigm shift during the mid-20th century as a result of the groundbreaking revelation concerning the molecular structure of DNA. The elucidation of the DNA double helix by

James Watson and Francis Crick in 1953 yielded a significant advancement in comprehending the mechanisms underlying the storage and replication of genetic information. This significant finding laid the foundation for the advancement of molecular genetics and marked the commencement of a novel phase in genetic investigation (Sadriyevna, 2020). During the latter part of the 20th century, researchers successfully deciphered the genetic code, which entails establishing the correspondence between DNA sequences and the amino acids that constitute proteins. The process of decoding the genetic code has greatly contributed to the comprehension of how genes encode the necessary information for the synthesis of proteins. As postulated by Francis Crick, the central dogma of molecular biology delineated the sequential transmission of genetic material from DNA to RNA to proteins (Lombardo, 2018).

The biotechnology and genomics fields experienced remarkable progress during the late 20th and early 21st centuries. The field of genetics research has been significantly transformed by the advent of various methodologies, including polymerase chain reaction (PCR), DNA sequencing, and genetic engineering. The Human Genome Project, which concluded in 2003, yielded a comprehensive depiction of the human genome, elucidating the precise positioning and arrangement of all human genes.

The emergence of the field of epigenetics in the 21st century has focused on investigating the impact of environmental factors and lifestyle choices on gene expression without inducing changes to the underlying DNA sequence. Epigenetic mechanisms are of paramount importance in the processes of development, disease manifestation, and hereditary transmission (Jacob, 2022).

The historical narrative surrounding the principles of heredity is characterized by a persistent pursuit of knowledge and notable breakthroughs. The comprehension of heredity has undergone substantial development throughout human history, progressing from ancient observations to the emergence of modern genetics and genomics. Currently, genetics occupies a prominent position in the realm of scientific advancement, exerting influence on diverse domains such as medical breakthroughs, agriculture, and biotechnology. Moreover, it possesses the capacity to potentially unravel the most profound enigmas surrounding the essence of life (Kevles, 1995).

Did You Know?

Mendel's selection of pea plants for his experiments was purposefully strategic. Pea plants possess a brief reproductive cycle, yield a substantial progeny, and exhibit readily discernible characteristics, rendering them highly suitable for investigating patterns of heredity.

3.3 MENDEL: THE FATHER OF GENETICS

During the year 1902, the fundamental principles of genetics, which were effectively employed by Archibald Garrod in studying the hereditary nature of alkaptonuria, had recently gained significant recognition within the field of biology. Remarkably, Johann Gregor Mendel (1822 – 1884) had actually discovered these principles approximately 35 years prior (Wolfe, 2012).

Mendel's birthplace is situated in the present-day territory of the Czech Republic. Despite his parents' modest financial means and occupation as farmers, he acquired a comprehensive education and gained acceptance into the Augustinian monastery in Brno in September 1843. Upon completion of his studies at the seminary, Mendel received ordination as a priest and was subsequently assigned to a pedagogical role within a nearby educational institution. The individual demonstrated exceptional proficiency in the field of education, leading to a commendation from the abbot of the monastery. As a result, he was recommended to pursue advanced studies at the University of Vienna, where he enrolled and completed his academic pursuits between 1851 and 1853 (Das, 2022). Mendel enrolled at the recently established Physics Institute, pursuing studies encompassing various disciplines. Following two years of study in Vienna, Mendel subsequently returned to Brno, where he engaged in educational instruction and initiated his empirical investigations involving pea plants. The individual in question engaged in a series of breeding experiments spanning from 1856 to 1863. Subsequently, the findings derived from these experiments were publicly disseminated during gatherings of the Brno Natural Science Society in 1865. The publication of Mendel's paper derived from the aforementioned lectures occurred in the year 1866. At the time, it appears that Mendel's discovery of the fundamental principles of inheritance went unnoticed by the scientific community (Schwarzbach et al., 2014).

In 1868, Mendel was appointed as the abbot of his monastery, which reduced his teaching responsibilities and ultimately led to the cessation of his genetics experiments due to the mounting administrative obligations. Regrettably, the individual in question passed away at the age of 61 on January 6, 1884, without receiving due recognition for his noteworthy contributions to the field of genetics. The true importance of Mendel's discovery remained unrecognized until the year 1900, when three botanists, namely Hugo de Vries, Erich von Tschermak, and Carl Correns, independently embarked on similar plant experiments and reached conclusions that aligned with Mendel's findings. Upon encountering Mendel's seminal paper, the researchers analyzed their findings within the framework of his principles and underscored the groundbreaking nature of his research (De Castro, 2016).

3.3.1 Mendel's Success

Mendel's methodology in investigating heredity proved to be efficacious due to various factors. The primary factor in his selection of the experimental subject was the pea plant *Pisum sativum*, which presented distinct advantages for genetic research (Figure 3.1). The cultivation process was straightforward, and Mendel had access to both the monastery garden and greenhouse facilities. Peas exhibit a notable rate of growth, culminating in the completion of a full life cycle within the span of a single growing season. In contemporary terms, the pace of one generation per year may appear remarkably sluggish. In comparison, fruit flies accomplish a full generation within a mere two weeks, while bacteria achieve the same feat in a mere 20 minutes. However, it is important to note that Mendel did not face any urgency to publish his findings promptly (Johnson & Stewart, 2002). Consequently, he had the luxury of meticulously observing the transmission of specific traits across multiple generations. If he had opted to conduct research on an organism characterized by a lengthier reproductive cycle, such as horses, it is plausible that he would not have been able to uncover the fundamental principles underlying the transmission of genetic traits. Pea plants exhibit a high fecundity, generating numerous offspring in the form of seeds. This characteristic enabled Mendel to discern significant mathematical ratios in the inherited traits that he observed in the subsequent generations (Charnley & Radick, 2013). The extensive assortment of pea varieties at Mendel's disposal played a pivotal role, as these varieties exhibited diverse traits and maintained genetic purity. Consequently, Mendel was able to initiate his experiments with plants that possessed diverse and well-documented genetic compositions (Jamieson & Radick, 2017).

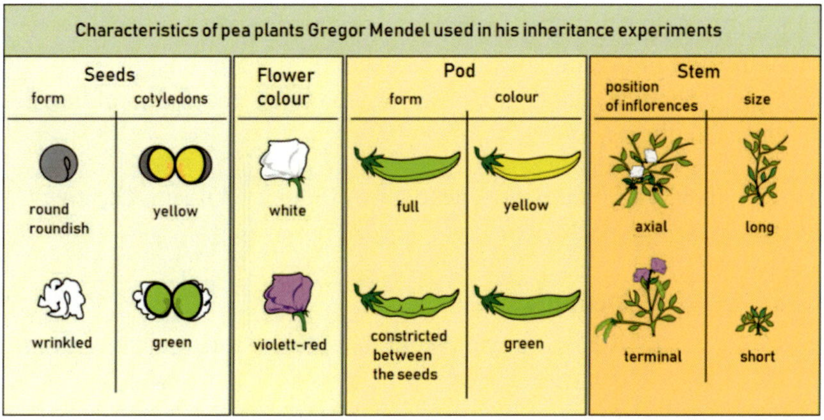

Figure 3.1. Mendel utilized the Pisum sativum plant in his investigations of heredity (Source: Https://commons.wikimedia.org/wiki/File:Gregor_Mendel_-_characteristics_of_pea_plants_-_english.png, Creative Commons License)

A significant portion of Mendel's achievements can be ascribed to the selection of seven specific features for his study, as depicted in Figure 3.1. The individual refrained from considering traits that exhibit a wide spectrum of variation. Instead, the individual directed their attention towards traits that manifest in two distinct forms, such as the contrast between white and gray seed coats, round and wrinkled seeds, and inflated and constricted pods (Vandervorst et al., 1998).

Ultimately, Mendel achieved success by employing an experimental methodology. In contrast to numerous preceding researchers who merely documented the outcomes of genetic crosses, Mendel developed hypotheses grounded in his preliminary observations and subsequently performed further crosses to experimentally validate his hypotheses. The individual maintained meticulous documentation of the quantities of offspring exhibiting each specific characteristic and calculated the proportions of the various types. The individual demonstrated a keen focus on meticulousness, exhibited proficiency in discerning patterns within intricate elements, and displayed attributes of patience and thoroughness. Notably, the individual dedicated a decade to conducting experiments prior to embarking on the task of documenting their findings (Bonneuil, 2006).

3.4 GENETIC TERMINOLOGY

Prior to analyzing Mendel's crosses and the subsequent deductions derived from them, it would be beneficial to revisit certain terminologies frequently employed in the field of genetics (refer to Table 3.1). Mendel was not acquainted with the term "gene" during his time. The term was not introduced until the year 1909 when Wilhelm Johannsen, a Danish geneticist, initially employed it. The conceptualization of a gene is subject to contextual variations, necessitating a dynamic definition that adapts to the exploration of diverse facets of hereditary phenomena (Dreimanis, 1989).

Table 3.1. Summary of important genetic terms (Source: William, Creative Commons License)

Term	Definition
Gene	A hereditary element that plays a role in determining a particular trait
Allele	Two or more alternative forms of a gene
Locus	An allele's specific location on a chromosome
Genotype	An individual's collection of alleles

Heterozygote	An Individual who has two distinct alleles at a locus
Homozygote	An Individual who carries two alleles from the same locus
Phenotype or trait	A character's external appearance
Character or characteristics	An attribute

Alleles, which are alternative forms of genes, are commonly observed (Figure 3.2). Mendel's experiments involved the crossing of plants, wherein the hereditary trait of seed shape was governed by a gene comprising two distinct alleles. These alleles corresponded to the production of either round or wrinkled seeds. The locus for a gene is a specific location on a chromosome where all alleles for that gene can be found (Edlin, 1987). The correct plural form of the term "locus" is "loci." In the field of genetics, it is considered improper and inaccurate to refer to multiple loci as "locuses." Hence, it can be observed that within the genetic makeup of pea plants, there exists a distinct location, referred to as a locus, on a chromosome, which is responsible for determining the shape of the seeds.

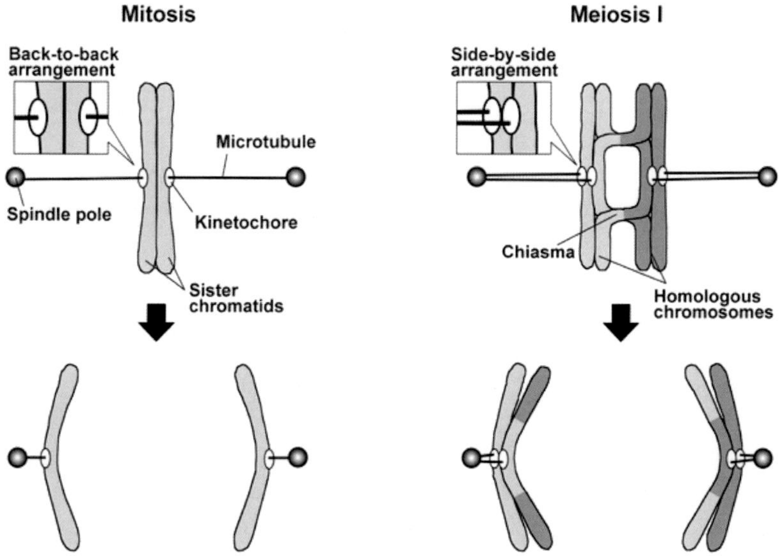

Figure 3.2. In the context of genetic loci, it is observed that diploid organisms possess a pair of alleles situated on distinct homologous chromosomes. (Source: Hirose Y, Suzuki R, Ohba T, Hinohara Y, Matsuhara H, Yoshida M, et al. (2011) Chiasmata Promote Monopolar Attachment of Sister Chromatids and Their Co-Segregation toward the Proper Pole during Meiosis I. PLoS Genet 7(3): e1001329. https://doi.org/10.1371/, Creative Commons License)

This particular genetic locus may potentially be occupied by either an allele associated with the presence of round seeds or an allele associated with the presence of wrinkled seeds. The term "allele" will be employed to denote a particular variant of a gene, while the term "gene" will be utilized in a broader sense to encompass any allele at a given locus (Samerski, 2006).

The genotype refers to the specific combination of alleles present within an individual organism. An organism that is diploid and has two alleles that are identical at a specific locus is considered to be homozygous for that locus. A person who carries two distinct alleles at a specific genetic locus is considered to be heterozygous (Lanie et al., 2004).

Phenotype, a significant concept, refers to the observable manifestation or physical expression of a particular trait. The term "phenotype" encompasses various types of characteristics, including physical, physiological, biochemical, and behavioral attributes. Therefore, the manifestation of round seeds, a body weight of 50 kg, and the presence of sickle-cell anemia are all examples of phenotypes. Within the context of this literary work, the terminology employed designates the concept of characteristic or character as a broad attribute, encompassing traits such as eye color. Conversely, the term trait is utilized to denote distinct expressions or manifestations of the said attribute, such as the specific hues of blue or brown eyes (Kessler et al., 2007).

The genotype plays a crucial role in shaping the potential for development by establishing specific boundaries and limitations on the developmental process. The phenotype development within these constraints is contingent upon the interactions of additional genes and environmental factors, with the relative contributions of these influences differing across various traits. In certain instances, the variations observed in phenotypes can be predominantly attributed to disparities in genotypes, indicating that the genetic boundaries governing the expression of the said phenotype are relatively constrained. The seed shape observed in Mendel's peas serves as a compelling illustration of a trait characterized by narrow genetic boundaries and predominantly genetic-based phenotypic variations. In the context of other characters, the significance of environmental disparities takes precedence, wherein the genotype's constraints are relatively extensive. The vertical extent attained by an oak tree upon reaching maturity is a phenotype that is significantly impacted by environmental factors, including the presence of water, sunlight, and nutrients. However, the height of a tree is inherently constrained by its genotype, as exemplified by the fact that an oak tree cannot attain a height of 300 meters, regardless of the abundance of sunlight, water, and fertilizer it receives. Hence, the genetic composition of an oak tree plays a role in determining its height to a certain extent. The determination of phenotypic differences is influenced by both genetic and environmental factors across various characteristics. A fundamental and noteworthy principle is that solely the genotype is passed down through inheritance. While the phenotype of an organism is influenced by its genotype to a certain degree, it is important to note that subsequent generations do not directly inherit phenotypes. The differentiation between genotype and phenotype constitutes a fundamental principle within the realm of contemporary genetics (Cas et al., 2008).

3.5 MONOHYBRID CROSSES

Mendel commenced his research by initially acquiring a total of 34 distinct types of pea plants, subsequently dedicating a span of two years to meticulously scrutinize and deliberate upon the selection of specific varieties to be employed in his experimental investigations. The researcher ensured the genetic purity of each variety by cultivating the plants for two successive generations and observing that all progeny exhibited identical traits to their respective progenitors. Subsequently, a series of controlled breeding experiments were conducted involving diverse varieties. In spite of peas typically exhibiting self-fertilization, Mendel performed crosses between distinct plants by manually manipulating the reproductive structures. This involved carefully exposing the immature anthers, eliminating them, and subsequently applying pollen from a separate plant onto the stigma (Stewart, 1982).

Mendel initiated his research by examining monohybrid crosses, which involved mating parents that exhibited dissimilarities in a single characteristic. In a particular experiment, Mendel conducted a cross between a pea plant that exhibited homozygosity for round seeds and another pea plant that displayed homozygosity for wrinkled seeds (refer to Figure 3.3). The initial generation resulting from a cross is the parental (P) generation (Browning & Lehman, 1988).

Upon conducting a cross between the two distinct varieties in the parental (P) generation, Mendel proceeded to carefully observe and analyze the progeny that arose as a consequence of this cross. With respect to seed characteristics, such as seed shape, the phenotype emerges upon seed maturation, as the seed traits are dictated by the developing embryo contained within the seed. In relation to traits pertaining to the plant organism, such as the length of the stem, the manifestation of the phenotype is not evident until the plant undergoes growth from the seed stage. Consequently, Mendel was required to exercise patience and wait until the subsequent spring season to sow the seeds and subsequently examine the phenotypes exhibited by the germinated plants (Otten et al., 1981).

The progeny resulting from the mating of the parental generation, referred to as the P generation, is designated as the F1 (first filial) generation. Upon analyzing the F1 generation, Mendel observed that they exclusively exhibited one of the phenotypes observed in the parental generation. Specifically, all the F1 seeds displayed a round shape. Mendel conducted a total of 60 crosses, consistently yielding the same outcome.

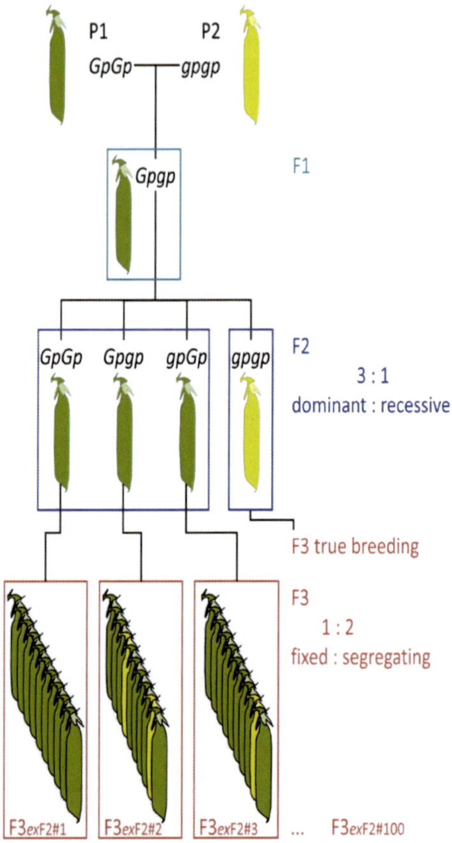

Figure 3.3. Mendel conducted monohybrid crosses (Source: Ellis, T.H.N., Hofer, J.M.I., Swain, M.T. et al. Mendel's pea crosses: varieties, traits and statistics. Hereditas 156, 33 (2019). https://doi.org/10.1186/s41065-019-0111-y, Creative Commons License)

Additionally, reciprocal crosses were performed, wherein one cross involved the extraction of pollen, the male gamete, from a plant exhibiting round seeds. In contrast, the reciprocal cross involved the extraction of pollen from a plant displaying wrinkled seeds. The reciprocal crosses yielded identical outcomes, as all the F1 progeny exhibited a round phenotype. Mendel's investigation extended beyond the analysis of solely the seeds resulting from these monohybrid crosses. In the subsequent spring, the individual proceeded to sow the F1 seeds, nurtured the resulting plants, and facilitated self-fertilization, thereby generating a subsequent generation known as the F2 generation (Ford,

2000). Both of the traits observed in the parental generation (P generation) were present in the second filial generation (F2). Mendel meticulously recorded a total of 5474 round seeds and 1850 wrinkled seeds in the F2 generation, as depicted in Figure 3.3. The observer observed the proportion of round and wrinkled seeds, noting that they occurred in a ratio of approximately 3 to 1. Specifically, it was observed that approximately the F2 seeds exhibited a round shape, while the remaining proportion displayed a wrinkled shape. Mendel performed monohybrid crosses on all seven characteristics under investigation in pea plants. In each cross, a consistent outcome was observed: the F1 generation exhibited resemblance solely to one of the two parental plants, while both parental traits manifested in the F2 generation at a ratio of approximately 3:1 (Manney & Manney, 1993).

3.5.1 What Monohybrid Crosses show

From the results of his monohybrid crosses, Mendel derived several significant conclusions. Initially, the author employed logical deduction to establish that while the F1 plants exclusively exhibit the phenotype of a single parent, they must nonetheless inherit genetic elements from both parents. The coexistence of round and wrinkled seeds in the F2 generation can be accounted for solely if the F1 plants harbored both round and wrinkled genetic traits that were inherited from the parental generation (P). The researcher reached the conclusion that it is necessary for each plant to possess two genetic factors that encode a particular trait (Lassiter, 2010).

The genetic factors that were initially identified by Mendel, known as alleles, are commonly denoted using letters as a convention. Specifically, the allele associated with round seeds is typically symbolized as R, while the allele linked to wrinkled seeds is represented as r. The plants in the parental generation (P generation) of Gregor Mendel's cross exhibited a homozygous genotype, with both alleles being identical. Specifically, the round-seeded parent possessed two copies of the dominant allele (RR), while the wrinkled-seeded parent possessed two copies of the recessive allele (rr) (see Figure 3.4a) (Caldwell & Teagarden, 2010).

Another inference derived by Mendel was the segregation of the two alleles in each plant during gamete formation, with one allele being transmitted to each gamete. The fusion of two gametes, one from each parent, forms a zygote. During this process, the allele contributed by the male parent combines with the allele contributed by the female parent, thereby determining the genotype of the offspring. Therefore, it can be observed that the F1 plants of Mendel inherited a R allele from the round-seeded plant and a r allele from the wrinkled-seeded plant. Nevertheless, it was noted that only the trait governed by the round allele (R) was present in the F1 generation, as all the offspring in the F1 generation exhibited round seeds. The characteristics exhibited no variation in the F1 generation's heterozygous offspring (Ka et al., 2019).

Mendel referred to traits that were expressed in the F1 heterozygous offspring as dominant, while traits that were not observed in the F1 generation were labeled as recessive. In cases where both dominant and recessive alleles coexist, the recessive allele is effectively concealed. The derivation of the concept of dominance from Mendel's monohybrid crosses constituted a significant conclusion. One of Mendel's key findings was that the two alleles of a given plant segregate into the gametes with equal likelihood. When the F1 plants (with the genotype Rr) underwent gamete production, it was observed that 50% of the gametes got the R allele, which is responsible for round seeds, while the remaining 50% received the r allele, responsible for wrinkled seeds. The gametes were subsequently paired in a random manner, resulting in the production of the following genotypes in equal proportions within the F2 generation: RR, Rr, rR, rr (Turner, 2007).

The principles of segregation and dominance have been derived and formalized based on the conclusions drawn by Mendel regarding inheritance through his monohybrid crosses. The principle of segregation postulates that every diploid organism has a pair of alleles for each specific trait. The two alleles undergo segregation during gamete formation, with each allele being allocated to a separate gamete. In addition, it is observed that the two alleles undergo segregation into gametes in a manner that results in equal proportions. The principle of dominance posits that in a genotype containing two distinct alleles, only the observable trait associated with the dominant allele is expressed in the phenotype. Mendel substantiated these principles through the process of self-fertilization in his F2 plants, resulting in the production of an F3 generation. The researcher observed that the F2 generation, derived from the cultivation of wrinkled seeds exhibiting the recessive trait (rr), yielded an F3 generation in which all plants created wrinkled seeds. The wrinkled-seeded plants in question were determined to be homozygous for the wrinkled allele (rr), resulting in the exclusive transmission of wrinkled alleles to their offspring (Morrison, 1935).

The F2 generation of plants derived from round seeds, which represent the dominant trait, can be categorized into two distinct types. In the context of self-fertilization, it was observed that a proportion of the F2 plants yielded a combination of round and wrinkled seeds in the subsequent F3 generation. The F2 generation consisted of plants that were heterozygous (Rr), resulting in the production of seeds with three different genotypes: RR (round), Rr (round), and rr (wrinkled). This led to a phenotypic ratio of 3:1, with three round seeds for every one wrinkled seed in the F3 generation. Approximately half of the F2 plants exhibited the second phenotype, exclusively manifesting the dominant round-seeded trait in the subsequent F3 generation. The F2 generation consisted of individuals that were homozygous for the round allele (RR), resulting in the exclusive production of round offspring in the subsequent F3 generation. Mendel proceeded to sow the seeds acquired from the F3 generation and subjected these plants to an additional three successive cycles of self-pollination. Within each successive cohort, approximately two-thirds of the plants exhibiting round seeds yielded progeny that displayed both round and wrinkled characteristics, while the remaining one-third exclusively produced offspring with round seeds. The obtained results align entirely with the principle of segregation (Awang-Kanak et al., 2016).

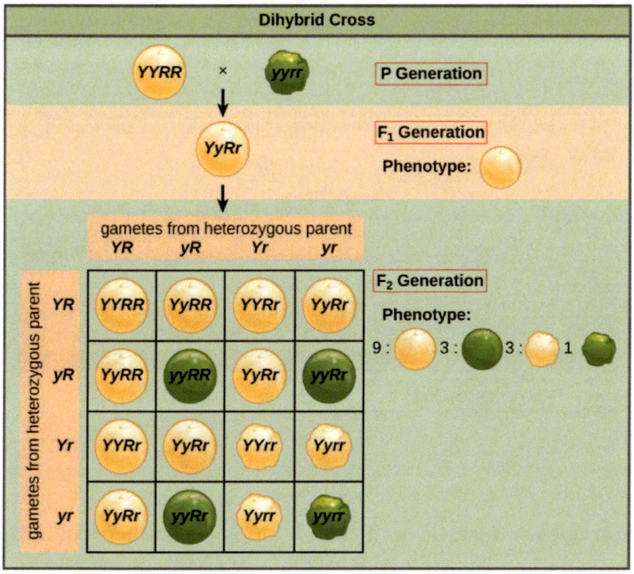

Figure 3.4. Mendel's experiments involving monohybrid crosses provided insights into the principle of segregation and dominance. (Source: By CNX OpenStax - http://cnx.org/contents/GFy_h8cu@10.53:rZudN6XP@2/Introduction, CC BY 4.0, https://commons.wikimedia.org/w/index.php?curid=49926375, Creative Commons License).

3.6 ANTICIPATING THE RESULTS OF GENETIC CROSSES

One of Gregor Mendel's primary objectives was to establish a methodology for accurately predicting the results of genetic crosses between plants exhibiting distinct phenotypic traits. This section will commence by introducing a concise and efficient technique, known as the Punnett square, for forecasting genetic cross outcomes. Subsequently, we will delve into the utilization of probability as a means to anticipate the results of such crosses (Bredahl, 2001).

The Punnett square is a tool used in genetics to predict the possible outcomes of a cross between two individuals. In order to elucidate the concept of the Punnett square, let us analyze an additional genetic cross conducted by Mendel. Mendel's experimental work involved the deliberate hybridization of two distinct pea varieties that exhibited contrasting heights, leading to the establishment of the dominance relationship between the tall (T) and short (t) traits. The individual conducted an experiment to investigate the transmission of dominant characteristics by crossing a first-generation tall plant with heterozygous genotype (Tt) with a short plant of homozygous genotype (tt), which served as the parental variety. The term "backcross" refers to the breeding technique involving the crossing of an F1 genotype with either of the parental genotypes (Darvasi & Soller, 1995).

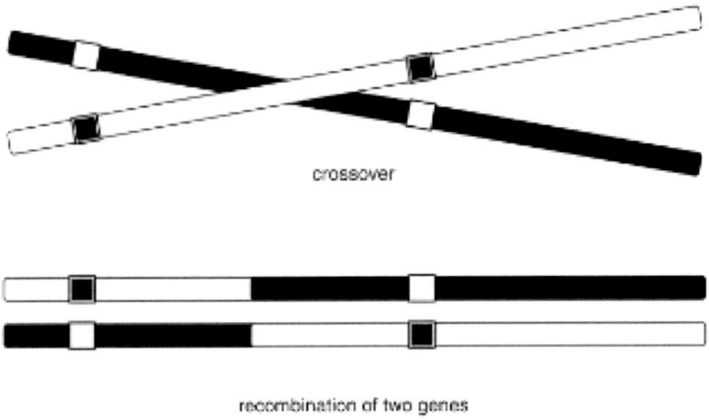

Figure 3.5. Segregation occurs because of the separation of homologous chromosomes during the process of meiosis. (Source: Mneimneh S (2012) Crossing Over…Markov Meets Mendel. PLoS Comput Biol 8(5): e1002462. https://doi.org/10.1371/journal.pcbi.1002462, Creative Commons License).

In order to make predictions regarding the offspring resulting from this cross, it is necessary to ascertain the gametes that will be generated by each parent. The principle of segregation elucidates that during gamete formation, the two alleles present in each parent segregate from one another, with one allele being transmitted to each gamete. All gametes derived from the homozygous tt genotype of a plant will inherit a single allele for short stature (t). The plant exhibiting height in this particular genetic cross possesses heterozygosity (Tt), resulting in an equal distribution of tall alleles (T) and short alleles (t) in its gametes, with each allele accounting for 50% of the total (Kasschau & Carrington, 1998). The construction of a Punnett square involves the delineation of a grid, wherein the gametes generated by one parent are positioned along the upper boundary, while the gametes produced by the other parent are arranged along the left side. Every individual cell, which is represented as a block within the Punnett square, contains one allele from each of the corresponding gametes. This combination of alleles determines the genotype of the offspring resulting from the fusion of those gametes. The progeny's genotype (Tt) is determined by the union of a gamete containing the dominant allele T from the tall plant and a gamete containing the recessive allele t from the short plant, as observed in the upper left-hand cell of the Punnett square. It is advantageous to document the phenotypic manifestation associated with each genotype. In this case, the offspring will exhibit a tall phenotype due to the dominance of the tall allele (Reiter et al., 2001).

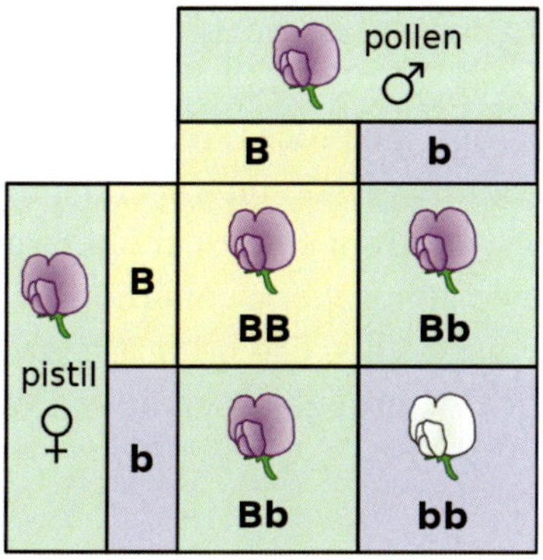

Figure 3.6. The Punnett square is a tool that can be employed to ascertain the outcomes of a genetic cross. (Source: Https://commons.wikimedia.org/wiki/File:Punnett_square_mendel_flowers.svg, Creative Commons License).

The procedure mentioned above is iteratively conducted for each individual cell within the Punnett square. Through the process of enumeration, it is possible to ascertain the various categories of offspring generated as well as their corresponding proportions. According to the data presented in Figure 3.6b, it can be observed that two cells exhibit tall (Tt) progeny, while two cells exhibit short (tt) progeny. Therefore, the anticipated genotypic ratio for this cross is 2 Tt to 2 tt, expressed as a 1:1 ratio. An alternative means of articulating this outcome is to assert that we anticipate the offspring to possess the genotype Tt, resulting in a tall phenotype. At the same time, the remaining progeny are projected to possess the genotype tt, leading to a short phenotype. In the context of this genetic cross, it is observed that the genotypic ratio and the phenotypic ratio exhibit equivalence (Wirtz et al., 1999).

Probability as a tool

An alternative approach to ascertain the result of a genetic cross involves the utilization of probability principles, akin to the methodology employed by Mendel in his crossbreeding experiments. Probability is a mathematical concept that quantifies the likelihood or chance of a specific event taking place. The frequency of a specific event is computed by dividing the total number of occurrences by the total number of possible outcomes (Zhaxybayeva & Gogarten, 2002).

The determination of the probability of a specific event can be facilitated by possessing knowledge regarding the manner or frequency of occurrence of said event. It is evident that the probability of obtaining a four when rolling a six-sided die is 1/6, as each side of the die has an equal likelihood of being the outcome. In this particular scenario, the comprehension of the event's characteristics, specifically the geometric attributes of the thrown die, facilitates the calculation of its probability (Shields et al., 2012). In alternative scenarios, the probability of an event is ascertained through the process of conducting a substantial number of observations. The probability of rain on a specific day, as communicated by a weather forecaster, is derived from empirical observations of numerous days characterized by comparable atmospheric conditions, wherein rain occurred on 40% of those days. The probability has been ascertained empirically in this particular instance, specifically through observation (Vahteristo et al., 2001).

Did You Know?

The seminal event in the field of genetics occurred in 1953 with the identification of the DNA double helix structure by James Watson and Francis Crick. Deoxyribonucleic acid (DNA) serves as the carrier of genetic instructions, playing a crucial role in the determination of an organism's traits and facilitating the transmission of hereditary information.

3.6.1. The Multiplication Rule

Two fundamental principles of probability prove valuable in predicting the ratios of offspring generated in genetic crosses. The initial concept to consider is the multiplication rule, which posits that the likelihood of multiple independent events transpiring concurrently is determined by the product of their individual probabilities. To demonstrate the application of the multiplication rule, let us once again examine the act of rolling dice. The probability of obtaining a four when rolling a single die is. In order to determine the likelihood of rolling a die twice and obtaining two instances of the number four, we can utilize the multiplication principle. The probability of obtaining a four on the first roll is represented by P(first roll = 4), while the probability of obtaining a four on the second roll is denoted as P(second roll = 4). Therefore, the probability of rolling a four on both the first and second rolls can be calculated by multiplying these individual probabilities, resulting in P(first roll = 4) x P(second roll = 4)

= 1/6 x 1/6 = 1/36 (as shown in Figure 3.7a). The primary determinant for the utilization of the multiplication rule is the presence of the conjunction "and" (Honeycutt & Pierce, 2007).

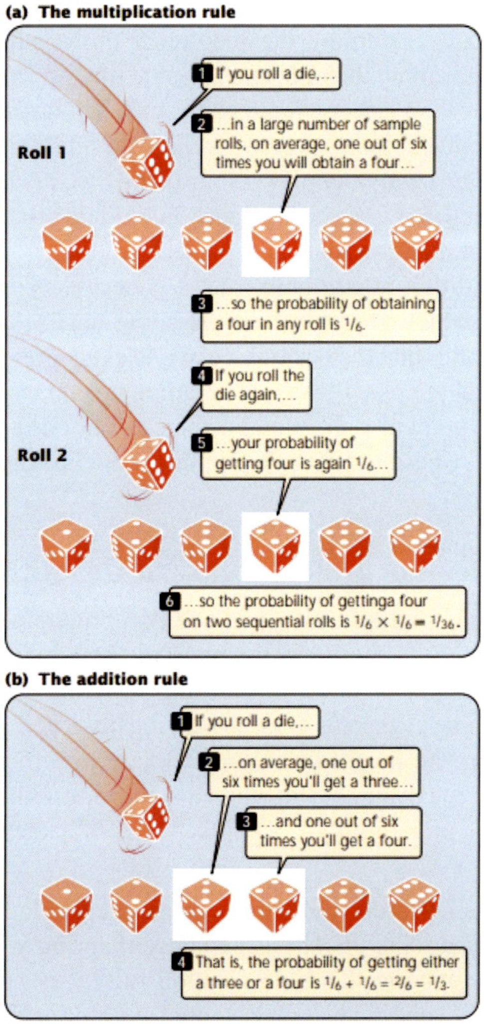

Figure 3.7. The utilization of the multiplication and addition rules enables the determination of the probability associated with combinations of events (Source: Ruddle, Creative Commons License).

In order for the multiplication rule to hold true, it is necessary for the events for which the joint probability is being computed to exhibit independence, meaning that the outcome of one event does not exert any effect on the outcome of the other event (Turelli & Orr, 1995).

3.6.2 The Addition Rule

The addition rule, a commonly employed principle in genetics, pertains to the second rule of probability. This rule posits that when dealing with two or more events that are mutually exclusive, the probability of any one of these events can be determined by summing the individual probabilities associated with each event. Let us examine this rule in a tangible manner. In order to determine the probability of rolling either a three or a four when throwing a die once, the addition rule is applied. This involves adding the individual probabilities of obtaining a three (1/6) and obtaining a four (also 1/6), resulting in a combined probability of 1/6 + 1/6 = 2/6 = 1/3 (Figure 3.7b). The primary determinant for the utilization of the addition rule is the presence of the terms "either" and "or." In order for the addition rule to hold true, it is necessary for the events under consideration, for which the probability is being determined, to be mutually exclusive. This implies that the occurrence of one event precludes the possibility of the other event taking place. As an illustration, it is impossible to roll a single die once and observe both a three and a four, as only one face of the die can be facing upwards. These events exhibit mutual exclusivity (Fernández et al., 2010).

3.6.3 The Application of Probability to Genetic Crosses

The utilization of the multiplication and addition principles of probability can serve as an alternative to the Punnett square methodology for the purpose of forecasting the anticipated ratios of offspring resulting from a genetic cross. First, let us examine a genetic cross involving two pea plants that possess different alleles for the height-determining locus, specifically Tt x Tt. Each plant produces an equal number of gametes carrying either a T allele or a t allele, resulting in a probability of ½ for each type of gamete (Tanriverdi et al., 2007).

he fusion of gametes derived from two parental individuals can result in four distinct combinations, thereby generating offspring with varying genetic compositions. By applying the multiplication rule, it is possible to ascertain the probability associated with each potential type. In order to determine the likelihood of producing TT offspring, it is necessary to multiply the probability of inheriting a T allele from the first parent (1/2) by the probability of inheriting a T allele from the second parent (1/2). The utilization of the multiplication rule is warranted in this context as it enables the determination of the probability associated with the occurrence of two independent events: the transmission of a T allele from the first parent and the transmission of a T allele from the second parent. Four distinct progeny result from this cross, each with their respective probabilities (Raj & Van Oudenaarden, 2008).

$$TT\left(T\ gamete, and\ T\ gamete\right) 1/2 x 1/2 = 1/4\ tall$$

$$Tt\left(T\ gamete, and\ t\ gamete\right) 1/2 x 1/2 = 1/4\ tall$$

$$tT\left(t\ gamete, and\ T\ gamete\right) 1/2 x 1/2 = 1/4\ tall$$

$$tt\left(t\ gamete, and\ t\ gamete\right) 1/2 x 1/2 = 1/4\ short$$

3.6.4 The binomial expansion and probability

When employing probability, it is crucial to acknowledge the existence of multiple potential outcomes for a given set of events. Let us contemplate a scenario wherein two parents possess heterozygosity for albinism, a hereditary trait in humans that manifests as diminished pigmentation in the skin, hair, and eyes (as depicted in Figure 3.8).

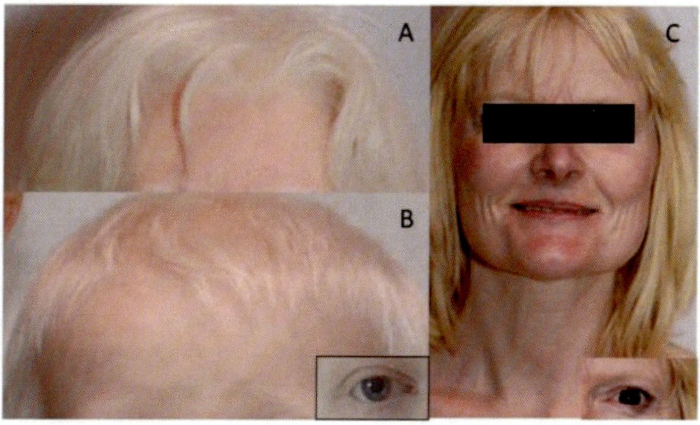

Figure 3.8. The occurrence of albinism in human beings is typically transmitted through the inheritance of a recessive trait (Source: Moreno-Artero E, Morice-Picard F, Lasseaux E, Robert MP, Coste V, Michaud V, Leclerc-Mercier S, Bremond-Gignac D, Arveiler B, Hadj-Rabia S. Oculo-Cutaneous Albinism Type 4 (OCA4): Phenotype-Genotype Correlation. Genes. 2022; 13(12):2198. https://doi.org/1, Creative Commons License).

When two parents who possess the heterozygous genotype for albinism (Aa x Aa) engage in reproduction, the likelihood of their offspring inheriting the homozygous recessive genotype for albinism (aa) is determined to be 1/4. In contrast, the probability of producing offspring with either the homozygous dominant genotype (AA) or the heterozygous genotype (Aa) resulting in normal pigmentation is determined to be 3/4. Consider a scenario in which we aim to determine the likelihood of a couple having three offspring with albinism. In this scenario, the only possible outcome for the couple to have three children with albinism is if their first child, second child, and third child all exhibit albinism. Here we simply apply the multiplication rule: ¼ x ¼ x ¼ = 1/64 (Chynoweth et al., 1986).

Suppose we pose the following inquiry: What is the likelihood of this couple conceiving three offspring, with one displaying albinism and the remaining two exhibiting normal pigmentation? The present scenario exhibits a higher degree of complexity. The initial offspring may exhibit albinism, while the subsequent two offspring remain unaffected. The likelihood of this particular sequence of events occurring is calculated as ¼ x ¾ x ¾ = 9/64 (Woo & Siu, 2004).

In an alternative scenario, the first and third offspring can exhibit typical pigmentation, while the second offspring displays albinism. The likelihood of this specific sequence occurring can be calculated by multiplying the probabilities of each event: ¾ x ¼ x ¾ = 9/64. Ultimately, it is possible that the initial pair of offspring will exhibit typical pigmentation, while the subsequent offspring may display albinism. The likelihood of this specific sequence occurring can be calculated by multiplying the probabilities of each event: ¾ x ¼ x ¾ = 9/64. Due to the occurrence of one child with albinism and two children with normal pigmentation in either the first, second, or third sequence, the addition rule is applied. The probabilities of these events are added together: 9/64 + 9/64 = 27/64 (Cifuentes & O'Connor, 1996).

Determining the probability of a couple having five children, specifically two with albinism and three with normal pigmentation, presents a challenging task as it involves the calculation of various combinations of children and their associated probabilities. The utilization of the binomial expansion facilitates the simplification of this task. The binomial distribution is represented by the expression $(a + b)^n$, where a represents the probability of one specific event, b represents the probability of the alternative event, and n represents the number of occurrences of the event. To calculate the probability of two out of five children having albinism (Janson, 2021):

$$a = the\ probability\ of\ a\ child\ having\ albinism = \frac{1}{4}$$

$$b = the\ probabilty\ of\ a\ child\ having\ normal\ pigmentation = \frac{3}{4}$$

The binomial for this situation is $(a + b)^5$ because there are five children in the family (n = 5). The expansion is:

$$(a+b)^5 = a^5 + 5a^4b + 10a^2b^3 + 5ab^4 + b^5$$

The initial coefficient in the binomial expansion, denoted as a^5, represents the probability of a specific event occurring, namely the birth of five children, all of whom possess the trait of albinism. This probability is contingent upon the value of a, which represents the probability of inheriting the trait of albinism. The second term, denoted as $5a^4b$, represents the probability of having four offspring with albinism and one offspring with normal pigmentation. Similarly, the third term, represented as $10a^3b^2$, corresponds to the probability of having three offspring with albinism and two offspring with normal pigmentation. This pattern continues for subsequent terms. In order to calculate the probability of a specific combination of events, the values of variables a and b are inserted. Thus, the probability of having two out of five children with albinism can be determined by this method (Frame, 1945):

$$10a^2b^3 \aleph 10\,(1/4)^2(3/4)^3 \quad 270/1024 \quad .26$$

The probability of obtaining a particular combination of albinism and pigmentation among a set of five children can be determined by utilizing the remaining terms in the expansion.

In this particular example, the expansion of the binomial was achieved through a specific method or technique. In a general context, the expansion of a binomial $(a + b)^n$ is comprised of a series of n + 1 terms. In the aforementioned example, the value of n is 5, resulting in a total of 6 terms: a^5, $5a^4b$, $10a^3b^2$, $10a^2b^3$, $5ab^4$, and b^5. In order to express the terms in written form, it is necessary to determine their respective exponents. The exponent of the variable "a" in the first term of a binomial expression is always equal to the power to which the binomial is raised, denoted as "n". In the given example, the value of n is 5, thus denoting the initial term as a^5. The exponent of the variable "a" decreases by one in each subsequent term. Therefore, the exponent of "a" is 4 in the second term (a^4), 3 in the third term (a^3), and so on. The first term of the sequence has an exponent of b equal to 0, indicating the absence of b. In each subsequent term, the exponent of b increases by 1, progressing from 0 to 5 in the given example (Thomas & Taub, 1982).

Subsequently, ascertain the coefficient associated with each term. The coefficient of the initial term is consistently 1. Therefore, in the given example, the initial term can be represented as $1a^5$. The coefficient of the second term is consistently equal to the exponent to which the binomial is raised. In the given example, this coefficient is 5, resulting in the term $5a^4b$. An alternative method for calculating the likelihood of a specific combination of events involves the utilization of the subsequent formula:

$$P = \frac{n!}{s!t!} a^s b^t$$

Let P denote the cumulative probability of event X, characterized by the probability a, transpiring s times, and event Y, characterized by the probability b, transpiring t times. In the context of albinism, event X can be defined as the manifestation of albinism in a child, while event Y represents the occurrence of a child with typical pigmentation. The variable 's' denotes the count of children with albinism, which in this case is 2, whereas t represents the count of children with normal pigmentation, which is 3. The factorial symbol, denoted by "!", represents the mathematical operation that calculates the product of all integers from n down to 1. In the given illustration, n = 5, so n! =5 x 4 x 3 x 2 x 1. By utilizing this formula, we can calculate the probability of two out of five children having albinism (Fisher, 1957).

$$P = 5!/2!3!(1/4)^2(3/4)^3$$

$$P = 5\times4\times3\times2\times1\,/2\times1\times3\times2\times1\ (1/4)^2(3/4)^3 = .26$$

3.7 THE TESTCROSS

The testcross is a valuable analytical mechanism for studying genetic crosses. The test cross, as well as the backcross, is depicted in Figure 3.6. A testcross is a method used to ascertain or uncover the genotype of the initial individual. Assume that a vertically elongated pea plant has been provided, devoid of any knowledge regarding its parental lineage (Schön et al., 2004). Due to the dominance of tallness as a genetic trait in peas, the plant in question may possess either a homozygous (TT) or heterozygous (Tt) genotype, although the specific genotype cannot be determined with certainty. The genotype of the organism can be ascertained through the execution of a testcross. In the scenario where the plant possesses a homozygous genotype (TT), conducting a testcross would result in the production of all tall offspring. Conversely, if the plant exhibits a heterozygous genotype (Tt), performing the testcross would yield an equal proportion of tall and short progeny (Tt x tt : Tt and tt) (Grattapaglia & Sederoff, 1994).

Did You Know?

Mutations are alterations in the DNA sequence that have the potential to result in genetic diversity. While certain mutations may have deleterious effects and lead to genetic disorders, others can have advantageous outcomes and contribute to the processes of evolution and adaptation.

3.8 INCOMPLETE DOMINANCE

Mendel's selection of seven characters in pea plants for extensive study was predicated on their manifestation of dominance. However, Mendel acknowledged the existence of characters whose traits do not conform to the principle of dominance. The researcher performed a series of experiments investigating the duration required for pea plants to reach the flowering stage. When the experimenter conducted a cross between two homozygous varieties that exhibited a disparity of 20 days in their flowering time, the F1 plants displayed an intermediate flowering period, falling between that of the two parental varieties. Incomplete dominance is observed when the heterozygote exhibits a phenotype that is intermediate between the phenotypes of the two homozygotes. The phenomenon of incomplete dominance is also observed in the expression of fruit color in eggplants. When a plant that is homozygous for the purple fruit trait (PP) is crossed with a plant that is homozygous for the white fruit trait (pp), all the resulting F1 plants that are heterozygous (Pp) exhibit the phenotype of violet fruit. When the first generation (F1) individuals are bred with each other, the second generation (F2) offspring exhibit three different phenotypes: purple (PP), violet (Pp), and white (pp), as depicted in (Yang et al., 2017).

Figure 3.9. The inheritance of fruit color in eggplant follows an incompletely dominant trait. (Source: Gramazio P, Yan H, Hasing T, Vilanova S, Prohens J and Bombarely A (2019) Whole-Genome Resequencing of Seven Eggplant (Solanum melongena) and One Wild Relative (S. incanum) Accessions Provides New Insights and Breeding Tools for Eggplant Enhancement. Front. Plant Sci. 10:1220. doi: 10.3389/fpls.2019.01220, Creative Commons License)

The observed 1:2:1 ratio deviates from the expected 3:1 ratio that would be anticipated if eggplant fruit color displayed dominant inheritance. In instances where incomplete dominance is observed in a trait, the genotypic ratios and phenotypic ratios of the progeny remain equivalent, as each genotype manifests a distinct phenotype. Acquiring eggplants that exhibit pure breeding for violet fruit is unattainable due to the inherent heterozygosity of all plants bearing violet fruit (Stone, 1968). Feather color in chickens serves as an additional illustration of incomplete dominance. The offspring resulting from the mating of a homozygous black chicken and a homozygous white chicken exhibit a phenotypic expression of gray coloration. When the gray F1 birds are crossed with each other, the resulting F2 birds exhibit a phenotypic ratio of 1 black: 2 grey: 1 white. The phenomenon of leopard white spotting in horses is characterized by an incomplete dominant inheritance pattern, whereby LL horses exhibit a white coat with numerous dark spots. Heterozygous Ll horses, on the other hand, display a reduced number of spots, while ll horses, which are homozygous recessive, do not possess any spots (refer to Figure 3.10) (Klymiuk et al., 2022).

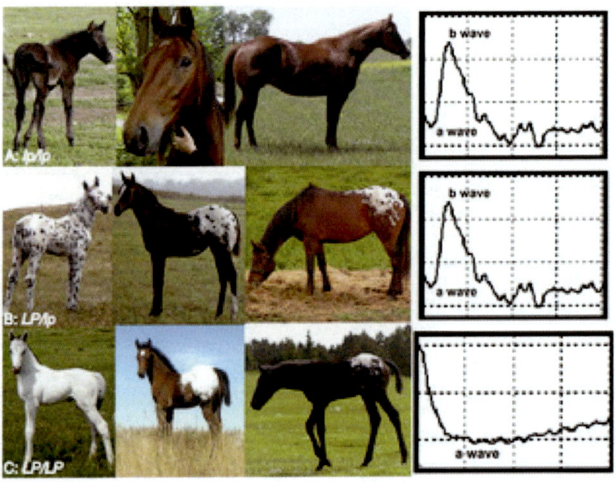

Figure 3.10. Leopard spotting in horses exhibits incomplete dominance (Source: Bellone RR, Holl H, Setaluri V, Devi S, Maddodi N, Archer S, et al. (2013) Evidence for a Retroviral Insertion in TRPM1 as the Cause of Congenital Stationary Night Blindness and Leopard Complex Spotting in the Horse. PLoS ONE 8(10): e78280. https://doi.or, Creative Commons License)

3.9 GENETIC SYMBOLS

Genetic crosses are commonly represented through the utilization of symbols for the purpose of denoting distinct alleles, as observed in previous studies. In the context of genetics, lowercase letters conventionally represent recessive alleles, while uppercase letters are employed to denote dominant alleles. In the context of genetic nomenclature, it is common practice to represent a single allele using two or three letters (Oh et al., 2020). The typical allele for a particular trait, commonly referred to as the wild type due to its prevalence in natural populations, is often represented by one or more letters accompanied by a plus sign (+). The observable characteristics of the mutant organism typically determine the selection of letter(s). In instances where the mutant phenotype is recessive, the initial letter is written in lowercase, while it is capitalized if the mutant phenotype is dominant. An instance of the recessive allele for yellow eyes in the Oriental fruit fly is denoted as ye, while the allele for the typical eye color, known as wild-type, is represented by ye+. Occasionally, the symbols denoting the wild-type allele are omitted, and a mere plus sign instead denotes the allele. Superscripts and subscripts are occasionally employed as a means of differentiation between genetic elements. The alleles Lfr1 and Lfr2 are indicative of dominant traits found at distinct loci in opium poppies, resulting in the manifestation of lacerate leaf margins (Proux et al., 1998).

On the other hand, the allele El^R is associated with goats and is responsible for limiting the length of their ears. Using a slash can serve as a means to differentiate between the alleles present within an individual's genotype (Clutterbuck, 1973). The genotype of a goat exhibiting heterozygosity for the restricted ears trait can be denoted as El^+/El^R. When multiple genotypes at different loci are presented concurrently, they may be separated by a space. The goat exhibits heterozygosity for a pair of alleles responsible for the expression of restricted ears, as well as heterozygosity for another pair of alleles responsible for the occurrence of goiter. This particular genotype can be denoted as El^+/El^R G/g (Berger et al., 2007).

3.10 MULTIPLE LOCI-CROSSES

In this section, we will proceed to expand upon Mendel's principle of segregation (Ohman & Chakrabarty, 1981).

3.10.1 Dihybrid Crosses

In addition to his investigations on monohybrid crosses, Gregor Mendel also conducted experiments involving the crossing of pea plant varieties that exhibited variations in two distinct characteristics, commonly referred to as dihybrid crosses. As an illustration, the individual possessed a homozygous strain of peas that yielded seeds with a round shape and endosperm of a yellow hue. Additionally, another homozygous strain was present, which generated seeds with a wrinkled texture and endosperm exhibiting a green coloration (Marshall, 2008). Upon performing a cross between the two, it was observed that all of the first-generation offspring (F1 progeny) exhibited round seeds and yellow endosperm. Subsequently, the individual proceeded to engage in self-fertilization of the F1 generation, resulting in the acquisition of the subsequent offspring observed in the F2 generation. There were a total of 315 round yellow seeds, 101 wrinkled yellow seeds, 108 round green seeds, and 32 wrinkled green seeds. Mendel observed that the expression of these traits followed an approximate ratio of 9:3:3:1. Specifically, 9/16 of the offspring exhibited a phenotype characterized by round shape and yellow color, 3/16 displayed a phenotype characterized by wrinkled shape and yellow color, 3/16 exhibited a phenotype characterized by round shape and green color, and 1/16 displayed a phenotype characterized by wrinkled shape and green color (Xing et al., 2006).

3.10.2. The Principle of Independent Assortment

Mendel conducted several dihybrid crosses involving pairs of characteristics, consistently observing a 9:3:3:1 ratio in the F2 generation. Including a third principle, namely the principle of independent assortment, provides a coherent explanation for the observed ratio concerning segregation and dominance, as recognized by Mendel in his dihybrid crosses (Peay et al., 2015). While the principle of segregation posits that during gamete formation, the two alleles of a specific locus separate from each other. On the other hand, the principle of independent assortment suggests that the separation of these two alleles is not influenced by the separation of alleles at other loci. Let us examine how the principle of independent assortment elucidates the outcomes observed by Mendel in his dihybrid cross experiment. Each plant contains a pair of alleles responsible for each characteristic, thus indicating that the parental plants must have possessed genotypes RRYY and rryy. The principle of segregation posits that alleles for each locus undergo separation, resulting in the transmission of one allele for each locus to each gamete. The round, yellow parent produces gametes with alleles RY, while the wrinkled, green parent produces gametes with alleles ry. The fusion of these two distinct types of reproductive cells results in the formation of the first filial generation (F1), each possessing the genotype RrYy (Sasaaman & Garthwalte, 1985). Due to the dominance of the round trait over the wrinkled trait and the yellow trait over the green trait, it can be inferred that the phenotype of the F1 generation will exhibit roundness and yellowness.

Upon self-fertilization of the F1 plants by Mendel, the alleles for each locus underwent segregation, wherein a single allele was transmitted to each gamete. The significance of the principle of independent assortment becomes evident in this context. There are two possible ways in which each pair of alleles can segregate: (1) The allele R segregates with the allele Y, while the allele r segregates with the allele y, producing gametes RY and ry. (2) Alternatively, the allele R segregates with the allele y, while the allele r segregates with the allele Y, leading to the production of gametes Ry and rY. The principle of independent assortment states that the alleles at each locus segregate independently. As a result, both types of segregation occur with equal frequency, producing all four types of gametes (RY, ry, Ry, and rY) in equal proportions. Combining the four types of gametes to generate the F2 generation, the resulting offspring exhibit various phenotypes, including round and yellow, wrinkled and yellow, round and green, and wrinkled and green. This phenotypic distribution follows a ratio of 9:3:3:1, as illustrated in Figure 3.11(Kabat et al., 1980).

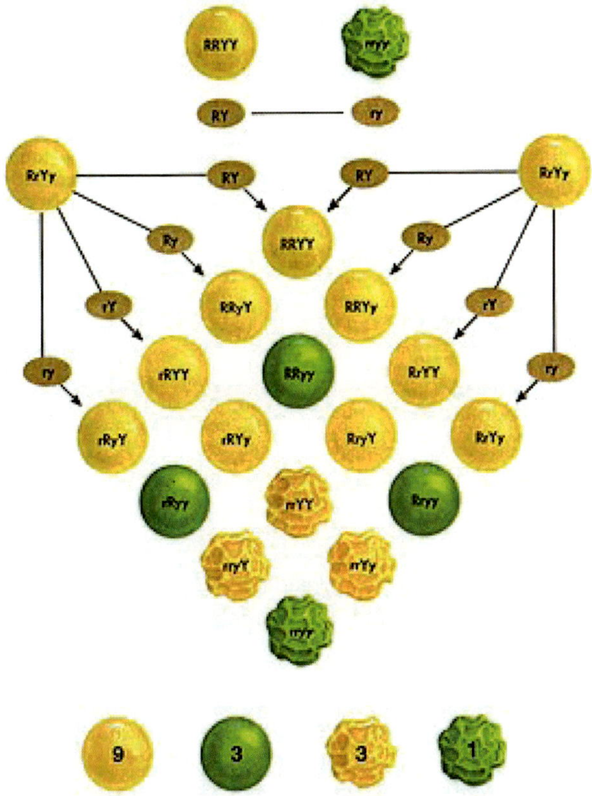

Figure 3.11. Mendel conducted dihybrid crosses (Source: By Archive - https://bs.wikipedia.org/wiki/Dihibridno_ukr%C5%A1tanje#/media/Datoteka:Pusum-dihybrid.jpg, CC BY-SA 4.0, https://commons.wikimedia.org/w/index.php?curid=116991009, Creative Commons License).

3.10.3. The Relation between Meiosis and the Principle of Independent Assortment

One crucial requirement for the principle of independent assortment is that it exclusively pertains to traits governed by genes located on separate chromosomes. This is because, in anaphase I of meiosis, the separation of homologous chromosomes occurs independently for each pair, resulting in an independent assortment of genes situated on different pairs of homologs (Ruch, 1998). Genes that are situated on the same chromosome will undergo co-segregation during anaphase I of meiosis, resulting in their joint transmission to the same gamete unless a genetic recombination event, known as crossing over, occurs. Genes that are situated on the same chromosome exhibit a lack of independent assortment unless they are positioned at a significant distance from each other, resulting in crossing-over occurring during each meiotic division (MacLeod et al., 2005).

Remember

The term "epigenetics" pertains to modifications in the expression of genes that do not entail modifications in the underlying DNA sequence. Epigenetic modifications can be influenced by environmental factors, thereby exerting an impact on the regulation and expression of genes.

3.10.4. Applying Probability & The Branch Diagram to Dihybrid Crosses

When the genes located at two different loci undergo independent assortment, a dihybrid cross can be conceptualized as the combination of two separate monohybrid crosses. In this analysis, we will investigate Mendel's dihybrid cross (RrYy x RrYy) by evaluating each characteristic individually (Figure 3.12a) (Preethi et al., 2020). When examining solely the morphology of the seeds, the genetic cross conducted was between individuals with the genotype Rr and Rr. This particular cross resulted in a phenotypic ratio of 3:1, with 75% of the progeny displaying a round shape and 25% exhibiting a wrinkled shape. Next, it is important to consider another characteristic, namely the color of the endosperm. The genetic cross involved the mating of individuals with the genotype Yy and Yy, resulting in a phenotypic ratio of 3:1. This ratio signifies that 3 out of 4 offspring exhibited the yellow phenotype. At the same time, 1 out of 4 displayed the green phenotype (Garimella & Sahin, 2022).

The monohybrid ratios can be combined using the multiplication rule to determine the proportion of offspring with varying seed shape and color combinations. The probability of obtaining progeny with both round and yellow seeds is: ¾ x ¼ = 3/16. The proportion of offspring exhibiting round and green seeds is calculated as ¾ x ¼ = 3/16. Similarly, the proportion of offspring displaying wrinkled and yellow seeds is determined by ¼ x ¾ = 3/16. Lastly, the proportion of progeny with wrinkled and green seeds is obtained as ¼ x ¼ = 1/16 (Shotwell, 2019). Branch diagrams serve as a practical method for systematically arranging and categorizing various combinations of characteristics, as depicted in Figure 3.12b. In the initial column, enumerate the proportions of phenotypes for a single character. In the subsequent column, please record the proportions of phenotypes for the second character, adjacent to each phenotype listed in the initial column. Place ¾ yellow and ¼ green next to both the round and wrinkled phenotypes. Connect the phenotypes listed in the first column with their respective phenotypes in the second column by drawing lines (Myers et al., 2014). Next, proceed to traverse each branch of the diagram, calculating the product of the probabilities associated with each trait along that particular branch. A single branch exhibits a genetic trait that results in a transition from a round shape to a yellow color, ultimately producing offspring with both a round shape and yellow coloration.

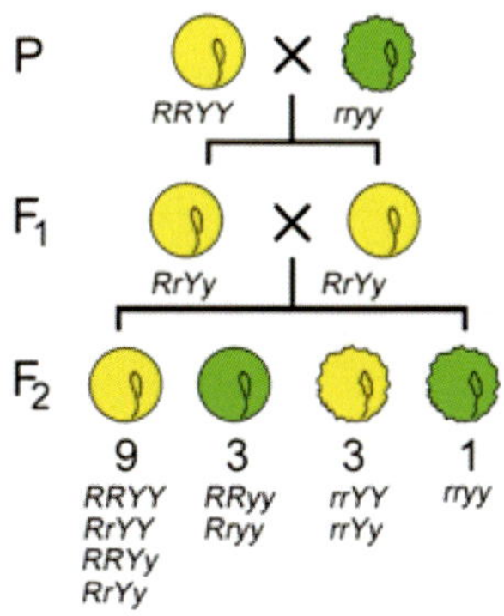

Figure 3.12. A branch diagram is a useful tool in the determination of phenotypes and the anticipated proportions of offspring resulting from a dihybrid cross (Source: Https://bio.libretexts.org/Bookshelves/Genetics/Online_Open_Genetics_%28Nickle_and_Barrette-Ng%29/06%3A_Genetic_Analysis_of_Multiple_Genes/6.01%3A_Dihybrid_Crosses, Creative Commons License)

Another branch emerges from circular to verdant, producing circular and verdant offspring, and so forth. The determination of the likelihood of offspring exhibiting a specific combination of characteristics is accomplished by utilizing the multiplicative principle. In this case, the probability of round seeds and yellow seeds is computed as ¾ x ¾ = 9/16. One benefit of utilizing a branch diagram is its ability to effectively document and organize the various potential combinations of traits that may manifest in the offspring. This tool has the capability to ascertain phenotypic or genotypic ratios for a multitude of characteristics. The utilization of probability proves to be significantly more efficient than the employment of the Punnett square when conducting crosses involving multiple loci (Thomson & Stewart, 1985). The determination of genotypic and phenotypic ratios can be efficiently achieved through the application of the multiplication rule, which involves combining relevant factors. The probability method is highly effective when determining the probability of a specific phenotype or genotype within the offspring resulting from a genetic cross. Assuming the necessity to determine the probability of acquiring the genotype Rryy in the second filial generation (F2) of the dihybrid cross depicted in Figure 3.11 (Patwardhan, 2022). The likelihood of acquiring the Rr genotype in a genetic cross between two individuals with the Rr genotype is 1/2, while the likelihood of producing offspring with the yy genotype in a genetic cross between two individuals with the Yy genotype is ¼. By applying the multiplication rule, the probability of the genotype Rryy can be determined as the product of the probabilities of each allele combination. In this case, the probability of obtaining the Rr genotype is ½, while the probability of obtaining the yy genotype is ¼. Therefore, the probability of the Rryy genotype is calculated as ½ x ¼ = 1/8 (Nath et al., 2021). To elucidate the merits of employing the probability method, let us contemplate the mating between individuals with genotypes AaBbccDdEe and AaBbCcddEe. Suppose there is a desire to ascertain the probability associated with the acquisition of offspring possessing the genotype aabbccddee. If a Punnett square were employed to ascertain this probability, it is plausible that the resolution of this inquiry could extend over a period of several months. Nevertheless, it is possible to expeditiously determine the probability of acquiring this particular genotype by deconstructing this cross into a sequence of individual locus crosses (Portillo, 2019).

Cross	Progeny Genotype	Probability
Aa x Aa	aa	¼
Bb x Bb	bb	¼
cc x Cc	cc	½
Dd x dd	dd	½
Ee x Ee	ee	¼

3.11 THE DIHYBRID TESTCROSS

In order to enhance our understanding of the branch diagram, we shall engage in an exercise involving the identification of phenotypic types and their respective proportions in a dihybrid testcross. This testcross involves the mating of the F1 plants, which possess round and yellow traits (Rr Yy) as a result of Mendel's dihybrid cross, with wrinkled and green plants (rryy), as depicted in Figure 3.13 (Venema, 2006). Disassemble the cross into a sequence of single-locus crosses. The cross between individuals with the genotype Rr and rr results in offspring with a 50% chance of having the round phenotype (Rr) and a 50% chance of having the wrinkled phenotype (rr) (Moemeka et al., 1998). The genetic cross between individuals with genotypes Yy and yy results in an equal proportion of offspring, with 50% exhibiting the yellow phenotype (Yy) and the remaining 50% displaying the green phenotype (yy). By applying the multiplication rule, the resulting proportion of offspring exhibiting both roundness and yellowness is determined as the product of the probability of roundness (½) and the probability of yellowness (½), yielding a value of ¼. The progeny exhibit four distinct trait combinations with the following frequencies: 25% RrYy, characterized by a round shape and yellow color; 25% Rryy, displaying a round shape and green color; 25% rrYy, exhibiting wrinkled shape and yellow color; and 25% rryy, characterized by wrinkled shape and green color (Marais et al., 1998).

	RY	***Ry***	***rY***	***ry***
RY	*RRYY*	*RRYy*	*RrYY*	*RrYy*
Ry	*RRYy*	*RRyy*	*RrYy*	*Rryy*
rY	*RrYY*	*RrYy*	*rrYY*	*rrYy*
ry	*RrYy*	*Rryy*	*rrYy*	*rryy*

Figure 3.13. Understanding the anticipated proportions of offspring resulting from a dihybrid testcross. (Source: Https://bio.libretexts.org/Bookshelves/Genetics/Online_Open_Genetics_%28Nickle_and_Barrette-Ng%29/06%3A_Genetic_Analysis_of_Multiple_Genes/6.01%3A_Dihybrid_Crosses, Creative Commons License)

3.12 TRIHYBRID CROSSES

The utilization of branch diagrams can also be extended to encompass trihybrid crosses, which involve the examination of three distinct characters (Vrijenhoek & Schultz, 1974). In a trihybrid cross, Mendel conducted a breeding experiment involving a pure-breeding variety exhibiting round seeds, yellow endosperm, and gray seed coats, which was crossed with another pure-breeding variety exhibiting wrinkled seeds, green endosperm, and white seed coats (Figure 3.14) (Khrustaleva et al., 2005). The branch diagram illustrates that the anticipated phenotypic ratio in the F2 generation is 27:9:9:9:3:3:3:1. It is noteworthy that the numerical outcomes obtained by Mendel from this particular cross align closely with the projected ratios. In the context of monohybrid crosses, it has been observed that the F2 generation yields three distinct genotypes, namely RR, Rr, and rr. In the context of dihybrid crosses, it is observed that a total of nine genotypes can be generated in the F2 generation (Shihab & Abood, 2019). This is derived from the combination of three genotypes for the first locus and three genotypes for the second locus, resulting in a multiplication of 3 x 3 = 9 possible genotypes.

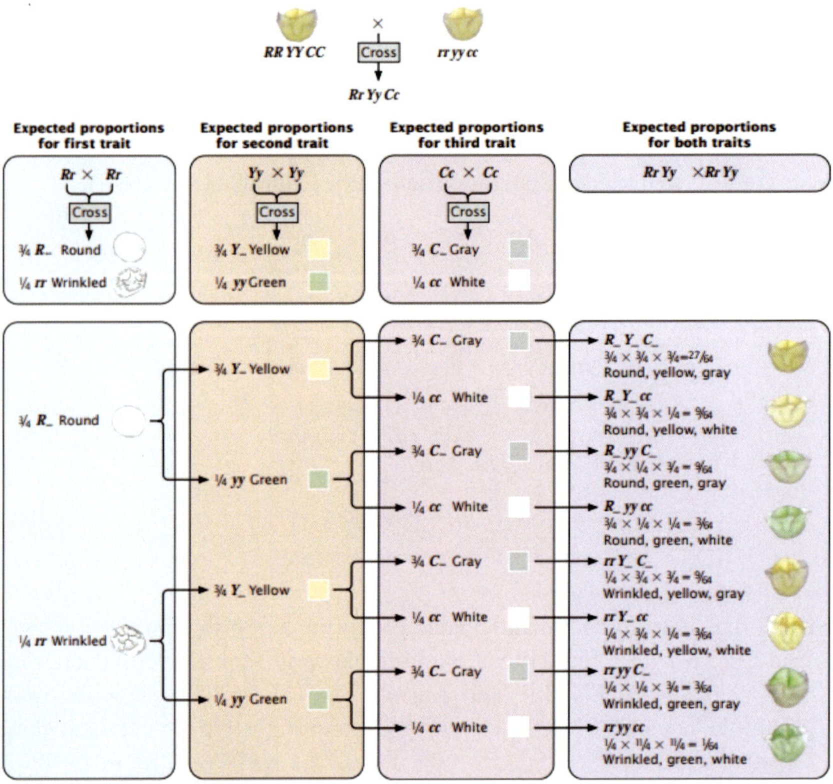

Figure 3.14. A branch diagram is a useful tool for the determination of phenotypes and the anticipated proportions of offspring resulting from a trihybrid cross(Source: Stern, Creative Commons License)

These genotypes include RRYY, RRYy, RRyy, RrYY, RrYy, Rryy, rrYY, rrYy, and rryy. In the case of two alternative alleles at each locus, it is observed that there exist three potential genotypes (Azis et al., 2022). Consequently, when considering the F2 generation resulting from a cross between individuals who are heterozygous for n loci, the total number of genotypes produced can be determined as 3n. In the case of incomplete dominance, it can be observed that the number of phenotypes will also be three times the number of genotypes (3n). This is due to the fact that incomplete dominance results in each genotype expressing a distinct phenotype. If the observed characteristics demonstrate dominance, the resulting number of distinct phenotypes will be twice the value of the number of alleles (2n) (Yan et al., 2020).

Did You Know?

An intriguing aspect of heredity lies in the observation that certain traits adhere to Mendelian inheritance patterns, which are governed by a solitary gene possessing two distinct alleles.

SUMMARY

The chapter titled "Principles of Heredity" delves into the realm of genetics, focusing on the significant contributions made by Gregor Mendel, widely recognized as the progenitor of the field. In this study, we explore fundamental genetic terminology and concepts that serve as the foundation for our comprehension of heredity.

The utilization of genetic terminology serves as a guiding tool in our exploration of the intricate mechanisms of inheritance. In the process of acquiring knowledge, we acquaint ourselves with essential terminologies such as alleles, genotypes, and phenotypes, which serve to elucidate the genetic constitution and observable attributes of individuals. The lexicon presented here plays a pivotal role in elucidating the complexities of genetic crosses and forecasting the results of diverse mating scenarios.

In the context of monohybrid cases, our attention is directed towards the transmission of a single characteristic governed by a pair of distinct alleles. The comprehension of the interplay between dominant and recessive alleles facilitates the ability to anticipate the inheritance patterns of traits in progeny arising from particular genetic crosses.

In order to address genetic crosses that involve multiple traits of increased complexity, the multiplication, and addition rules are utilized. The multiplication rule enables the determination of the probability of concurrent occurrences of specific combinations of traits, whereas the addition rule facilitates the calculation of the probability of inheriting any one of multiple alternative traits.

Probability and the binomial expansion are fundamental concepts in genetics, as they allow for precise predictions to be made regarding the transmission of traits. Through the utilization of mathematical principles, valuable insights are acquired regarding the probabilities associated with various outcomes in genetic crosses, thereby augmenting our comprehension of the fundamental principles governing heredity.

The testcross is a valuable method utilized in deducing an unknown genotype by mating an individual with a dominant phenotype but an uncertain genotype with another individual that is homozygous recessive. The conducted cross facilitates the elucidation of the genetic constitution of the subject under examination, thereby providing insights into the fundamental patterns of inheritance.

As we delve further into the intricacies of heredity, we are confronted with instances of incomplete dominance. In these instances, neither allele exhibits complete dominance, resulting in the manifestation of intermediate phenotypic traits. The comprehension of the extensive range of genetic inheritance is enhanced through the understanding of the patterns of incomplete dominance.

In order to broaden our investigation, we proceed to examine multiple loci crosses, which entail the interplay of genes situated on distinct chromosomes. The presence of complex genetic interactions plays a significant role in the extensive variability observed among populations, thereby influencing the range of characteristics exhibited by individuals.

In summary, the chapter titled "Principles of Heredity" provides valuable insights into the lasting impact of Gregor Mendel, who is widely recognized as a seminal figure in the field of genetics. We acquire knowledge of fundamental genetic terminology, which empowers us to navigate the intricate aspects of genetic inheritance with assurance. By utilizing monohybrid cases and employing probability and mathematical principles, we are able to make accurate predictions regarding the outcomes of genetic crosses. The testcross, incomplete dominance, and multiple loci crosses contribute to the advancement of our comprehension regarding the various mechanisms that regulate heredity. These concepts shed light on the complex interactions among genes, which play a crucial role in shaping the intricate fabric of life.

REFERENCES

1. Awang-Kanak, F., Masnoddin, M., Matawali, A., Daud, M. A., & Jumat, N. R. (2016). Difficulties experience by science foundation students on basic mendelian genetics topic: A preliminary study. *Transactions on Science and Technology*, *3*(1-2), 283-290.
2. Azis, D., Syanur, M. N., Amanto, A., & Zakaria, L. (2022). An Analysis Genotype Inheritance in a Trihybrid Cross by Applying a Diagonalization of Matrix Method. *Desimal: Jurnal Matematika*, *5*(3), 305-314.
3. Bateson, W. (1903). Mendel's principles of heredity in mice. *Nature*, *68*(1750), 33-34.
4. Berger, S. I., Posner, J. M., & Ma'ayan, A. (2007). Genes2Networks: connecting lists of gene symbols using mammalian protein interactions databases. *BMC Bioinformatics*, *8*, 1-6.
5. Berry, R. J. (1967). Genetics: a survey of the principles of heredity. *The Eugenics Review*, *59*(2), 117.
6. Bonduriansky, R. (2012). Rethinking heredity, again. *Trends in ecology & evolution*, *27*(6), 330-336.
7. Bonneuil, C. (2006). Mendelism, plant breeding and experimental cultures: Agriculture and the development of genetics in France. *Journal of the History of Biology*, *39*, 281-308.
8. Bredahl, L. (2001). Determinants of consumer attitudes and purchase intentions with regard to genetically modified food–results of a cross-national survey. *Journal of consumer policy*, *24*(1), 23-61.
9. Browning, M. E., & Lehman, J. D. (1988). Identification of student misconceptions in genetics problem solving via computer program. *Journal of Research in Science Teaching*, *25*(9), 747-761.
10. Caldwell, J., & Teagarden, K. (2010). What caused the ghostly leaves? inquiry-based investigation of the genetics and molecular biology of corn albinism. In *Tested Studies for Laboratory Teaching, Proceedings of The 31st Workshop/Conference of The Association for Biology Laboratory Education (ABLE), Oklahoma* (pp. 50-65).
11. Cas, R., Porritt, L., Pittari, A., & Hayman, P. (2008). A new approach to kimberlite facies terminology using a revised general approach to the nomenclature of all volcanic rocks and deposits: descriptive to genetic. *Journal of Volcanology and Geothermal Research*, *174*(1-3), 226-240.
12. Charnley, B., & Radick, G. (2013). Intellectual property, plant breeding and the making of Mendelian genetics. *Studies in History and Philosophy of Science Part A*, *44*(2), 222-233.

13. Chynoweth, G. H., Blankinship, D. A., & Parker, M. W. (1986). The binomial expansion: Simplifying the evaluation process. *Journal of Counseling & Development*, *64*(10), 645-647.
14. Cifuentes, A., & O'Connor, G. (1996). The binomial expansion method applied to CBO/CLO analysis. *Moody's Investor Services, December*, *13*, 584.
15. Clutterbuck, A. J. (1973). Gene symbols in *Aspergillus nidulans*. *Genetics Research*, *21*(3), 291-296.
16. Cobb, M. (2006). Heredity before genetics: a history. *Nature Reviews Genetics*, *7*(12), 953-958.
17. Darvasi, A., & Soller, M. (1995). Advanced intercross lines, an experimental population for fine genetic mapping. *Genetics*, *141*(3), 1199-1207.
18. Das, S. S. (2022). Mendel paved the path toward understanding genetic diseases. *Egyptian Journal of Medical Human Genetics*, *23*(1), 1-3.
19. De Castro, M. (2016). Johann Gregor Mendel: paragon of experimental science. *Molecular Genetics & Genomic Medicine*, *4*(1), 3.
20. Delhommeau, F., Cynober, T., Schischmanoff, P. O., Rohrlich, P., Delaunay, J., Mohandas, N., & Tchernia, G. (2000). Natural history of hereditary spherocytosis during the first year of life. *Blood, The Journal of the American Society of Hematology*, *95*(2), 393-397.
21. Dreimanis, A. (1989). Tills: their genetic terminology and classification. *Genetic classification of glacigenic deposits*, 58, 17-83.
22. Edlin, G. J. (1987). Inappropriate use of genetic terminology in medical research: a public health issue. *Perspectives in Biology and Medicine*, *31*(1), 47-56.
23. Fernández, A., García, S., Luengo, J., Bernadó-Mansilla, E., & Herrera, F. (2010). Genetics-based machine learning for rule induction: state of the art, taxonomy, and comparative study. *IEEE Transactions on Evolutionary Computation*, *14*(6), 913-941.
24. Fisher, R. A. (1957). The underworld of probability. *Sankhyā: The Indian Journal of Statistics (1933-1960)*, *18*(3/4), 201-210.
25. Ford, R. H. (2000). Inheritance of kernel color in corn: explanations & investigations. *The American Biology Teacher*, 102, 181-188.
26. Frame, J. S. (1945). Mean deviation of the binomial distribution. *The American Mathematical Monthly*, *52*(7), 377-379.
27. Froggatt, P., & Nevin, N. C. (1971). Galton's 'Law of Ancestral Heredity': its influence on the early development of human genetics. *History of Science*, *10*(1), 1-27.

28. Garimella, U., & Sahin, N. (2022). Inheritance Patterns: Probability Rules & Probability Trees. *The American Biology Teacher*, *84*(1), 22-27.

29. Grattapaglia, D., & Sederoff, R. (1994). Genetic linkage maps of *Eucalyptus grandis* and *Eucalyptus urophylla* using a pseudo-testcross: mapping strategy and RAPD markers. *Genetics*, *137*(4), 1121-1137.

30. Hirsch, J. (1997). Some history of heredity-vs-environment, genetic inferiority at Harvard (?), and The (incredible) Bell Curve. *Genetica*, *99*, 207-224.

31. Honeycutt, B. B., & Pierce, B. A. (2007). Illustrating probability in genetics with hands-on learning: making the math real. *The American Biology Teacher*, *69*(9), 544-551.

32. Horin, P. (1998). Biological principles of heredity of and resistance to disease. *Revue Scientifique et Technique (International Office of Epizootics)*, *17*(1), 302-314.

33. Jacob, F. (2022). *The logic of life: A history of heredity*. Princeton University Press. (Vol. 62, pp. 987-990).

34. Jamieson, A., & Radick, G. (2017). Genetic determinism in the genetics curriculum: An exploratory study of the effects of Mendelian and Weldonian emphases. *Science & Education*, *26*(10), 1261-1290.

35. Janson, S. (2021). On the probability that a binomial variable is at most its expectation. *Statistics & Probability Letters*, *171*, 109020.

36. Johnson, S. K., & Stewart, J. (2002). Revising and assessing explanatory models in a high school genetics class: A comparison of unsuccessful and successful performance. *Science Education*, *86*(4), 463-480.

37. Ka, K., Ryoo, R., Jang, Y., Park, Y., Jeong, Y., Kang, J., ... & Jeon, S. (2019). Characteristics of fruiting bodies formed upon monohybrid cross of *Lentinula edodes* strains. *Korean Journal of Mycology*, *47*(2), 173-179.

38. Kabat, E. A., Wu, T. T., & Bilofsky, H. O. W. A. R. D. (1980). Evidence indicating independent assortment of framework and complementarity-determining segments of the variable regions of rabbit light chains. Delineation of a possible J minigene. *The Journal of Experimental Medicine*, *152*(1), 72-84.

39. Kasschau, K. D., & Carrington, J. C. (1998). A counterdefensive strategy of plant viruses: suppression of posttranscriptional gene silencing. *Cell*, *95*(4), 461-470.

40. Kessler, L., Collier, A., & Halbert, C. H. (2007). Knowledge about genetics among African Americans. *Journal of Genetic Counseling*, *16*, 191-200.

41. Kevles, D. J. (1995). *In the name of eugenics: Genetics and the uses of human heredity*. Harvard University Press. (Vol. 2, pp. 95-98).

42. Khrustaleva, L. I., De Melo, P. E., Van Heusden, A. W., & Kik, C. (2005). The integration of recombination and physical maps in a large-genome monocot using haploid genome analysis in a trihybrid Allium population. *Genetics*, *169*(3), 1673-1685.
43. Klymiuk, V., Chawla, H. S., Wiebe, K., Ens, J., Fatiukha, A., Govta, L., ... & Pozniak, C. J. (2022). Discovery of stripe rust resistance with incomplete dominance in wild emmer wheat using bulked segregant analysis sequencing. *Communications Biology*, *5*(1), 826.
44. Lanie, A. D., Jayaratne, T. E., Sheldon, J. P., Kardia, S. L., Anderson, E. S., Feldbaum, M., & Petty, E. M. (2004). Exploring the public understanding of basic genetic concepts. *Journal of Genetic Counseling*, *13*(4), 305-320.
45. Lassiter, C. S. (2010). An Inexpensive Method to Simulate a Monohybrid Cross Using Wild-Type Zebrafish. *Bioscene: Journal of College Biology Teaching*, *36*(1), 20-25.
46. Lombardo, P. A. (2018). The power of heredity and the relevance of eugenic history. *Genetics in Medicine*, *20*(11), 1305-1311.
47. Lynch, H. T., Smyrk, T. C., Watson, P., Lanspa, S. J., Lynch, J. F., Lynch, P. M., ... & Boland, C. R. (1993). Genetics, natural history, tumor spectrum, and pathology of hereditary nonpolyposis colorectal cancer: an updated review. *Gastroenterology*, *104*(5), 1535-1549.
48. MacLeod, A., Tweedie, A., McLellan, S., Taylor, S., Cooper, A., Sweeney, L., ... & Tait, A. (2005). Allelic segregation and independent assortment in *T. brucei* crosses: proof that the genetic system is Mendelian and involves meiosis. *Molecular and biochemical parasitology*, *143*(1), 12-19.
49. Mange, A. P. (1959). Genetics, A Survey of the Principles of Heredity. *American Journal of Human Genetics*, *11*(1), 86.
50. Manney, T. R., & Manney, M. L. (1993). Using yeast genetics to generate a research environment. *Genetics*, *134*(1), 387-391.
51. Marais, GF*, Wesselsa, WG*, Horn, M.* and Du Toit, F. (1998). Association of a stem rust resistance gene (Sr45) and two Russian wheat aphid resistance genes (Dn5 and Dn7) with mapped structural loci in common wheat. *South African Journal of Plant and Soil*, *15*(2), 67-71.
52. Marsh, M. D. (1911). Mendel's principles of heredity. *East African Geographical Review*, *2*(3), 52-60.
53. Marshall, P. A. (2008). Mapping linked genes in *Drosophila melanogaster* using data from the F2 generation of a dihybrid cross. *The American Biology Teacher*, *70*(9), 554-556.
54. Moemeka, A. N., Hildebrandt, A. L., Radaskiewicz, P., & King, T. R. (1998). Brief communication. Shorn (shn): a new mutation causing

hypotrichosis in the Norway rat. *Journal of Heredity*, *89*(3), 257-260.

55. Morrison, T. F. (1935). A Demonstration Model for Genetics. *School Science and Mathematics*, *35*(6), 571-575.
56. Müller-Wille, S., & Rheinberger, H. J. (Eds.). (2007). *Heredity produced: at the crossroads of biology, politics, and culture.* Mit Press. 4(54), *1500-1870.*
57. Myers, R. B., Millman, B., & Noor, M. A. (2014). Genetics and evolution: an iOS application to supplement introductory courses in transmission and evolutionary genetics. *G3: Genes, Genomes, Genetics*, *4*(5), 779-781.
58. Nath, B. B., Roy, J. K., & Lakhotia, S. C. (2021). Genetic crosses with *Drosophila melanogaster* to understand Mendelian inheritance. *Academy of Sci*, 67, 124-126.
59. Oh, S., Abdelnabi, J., Al-Dulaimi, R., Aggarwal, A., Ramos, M., Davis, S., ... & Waldron, L. (2020). HGNChelper: identification and correction of invalid gene symbols for human and mouse. *F1000Research*, *9*, 8722-8725.
60. Ohman, D. E., & Chakrabarty, A. M. (1981). Genetic mapping of chromosomal determinants for the production of the exopolysaccharide alginate in a *Pseudomonas aeruginosa* cystic fibrosis isolate. *Infection and immunity*, *33*(1), 142-148.
61. Otten, L. A. B. M., De Greye, H., Hernalsteens, J. P., Van Montagu, M., Schieder, O., Straub, J., & Schell, J. (1981). Mendelian transmission of genes introduced into plants by the Ti plasmids of Agrobacterium tumefaciens. *Molecular and General Genetics MGG*, *183*, 209-213.
62. Patwardhan, D. (2022). Mendelian Principle of Inheritance. In *Genetics Fundamentals Notes* Singapore: Springer Nature Singapore. (Vol. 3, pp. 53-83).
63. Peay, K. G., Russo, S. E., McGuire, K. L., Lim, Z., Chan, J. P., Tan, S., & Davies, S. J. (2015). Lack of host specificity leads to independent assortment of dipterocarps and ectomycorrhizal fungi across a soil fertility gradient. *Ecology Letters*, *18*(8), 807-816.
64. Portillo, B. I. (2019). "Finding Probability": A Tool to Develop Mastery in Solving Mendelian Genetics Dihybrid Cross. *Ascendens Asia Journal of Multidisciplinary Research Abstracts*, *3*(2), 98-100.
65. Preethi, D., Vimala, J., & Rajareega, S. (2020). A systematic study in the applications of fuzzy hyperlattice. *AIMS Mathematics*, *6*(2), 1695-1705.
66. Proux, D., Rechenmann, F., Julliard, L., Pillet, V., & Jacq, B. (1998). Detecting gene symbols and names in biological texts a first step toward pertinent information extraction. *Genome Informatics*, *9*, 72-80.

67. Raj, A., & Van Oudenaarden, A. (2008). Nature, nurture, or chance: stochastic gene expression and its consequences. *Cell*, *135*(2), 216-226.
68. Reiter, L. T., Potocki, L., Chien, S., Gribskov, M., & Bier, E. (2001). A systematic analysis of human disease-associated gene sequences in *Drosophila melanogaster*. *Genome Research*, *11*(6), 1114-1125.
69. Riggs, L., Collins, B. C., Kleinert, H., & Knight, V. F. (2013). Teaching principles of heredity to high school students with moderate and severe disabilities. *Research and Practice for Persons with Severe Disabilities*, *38*(1), 30-43.
70. Ruch, D. G. (1998). A cookie model for the development of the concept of independent assortment. *The American Biology Teacher*, *60*(9), 696-698.
71. Sadriyevna, A. M. (2020). Science of Genetics and a Brief History of Its Creation. the Creation of the Laws of Heredity. *European Scholar Journal*, *1*(3), 14-15.
72. Samerski, S. (2006). The unleashing of genetic terminology: how genetic counselling mobilizes for risk management. *New Genetics and Society*, *25*(2), 197-208.
73. Sasaaman, C., & Garthwalte, R. L. (1985). Independent assortment of four loci in the Isopod, *Porcellio dilatatus*. *Journal of Heredity*, *76*(6), 470-471.
74. Schmale, G. A., Conrad 3rd, E. U., & Raskind, W. H. (1994). The natural history of hereditary multiple exostoses. *Jbjs*, *76*(7), 986-992.
75. Schön, C. C., Utz, H. F., Groh, S., Truberg, B., Openshaw, S., & Melchinger, A. E. (2004). Quantitative trait locus mapping based on resampling in a vast maize testcross experiment and its relevance to quantitative genetics for complex traits. *Genetics*, *167*(1), 485-498.
76. Schwarzbach, E., Smýkal, P., Dostál, O., Jarkovská, M., & Valova, S. (2014). Gregor J. Mendel-genetics founding father. *Czech Journal of Genetics and Plant Breeding*, *50*(2), 43-51.
77. Shields, B. M., McDonald, T. J., Ellard, S., Campbell, M. J., Hyde, C., & Hattersley, A. T. (2012). The development and validation of a clinical prediction model to determine the probability of MODY in patients with young-onset diabetes. *Diabetologia*, *55*, 1265-1272.
78. Shihab, K. M., & Abood, I. D. (2019). Genetic segregation of tomato trihybrid, double cross and detection of Mi1. 2, Mi-3 resistance genes against root-knot nematode (Meloidogyne spp.). *International Journal of Agricultural and Statistical Sciences*, *15*(1), 153-162.
79. Shotwell, M. (2019). The Misuse of Genetics: The Dihybrid Cross & the Threat of "Race Crossing". *The American Biology Teacher*, *81*(1), 3-10.

80. Stewart, J. H. (1982). Difficulties experienced by high school students when learning basic Mendelian genetics. *The American Biology Teacher*, *44*(2), 80-89.
81. Stone, B. (1968). A formula for determining degree of dominance in cases of monofactorial inheritance of resistance to chemicals. *Bulletin of the World Health Organization*, *38*(2), 325.
82. Tanriverdi, S., Blain, J. C., Deng, B., Ferdig, M. T., & Widmer, G. (2007). Genetic crosses in the apicomplexan parasite *Cryptosporidium parvum* define recombination parameters. *Molecular Microbiology*, *63*(5), 1432-1439.
83. Thomas, M. A., & Taub, A. E. (1982). Calculating binomial probabilities when the trial probabilities are unequal. *Journal of Statistical Computation and Simulation*, *14*(2), 125-131.
84. Thomson, N., & Stewart, J. (1985). Secondary school genetics instruction: making problem solving explicit and meaningful. *Journal of Biological Education*, *19*(1), 53-62.
85. Turelli, M., & Orr, H. A. (1995). The dominance theory of Haldane's rule. *Genetics*, *140*(1), 389-402.
86. Turner, H. M. (2007). Analogy & demonstration: what is the chromosomal basis for Mendel's second theory? *The American Biology Teacher*, *69*(1), 8-8.
87. Vahteristo, P., Eerola, H., Tamminen, A., Blomqvist, C., & Nevanlinna, H. (2001). A probability model for predicting BRCA1 and BRCA2 mutations in breast and breast-ovarian cancer families. *British journal of cancer*, *84*(5), 704-708.
88. Vandervorst, M., Liebaers, I., Sermon, K., Staessen, C., De Vos, A., Van de Velde, H., ... & Devroey, P. (1998). Successful preimplantation genetic diagnosis is related to the number of available cumulus-oocyte complexes. *Human reproduction (Oxford, England)*, *13*(11), 3169-3176.
89. Venema, D. R. (2006). Enhancing undergraduate teaching and research with a Drosophila virginizing system. *CBE—Life Sciences Education*, *5*(4), 353-360.
90. Vrijenhoek, R. C., & Schultz, R. J. (1974). Evolution of a trihybrid unisexual fish (Poeciliopsis, Poeciliidae). *Evolution*, *28*(2), 306-319.
91. Wirtz, E., Leal, S., Ochatt, C., & Cross, G. M. (1999). A tightly regulated inducible expression system for conditional gene knock-outs and dominant-negative genetics in *Trypanosoma brucei*. *Molecular and biochemical parasitology*, *99*(1), 89-101.

92. Wolfe, A. J. (2012). The Cold War context of the Golden Jubilee, or, why we think of Mendel as the father of Genetics. *Journal of the History of Biology*, *45*(3), 389-414.
93. Woo, W. H., & Siu, T. K. (2004). A dynamic binomial expansion technique for credit risk measurement: a Bayesian filtering approach. *Applied Mathematical Finance*, *11*(2), 165-186.
94. Xing, Y., Ingvardsen, C., Salomon, R., & Lübberstedt, T. (2006). Analysis of sugarcane mosaic virus resistance in maize in an isogenic dihybrid crossing scheme and implications for breeding potyvirus-resistant maize hybrids. *Genome*, *49*(10), 1274-1282.
95. Yan, X., Cheng, M., Li, Y., Wu, Z., Li, Y., Li, X., ... & Tang, Q. (2020). Tripsazea, a novel trihybrid of *Zea mays*, *Tripsacum dactyloides*, and *Zea perennis*. *G3: Genes, Genomes, Genetics*, *10*(2), 839-848.
96. Yang, J., Mezmouk, S., Baumgarten, A., Buckler, E. S., Guill, K. E., McMullen, M. D., ... & Ross-Ibarra, J. (2017). Incomplete dominance of deleterious alleles contributes substantially to trait variation and heterosis in maize. *PLoS Genetics*, *13*(9), e1007019.
97. Zhaxybayeva, O., & Gogarten, J. P. (2002). Bootstrap, Bayesian probability and maximum likelihood mapping: exploring new tools for comparative genome analyses. *BMC Genomics*, *3*(1), 1-15.

CHAPTER 4

PRINCIPLES OF GENETIC TRANSMISSION

Contents

4.1 INTRODUCTION

The field of genetics, an intriguing and complex discipline within the realm of biology, explores the mechanisms of heredity and the diversity of characteristics observed in organisms. Genetic transmission is a fundamental mechanism by which genetic information is transferred from one generation to the succeeding one, influencing the traits and biodiversity observed in the Earth's biosphere. The principles of genetic transmission have undergone significant development, from Gregor Mendel's discoveries in the 19th century and the revolutionary advancements in molecular biology. These advancements have established genetic transmission as a fundamental pillar of contemporary scientific knowledge (Lanigan et al., 2020).

To comprehend the fundamental concepts of genetic transmission, it is imperative to embark on a historical exploration that traces its origins to the 19th century, during which the fundamental tenets of genetics were established. Gregor Mendel, a member of the Augustinian monastic order, conducted pioneering experiments involving pea plants, which led to the discovery of distinct heritable units that dictate the transmission of traits. Mendel's laws of segregation and independent assortment, derived from his meticulously conducted experiments, revealed the fundamental principles of genetic transmission in a straightforward and refined manner (Yang & Kim, 2018).

The genetic material's enigmatic nature was unraveled through developments in molecular biology during the 20th century. The seminal revelation of the double-helical configuration of deoxyribonucleic acid (DNA) by James Watson and Francis Crick in 1953 brought about a profound transformation within genetics. This significant discovery yielded an essential understanding regarding the storage, replication, and transmission of genetic information across successive generations. Scientists facilitated the comprehension of genetic transmission at the molecular level by understanding the principles of complementary base pairing and the central dogma of molecular biology (Emerson et al., 2017).

In light of Mendel's pioneering research, contemporary geneticists have significantly advanced our understanding of inheritance patterns, surpassing the conventional notions of simple dominance and recessiveness. This chapter comprehensively reviews Mendelian genetics, delving into more intricate concepts, including numerous alleles, incomplete dominance, codominance, and gene interactions. The author presents numerous instances from the realms of flora and fauna to enhance readers' comprehension of the intricate interplay between genes, ultimately resulting in the wide array of observable phenotypic variations found in the natural world (Li & Selloni, 2014).

During the early 20th century, a significant correlation between genetics and cytology emerged, marking a groundbreaking development. The seminal work conducted by Thomas Hunt Morgan and his colleagues led to the discovery that genes are situated on chromosomes, thereby establishing the fundamental framework for comprehending the chromosomal underpinnings of inheritance. This chapter delves into examining the physical association between genes and chromosomes, elucidating how specific traits are interconnected due to their proximity to a shared chromosome. Moreover, this study delves into the intriguing realm of sex-linked inheritance and investigates the impact of sex chromosomes on the manifestation of particular characteristics (Fasoula & Fasoula, 2002).

Meiosis, a specialized form of cellular division that takes place during the formation of gametes, is of paramount importance in the transmission of genetic information. The present chapter provides an in-depth analysis of the process of meiosis, examining the complex mechanisms by which homologous chromosomes align, engage in genetic recombination through crossing-over, and subsequently segregate to generate gametes that exhibit a high degree of genetic diversity. Examining independent assortment during metaphase I and II also focuses on its significance in contributing to genetic variation and the extensive diversity of offspring in sexually reproducing organisms (Altenberg & Feldman, 1987).

Although Mendel's laws offer a robust framework for comprehending inheritance, it is important to acknowledge that the principles of probability also influence genetic transmission. This chapter examines the application of probability theory in predicting the result of genetic crosses and determining the probability of specific phenotypes in offspring. Punnett squares and other analytical instruments facilitate the comprehension of genetic data, thereby illustrating genetic transmission's intrinsic unpredictability and stochastic nature (Piaget, 2013).

The principles governing genetic transmission are not solely of scientific interest but also bear significant consequences for human health and inheritance. This chapter provides an in-depth exploration of human genetics, specifically focusing on transmitting diverse traits and disorders within the human population. The investigation of pedigree analysis, genetic counseling, and advancements in techniques such as DNA sequencing contributes to our understanding of human genetic inheritance and the influence of genetics on the development of human traits and diseases (Hain et al., 1985).

In addition to traditional Mendelian inheritance, epigenetics has become a captivating investigation realm. Epigenetic mechanisms, such as DNA methylation and histone modification, can influence gene expression and heredity without causing any alterations to the fundamental DNA sequence. This chapter elucidates the intricate relationship between genetics and epigenetics, emphasizing the influence of environmental factors on the development of an organism's observable characteristics, known as its phenotype. Additionally, it explores the possibility of epigenetic marks being inherited across generations, leading to transgenerational effects (Teem et al., 2019).

The principles governing genetic transmission offer a captivating exploration of the intricate mechanisms and marvels of heredity. This chapter provides a fundamental basis for comprehending the mechanisms that regulate the transfer of genetic information, spanning from Mendel's experiments with peas to elucidate DNA's molecular structure and encompassing chromosomal associations and the intricate process of meiosis. Genetic transmission not only influences the characteristics exhibited by individual organisms but also plays a crucial role in promoting diversity and facilitating the adaptation of entire populations and species. As our understanding of genetic transmission deepens, we acquire significant knowledge regarding the interdependence of various organisms and the intricate mechanisms through which genetic material is inherited, guaranteeing life's persistence and perpetuation (Hallowell et al., 2006).

A mere gram of DNA possesses an astonishing capacity for information storage, estimated to be approximately 215 petabytes (equivalent to 215 million gigabytes) of data.

4.2 THE MONOHYBRID CROSSES

The narrative of Gregor Mendel stands as a captivating legend within the annals of contemporary scientific history. Residing as a monastic member within the esteemed St. Thomas monastery in the town of Brno, formerly known as Brünn, presently located in the Czech Republic, Mendel imparted scientific knowledge at a nearby vocational institution while concurrently engaging in the pursuit of biological experimentation. The pivotal experiments involved the crossbreeding of sweet peas, conducted between 1856 and 1863, within a modest garden plot located discreetly within the monastery premises' confines. The individual in question communicated the findings of his experiments to a regional natural history society, subsequently disseminating the results and his corresponding analysis through publication in the society's scientific journal in the year 1866. Additionally, he initiated a correspondence with Carl Nägeli, a prominent botanist based in Munich, who held a distinguished position within the field during that era (Ka et al., 2019).

Nevertheless, the importance of Mendel's work went unrecognized by all. In 1868, Mendel was elected as the monastery's abbot, ceasing his scientific endeavors. In the period preceding his demise in 1884, Mendel purportedly conveyed to a junior member of the monastic community his sentiment that his scientific endeavors had yielded considerable gratification, and he harbored a conviction that his contributions would soon garner recognition on a global scale. The prophecy came to fruition after 16 years when Hugo de Vries, Carl Correns, and Erich von Tschermak, operating autonomously and in distinct European nations, conducted experiments akin to those of Mendel (Stewart, 1982). Their endeavors shed light on Mendel's publication and acknowledged him as the original discoverer.

While there is an ongoing debate among contemporary historians of science regarding Mendel's underlying motivations for conducting his research, it is widely acknowledged that Mendel possessed exceptional skills as an experimenter. His meticulous and extensively recorded experiments are universally recognized for their high quality. The research paper provides an initial comprehensive elucidation of transmission genetics, which pertains to the statistical principles governing hereditary traits' inheritance across successive generations. The experiments conducted by Mendel were widely accepted as the cornerstone of genetics due to their inherent elegance. Consequently, the principles of hereditary transmission derived from his findings have come to be known as Mendelian genetics (Ford, 2000).

One notable distinction between Mendel's methodology and that employed by other plant hybridizers during his time lies in Mendel's inclination toward quantitative analysis of traits that could be dichotomously classified, such as round seeds instead of wrinkled seeds. He conducted straightforward crossing experiments and sought statistical regularities that could indicate overarching principles (Belyaev et al., 1986).

The researcher expressed his intention to investigate the various manifestations of hybrid offspring and to establish their quantitative relationships. Mendel opted to utilize peas as the subject of his experiments due to two distinct rationales. Initially, the individual possessed the opportunity to examine a range of options that exhibited discernible contrasting attributes, including seeds that varied in shape (round or wrinkled) and color (yellow or green). Furthermore, his previous research findings have demonstrated that peas typically engage in self-pollination, a reproductive process wherein pollen generated within a flower is utilized to fertilize the eggs within the same flower (Figure 4.1) (Nelson et al., 1982).

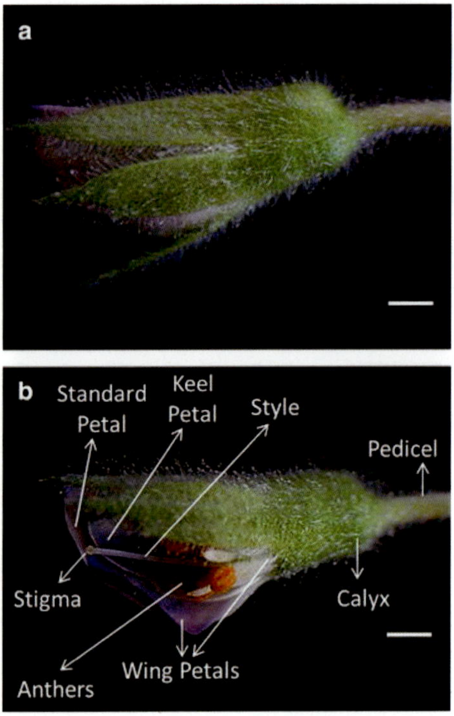

Figure 4.1. Crossing pea plants requires a minor technique in which the anthers of a flower are removed before they produce pollen. (Source: Kalve, S., Tadege, M. A comprehensive technique for artificial hybridization in Chickpea (Cicer arietinum). Plant Methods 13, 52 (2017). https://doi.org/10.1186/s13007-017-0202-6, Creative Commons License)

To generate hybrids through the process of cross-pollination, it was sufficient to expose the keel petal, which encloses the reproductive structures, eliminate the premature anthers, which are responsible for producing pollen, before their pollen release, and subsequently apply pollen from a different plant's flower onto the stigma, the female structure. Mendel acknowledged the necessity of investigating heritable traits that exhibited consistency within a particular pea variety yet displayed dissimilarity across different varieties (Davvaz et al., 2013). For instance, he observed a consistent occurrence of round seeds within one variety and a consistent occurrence of wrinkled seeds within another variety. Due to this rationale, he initiated his experiments by cultivating true-breeding varieties, wherein the plants exhibited the exclusive ability to yield offspring identical to themselves through regular self-pollination. The various cultivars, which exhibited consistent inheritance patterns for traits such as seed shape, seed color, color of flowers, pod shape, and other distinct characteristics chosen by Mendel for his study (see chapter 3), were used as the parental plants for subsequent crossbreeding experiments (Otten et al., 1981). A hybrid refers to the progeny resulting from a mating between parents exhibiting contrasting characteristics in one or multiple traits. A monohybrid refers to a hybrid offspring resulting from mating two individuals that exhibit variation in only one specific trait of interest. (Although there may be additional differences in other traits, these other differences are disregarded for the experiment) (Lumme & Pohjola, 1980).

Figure 4.2. Mendel studied the seven character differences in peas. (Source: Quilichini TD, Gao P, Yu B, Bing D, Datla R, Fobert P, Xiang D. The Seed Coat's Impact on Crop Performance in Pea (Pisum sativum L.). Plants. 2022; 11(15):2056. https://doi.org/10.3390/plants11152056, Creative Commons License)

It is of value to scrutinize a selection of Mendel's initial experiments to gain insight into his methodologies and his interpretations of the outcomes. One set of characters that the researcher examined was the comparison between round and wrinkled seeds.

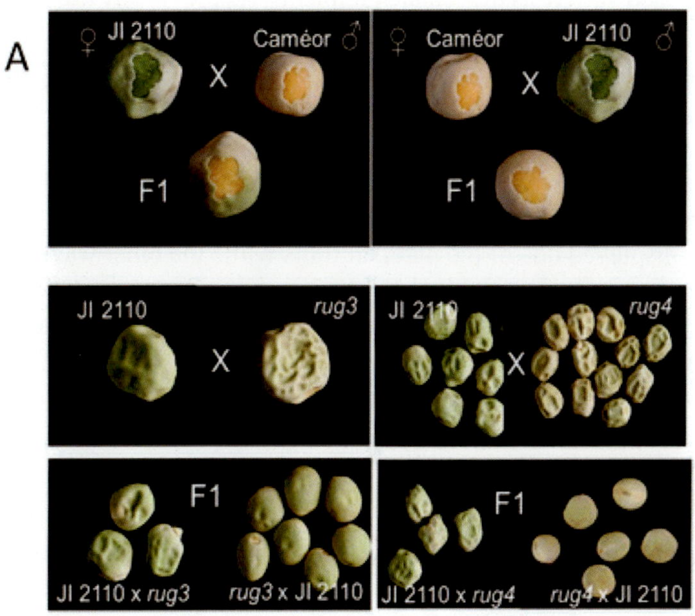

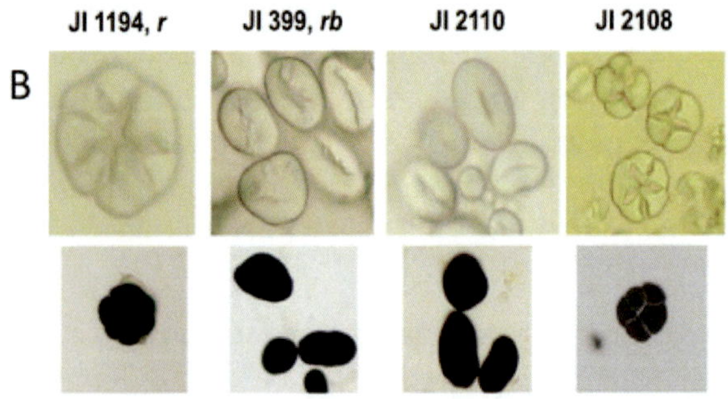

Figure 4.3. Mendel was the first to show that the characteristics of the progeny produced by a cross do not depend on which parent is the male and which is the female: this image is of a recent experiment (Source: Rayner T, Moreau C, Ambrose M, Isaac PG, Ellis N, Domoney C. Genetic Variation Controlling Wrinkled Seed Phenotypes in Pisum: How Lucky Was Mendel? International Journal of Molecular Sciences. 2017; 18(6):1205. https://doi.org/10.3390/ijms18061205, Creative Commons License)

The hybridization experiment involving the cross-pollination of plants with wrinkled seeds from various plant species with plants possessing round seeds resulted in the production of exclusively round hybrid seeds (Matthews & Craig, 1989). The F1 generation is a term geneticists employ to differentiate hybrid seeds or plants from their pure-breeding parents, known as the P1 generation. Mendel conducted the reciprocal cross experiment, wherein he employed plants from the round-seeded variety as the pollen donors and plants from the wrinkled-seeded variety as the female recipients. As previously observed, all F1 seeds exhibited a spherical shape (Lewis et al., 2000).

The result of a genetic cross is independent of the specific traits exhibited by the male and female individuals involved, as reciprocal crosses yield identical results (Kumar & Patra, 2010). Consistent outcomes were achieved by Mendel when he conducted crosses between plants that exhibited variations in any of the pairs of alternative characteristics. In every instance, the F1 offspring exclusively displayed one of the parental traits while the other was absent. The phenotypic characteristic exhibited by the F1 generation in each monohybrid cross is depicted on the right side in Figure 4.2. Mendel designated the trait exhibited in the hybrids as the dominant trait, while he referred to the trait not manifested in the hybrids as the recessive trait (Awang-Kanak et al., 2016).

Did You Know?

In 2003, scientists successfully engineered the world's inaugural genetically modified pet, specifically a zebrafish that exhibits bioluminescent properties.

4.3 TRAITS PRESENT IN THE PROGENY OF THE HYBRIDS

When monohybrid offspring self-fertilize, the recessive characteristic resurfaces in the following generations. For instance, after producing the round hybrid seeds from the cross between round and wrinkled kinds and self-fertilizing them, some were round, and some were wrinkled. Both kinds were present in exact numbers. Mendel quantified 5474 spherical seeds and 1850 wrinkled seeds. The author noted a 3:1 ratio. F1 self-fertilization produces F2 offspring seeds. Three F2 offspring had dominant traits, and one had recessive traits, according to Mendel. The following figure summarizes the hybridization of round and wrinkled variants (Verhaegen et al., 1997).

Table 4.1. Results of Mendel's monohybrid experiments (Source: Albert, Creative Commons License)

Parental Traits	F1 Trait	Number of F2 progeny	F2 Ratio
round × wrinkled (Seeds)	Round	5474 round, 1850 wrinkled	2.96: 1
yellow × green (seeds)	Yellow	6022 yellow, 2001 green	3.01: 1
purple × white (flowers)	Purple	705 purple, 224 white	3.15: 1
inflated ×constricted (pods)	Inflated	882 inflated, 299 con-stricted	2.95:1
green × yellow (unripe pods)	Green	428 green, 152 yellow	2.82:1
axial × terminal (flower position)	Axial	651 axial, 207 terminal	3.14: 1
long × short (stems)	Long	787 long, 277 short	2.84: 1

The F2 generation of plant crosses, involving variations in any of the pairs of alternative characteristics, yielded comparable outcomes, as shown in Table 4.1. It is important to acknowledge that the initial two characteristics, round versus wrinkled seeds and yellow versus green seeds, have a significantly higher number of observations than the remaining traits. This disparity can be attributed to the fact that these particular traits can be directly classified in the seeds. In contrast, the remaining traits can only be classified in mature plants. The primary findings derived from the data presented in Table 4.1 were (Dudits et al., 1987):

- F1 hybrids only express dominant traits.
- F2 plants have dominant or recessive traits.

- In the second filial generation (F2), dominant plants outnumber recessive plants threefold. The F2 ratio, or dominant-to-recessive allele ratio, is 3:1 (Bison et al., 2006).

4.3.1 Mendel's Genetic Hypothesis & its Experiment Tests

Mendel's monohybrid crosses demonstrated the reappearance of the recessive trait in its unaltered form in the F2 generation without any observable distinctions from the trait exhibited by the initial P1 recessive plants. In his correspondence, Mendel observed that in the F2 generation, the two parental traits manifest independently and remain unaltered, lacking any discernible evidence of inheritance or assimilation between them (Davey Smith & Ebrahim, 2003). Based on this discovery, Mendel concluded that the hereditary factors responsible for the traits observed in the parental lines were transmitted as two distinct elements that maintained their genetic integrity in the hybrid offspring. But, the hereditary determinants do not undergo a process of "mixing" or "contamination" with one another. Therefore, it is possible for a plant exhibiting the dominant trait to possess, without alteration, the genetic factor responsible for the recessive characteristic (Rabinowitz & Laird, 2000).

To elucidate his findings, Mendel formulated a genetic hypothesis that can be comprehended by consulting Figure 4.4. The individual presumed that every gamete, a reproductive cell, encompasses a single instance of each hereditary determinant category within the plant. The genetic factor responsible for the inheritance of round seeds was denoted as 'A', while the genetic factor responsible for the inheritance of wrinkled seeds was denoted as 'a'. Mendel postulated that in the true-breeding variety characterized by round seeds, every reproductive cell would possess the genetic factor 'A', while in the true-breeding variety exhibiting wrinkled seeds, every reproductive cell would possess the genetic factor 'a'. When the different varieties are subjected to crossbreeding, the resulting F1 hybrid is expected to inherit one allele from 'A' and 'a', resulting in Aa's genetic composition (Bowden et al., 2015).

If allele 'A' demonstrates dominance over allele 'a', it is anticipated that the F1 generation seeds will exhibit a spherical shape. When a first-generation (F1) plant engages in self-fertilization, the alleles 'A' and a segregate independently and are evenly distributed among the gametes. Hence, as depicted in in research, the stochastic segregation of gametes is anticipated to yield an F2 progeny with a genetic constitution consisting of 1/4 AA, 1/2 Aa, and 1/4 aa. Based on the principles of Mendelian genetics, individuals possessing the AA and Aa genotypes are anticipated to produce round seeds. In contrast, those with the aa genotype will exhibit wrinkled seeds. As a result, the expected proportion of round seeds to wrinkled seeds is 3:1 (Kalina, 2014).

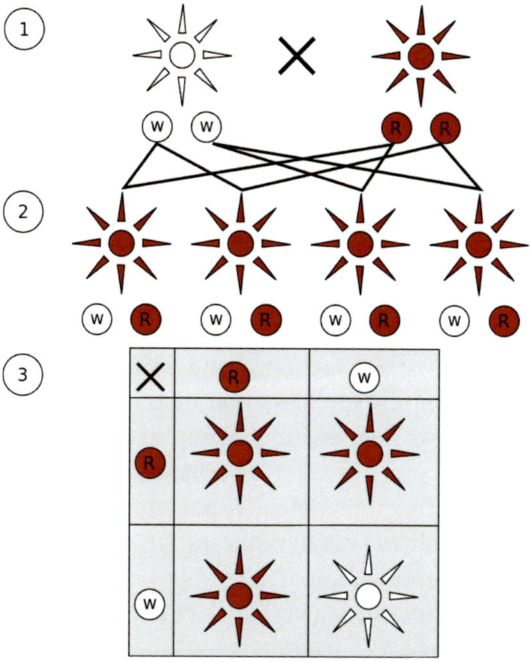

Figure 4.4. A diagrammatic explanation of Mendel's genetic hypothesis to explain the 3: 1 ratio of dominant: recessive phenotypes observed in the F2 generation of a monohybrid cross.Source: By Benutzer:Magnus Manske - http://de.wikipedia.org/wiki/Datei:Mendelian_inheritance_3_1.png, Public Domain, https://commons.wikimedia.org/w/index.php?curid=10396755, Creative Commons License)

The genetic variations AA, Aa, and aa can be denoted alternatively as A/A, A/a, and a/a using slashes. It is imperative to acknowledge that both manifestations of symbolism are mutually substitutable and effectively communicate identical information. The genetic hypothesis illustrated in Figure 4.4 further demonstrates one of Mendel's notable deductions. Plants with similar phenotypic traits, such as spherical seeds, can possess divergent genotypic characteristics, as indicated by the variations in their progeny resulting from crossbreeding. For instance, in the context of true breeding, the seeds of the round variety exhibit a genetic composition of AA (Laird et al., 2000).

Conversely, when round and wrinkled varieties are crossed to produce F1 hybrid seeds, the genetic composition of the resulting seeds is Aa. Nevertheless, which methodologies can be utilized to evaluate the genetic hypothesis? Mendel observed that a significant consequence of his hypothesis was related to the genetic makeup of the round seeds in the second filial generation (F2). If the hypothesis holds, it can be deduced that around 33.33% of the round seeds have the genetic composition AA. In contrast, the remaining 66.67% of the round seeds display the genetic composition Aa (Sanderson et al., 2019). The principle mentioned earlier is illustrated in Figure 4.5. The ratio of genotypes AA: Aa: aa in the F2 generation was observed to be 1: 2: 1. Nevertheless, upon excluding the recessive genotype aa from analysis, the proportion of genotypes AA: Aa is observed to be 1:2. To clarify, it can be stated that roughly 33% of the observed round seeds exhibit the AA genotype, whereas the remaining 67% exhibit the Aa genotype (Zhang et al., 2017). It is anticipated that plants resulting from self-fertilization and possessing AA genotypes will demonstrate the phenomenon of true breeding, specifically concerning the trait of round seed morphology. In contrast, plants derived from the Aa genotypes are expected to exhibit a phenotypic ratio of 3:1 for round and wrinkled seeds. Furthermore, it is anticipated that all the seeds in the second filial (F2) generation that display wrinkled traits will possess the genetic makeup of aa. As a result, these seeds are expected to demonstrate consistent inheritance of wrinkled seed traits due to the process of self-fertilization. Mendel performed self-fertilization on the F2 generation plants to validate his hypotheses for multiple traits. The diagram below offers a concise representation of the results obtained from the comparison between round and wrinkled seeds (Verbanck et al., 2018).

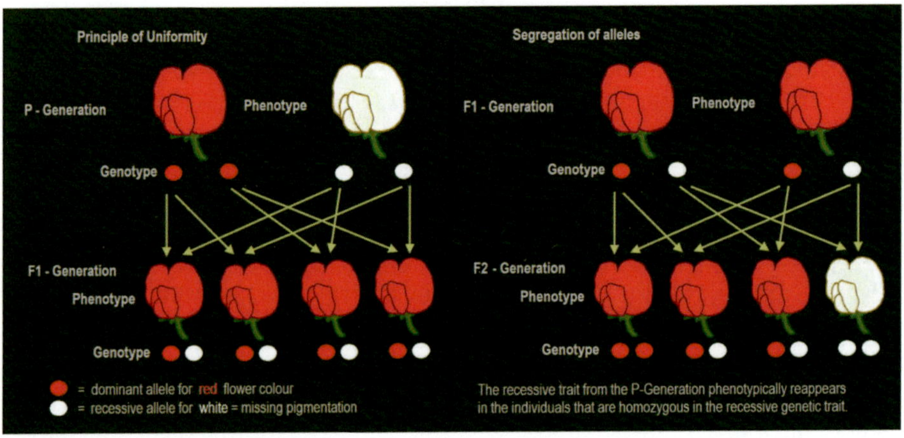

Figure 4.5. In the F2 generation, the ratio of AA: Aa is 1: 2. (Source: By Sciencia58 - Own work, CC0, https://commons.wikimedia.org/w/index.php?curid=82825440, Creative Commons License)

According to Mendel's genetic concept, wrinkled seeds: F2 generation plants have real breeding qualities. Wrinkled seeds were specific to F3. 193 of 565 F2 round seed plants produced round F3 seeds. The remaining 372 plants produced round and wrinkled seeds in a 3:1 ratio. The genetic theory in Figure 4.4 predicts a ratio of 1: 2 for AA: Aa types, which is 1: 1.93 for 193: 372. The F2 progeny and its genetic phenotypic distribution matched Figure 4.4's predicted 1: 2: 1 ratio of AA: Aa : aa (Xue et al., 2015).

4.4 THE PRINCIPLE OF SEGREGATION

The diagram in Figure 4.4 is the heart of Mendelian genetics. You should master it and be able to use it to deduce the progeny types produced in crosses. Be sure you thoroughly understand the meaning, and the biological basis, of the ratios 3: 1 and 1 : 2 and 1: 2 : 1. The following list highlights Mendel's key assumptions in formulating his inheritance model (Bhattacharyya et al., 2007).

1. For each of the traits that Mendel studied, a pea plant contains two hereditary determinants.
2. For each pair of hereditary determinants present in a plant, the members may be identical (AA) or different (for example, Aa) (Champagne, 1973).
3. Each reproductive cell (gamete) produced by a plant contains only one of each pair of hereditary determinants (that is, either A or a).
4. In the formation of gametes, any particular gamete is equally likely to include either hereditary determinant (hence, from an Aa plant, half the gametes contain A and the other half contain a) (James & Taeuber, 1985).
5. The union of male and female reproductive cells is a random process that reunites the hereditary determinants in pairs.

The essential feature of Mendelian genetics is the separation, technically called segregation, in unaltered form, of the two hereditary determinants in a hybrid plant in the formation of its reproductive cells (points 3 and 4 in the preceding list). The principle of segregation is sometimes called Mendel's first law, although Mendel never used this term (Reardon & Firebaugh, 2002):

Principle Of Segregation:

During gamete formation, the paired hereditary determinants undergo segregation, resulting in an equal likelihood for each gamete to contain either pair member (Milligan, 2018).

In addition to the principle of segregation, another significant assumption, implied by points 1 and 5 in the list, is that the hereditary determinants exist in pairs within both the parental and the progeny organisms but as single copies within the reproductive cells (Brown, 1939).

4.4.1 Important Genetic Terminology

Mendel struggled to find a language that sufficiently described his concepts. Thus, he occasionally confused an organism's genetics with its outward traits. Following Mendel's rediscovery, the necessary lexicon and essential terminologies were acquired (Samerski, 2006).

- A hereditary determinant of a trait is called a gene.
- The different forms of a particular gene are called alleles. In Figure 4.4, the gene alleles for seed shape are A for round seeds and a for wrinkled seeds. A and a are alleles because they are alternative forms of the gene for seed shape. Alternative alleles are typically represented by the same letter or combination of letters, distinguished by uppercase and lowercase, superscripts and subscripts, or some other typographic identifier (Kessler et al., 2007).
- The genotype is the genetic constitution of an organism or cell. For seed shape in peas, AA, Aa, and aa are possible genotypes for the A and A alleles. Because gametes contain only one allele of each gene, 'A' and 'a' are examples of genotypes of gametes (Lanie et al., 2004).
- A genotype in which the members of a pair of alleles are different, as in the Aa hybrids in Figure 4.4, is said to be heterozygous; a genotype in which the two alleles are alike is said to be homozygous. Homozygous organisms may be homozygous dominant (AA) or recessive (aa). The terms homozygous and heterozygous can-not apply to gametes, which contain only one allele of each gene.
- The observable properties of an organism constitute its phenotype. Round, wrinkled, yellow, and green seeds are all phenotypes. The phenotype of an organism does not necessarily tell you anything about its genotype. For example, a seed with the phenotype "round" could have the genotype AA or Aa (Edlin, 1987).

4.5 VERIFICATION OF MENDELIAN SEGREGATION BY THE TESTCROSS

Mendel also crossed F1 heterozygous genotypes with plants with homozygous recessive characteristics to test the genetic theory in Figure 2.4. A testcross is mating an organism heterozygous for one or more genes (e.g., Aa) with an organism homozygous for recessive alleles (e.g., aa). Figure 4.6 shows the testcross result (Miedaner et al., 2012). The heterozygous parent produces 'A' and gametes equally, unlike the homozygous recessive parent; hence the expected children will be half Aa and half aa. The former genotype has the dominant phenotype because allele 'A' is dominant over allele 'a', whereas the later genotype has the recessive phenotype. Due to its usefulness, genetic analysis uses test crosses (Schön et al., 2004).

Examining a testcross allows for directly observing the relative frequencies of various gametes produced by the heterozygous parent. This is possible as the recessive parent solely contributes recessive alleles, which can be discerned through the phenotypes of the offspring. Mendel conducted test crosses comparing round against wrinkled seeds, yellow versus green seeds, purple versus white blooms, and long versus short stems. Table 4.2 shows the results. The offspring's observable qualities always match the expected 1:1 ratio from the parent's heterozygosity (Grattapaglia & Sederoff, 1994).

Figure 4.6. In a testcross of a Yy heterozygous parent with a yy homozygous recessive, the progeny are Yy and yy in the ratio of 1 : 1. (Source: By KatieAnn127 - Own work, CC BY-SA 4.0, https://commons.wikimedia.org/w/index.php?curid=37309605, Creative Commons License)

Backcrossing hybrid creatures with one of the parental genotypes is another important genetic crossing. In later chapters, geneticists and plant and animal breeders use backcrosses. Table 4.2 test crosses are backcrosses. Each F1 heterozygous parent was obtained from a mating between a homozygous dominant and a recessive individual (Werner et al., 2020).

Table 4.2. Mendels' Testcross results (Source: Albert, Creative Commons License)

Testcross (F1 heterozygote × homozygous recessive)	**Progeny from testcross**	**Ratio**
Round × wrinkled seeds	193 round, 192 wrinkled	1.01: 1
Yellow × green seeds	196 yellow, 189 green	1.04:1
Purple × white flowers	85 purple, 81 white	1.05:1
Long × short stems	87 long 79 short	1.10:1

4.6 SEGREGATION OF TWO OR MORE GENES

Mendel performed studies to see if the same inheritance pattern applied to each pair of alleles when several allelic pairings were segregating in hybrids. For example, plants from a genetically stable lineage with round, yellow seeds were crossed with wrinkled, green-seeded plants. The F1 offspring have dihybrid characteristics. The seeds were spherical and golden. Since roundness dominates wrinkledness and yellowness dominates greenness, the F1 phenotype was round and yellow (Sachs et al., 1998).

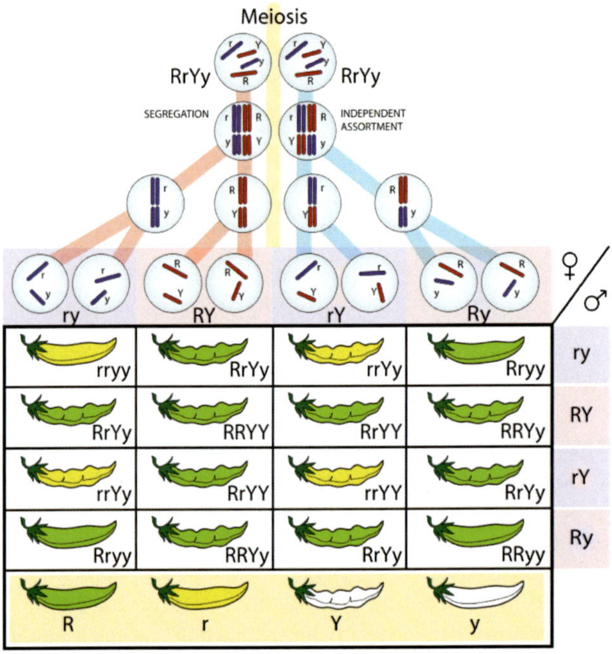

Figure 07. When combined at random with the 3: 1 ratio of round : wrinkled, the 3: 1 ratio of yellow: green yields the 9 : 3 : 3 : 1 ratio that Mendel observed in the F2 progeny of the dihybrid cross. (Source: By LadyofHats - Own work based on:[1], [2], [3], [4], [5], Public Domain, https://commons.wikimedia.org/w/index.php?curid=4770847, Creative Commons License)

Subsequently, Mendel engaged in self-fertilizing the F1 offspring to procure seeds for the F2 generation. The researcher observed four distinct seed phenotypes in the offspring and recorded the following numerical values while quantifying the seeds (Muller, 1914).

round, yellow	315
round, green	108
wrinkled, yellow	101
wrinkled, green	32
Total	556

In the provided data, Mendel observed the anticipated monohybrid 3:1 ratio for each trait. To each characteristic, the offspring were (Lorieux et al., 1995).

$$Round : wrinkled$$
$$=(315+108) :(101$$
$$=423 :133$$
$$=3.18 :1$$
$$Yellow : green$$
$$\aleph(315 \quad 101) :(108 \quad 32)$$
$$=416+140$$
$$=2.97 :1$$

Moreover, in the second filial generation (F2) resulting from the dihybrid cross, the individual 3 : 1 ratio for each trait was randomly combined, as illustrated in Figure 4.7. When the phenotypes of two traits are randomly combined, it is observed that 3/4 of the offspring exhibiting a round phenotype 3/4 will display a yellow phenotype. In contrast, the remaining 1/4 will exhibit a green phenotype. Similarly, among the 1/4 of the offspring that exhibit a wrinkled phenotype, 3/4 will display a yellow phenotype while the remaining 1/4 will exhibit a green phenotype. The anticipated ratios of round yellow to round green to wrinkled yellow to wrinkled green are calculated as 3/4 × 3/4 to 3/4 × 1/4 to 1/4 × 3/4 to 1/4 × 1/4, resulting in the overall proportions (Liu et al., 1997).

The ratio of 315 : 108 : 101 : 32, as observed, is approximately 9.84: 3.38 : 3.16 : 1. This observed ratio is reasonably close to the expected ratio of 9 : 3 : 3 : 1, which is obtained by cross-multiplying the individual 3 : 1 ratio depicted in Figure 4.7 (Wu, 1991).

4.7 THE PRINCIPLE OF INDEPENDENT ASSORTMENT

Mendel found a ratio of 9:3:3:1 for all pairs of traits in his studies. He also deduced the biological explanation. Using the dihybrid round × wrinkled cross, we may denote the pair's dominant and recessive alleles that affect seed shape as W and w and the allelic pair that affect seed color as G and g. Mendel hypothesized that the 9 : 3 : 3 : 1 ratio in the F2 generation was caused by the separate segregation of the alleles W and w, which define round or wrinkled seeds, and G and g, which determine yellow or green seeds. Each pair of alleles segregates into gametes separately. WW GG (round, yellow seeds) and ww gg (wrinkled, green seeds) are the parental genotypes in the P1 generation. The F1 genotype is double heterozygote Ww Gg. The genotype might be symbolized (Ray, 2008).

$W / w ; G / g \, or \, W / w G / g$

The slash, commonly known as a virgule, is substituted with a brief horizontal line.

The outcome of independent assortment in the F1 generation plants demonstrates that the 'W' allele has an equal probability of being present in a gamete alongside either the 'G' or 'g' allele. At the same time, the 'w' allele also has an equal likelihood of being included in a gamete alongside either the 'G' or 'g' allele. The concept of independent segregation is visually depicted in Figure 4.8. When two pairs of alleles undergo independent assortment, the gametes the double heterozygote produces are (Ruch, 1998).

$1/4 WG 1/4 Wg 1/4 wG 1/4 wg$

The four gametes of the following generation are randomly combined to generate zygotes. The Punnett square shows F2 genotypes from F1 female and male gametes in a cross-multiplication-like arrangement. The Punnett square shows F2 offspring traits. Note the 9:3:3:1 phenotypic ratio. A Punnett square shows that the F2 generation's genotypic ratio is not 9:3:3:1 (Yang et al., 2009).

Independent assortment determines the F2 genotype ratio.

$1 : 2 : 1 : 2 : 4 : 2 : 1 : 2 : 1$

The reason for this ratio is shown in Figure 4.10.

In seeds possessing the WW genotype, the ratio of GG : Gg : gg is observed to be 1 : 2 : 1. In the context of seeds possessing the Ww genotype, the observed ratio is 2 : 4 : 2. This ratio is derived by multiplying the standard 1 : 2 : 1 ratio by 2, as there exists a twofold greater frequency of Ww genotypes compared to both WW and ww genotypes. Within the subset of seeds possessing the ww genotype, the distribution of genotypes GG, Gg, and gg adheres to a ratio of 1:2:1 (Peay et al., 2015). The phenotypic characteristics of the seeds are displayed below the corresponding genotypic compositions. The phenotypic ratio is observed to be 9 : 3 : 3 : 1. According to the data presented in Figure 4.10, it is evident that the ratio of WW : Ww : ww among seeds with the GG genotype is 1 : 2 : 1. Similarly, among seeds with the Gg genotype, the ratio is 2

: 4 : 2. Lastly, among seeds with the gg genotype, the ratio is 1 : 2 : 1. Hence, the principle of independent segregation posits that within each potential genotype resulting from a single allele pair, the proportion of homozygous dominant, heterozygous, and homozygous recessive genotypes for the other allele pair follows a ratio of 1:2:1 (Kabat et al., 1980).

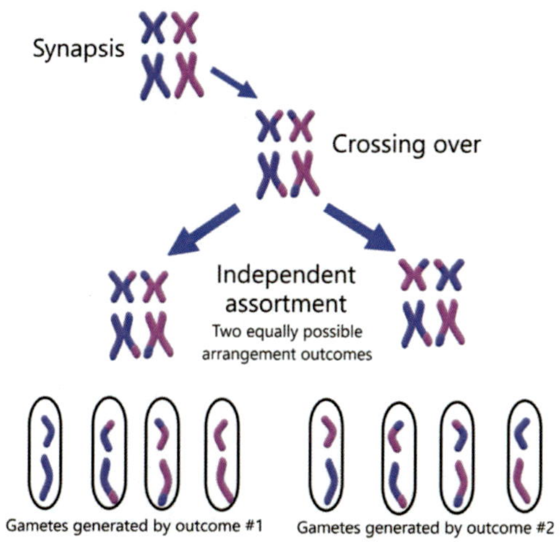

Figure 4.8. Independent segregation of the allele pairs means that, among each of the classes, the ratio is 1 : 1. (Source: https://commons.wikimedia.org/wiki/File:Independent_assortment.png, Creative Commons License)

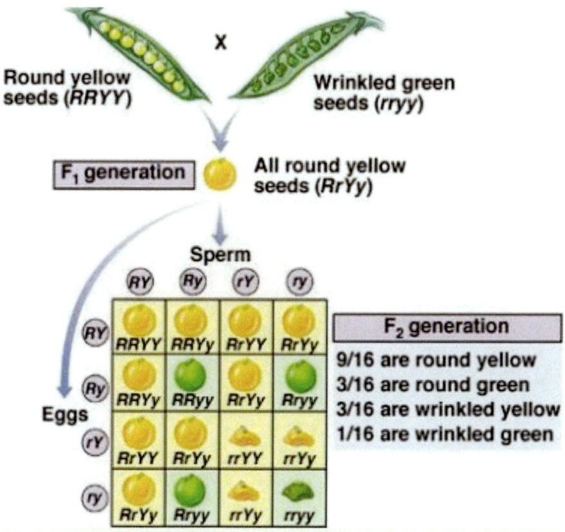

Figure 4.9. Diagram showing the basis for the 9 : 3 : 3 : 1 ratio of F2 phenotypes resulting from a cross in which the parents differ in two traits determined by genes that undergo independent assortment (Source: Susan, Creative Commons License)

Mendel conducted an experiment to examine the independent segregation hypothesis by verifying the anticipated genotypes' presence in the projected proportions (Buchanan, 2011). The tests were conducted by cultivating plants derived from the F2 seeds and obtaining F3 offspring through self-pollination. To illustrate the experiments conducted, let us consider a specific set of crosses wherein the experimenter cultivated plants derived from F2 seeds exhibiting the phenotypic traits of being round and green. It is worth noting that Figures 4.9 and 4.10 illustrate the expected genotypes of round, green F2 seeds as Ww gg and WW gg in a ratio of 2:1 (Feng et al., 2022) Mendel cultivated a total of 102 plants derived from these seeds, wherein 67 of them exhibited the phenotypic combination of round, green seeds alongside wrinkled, green seeds, thereby suggesting the presence of the parental genotype Ww gg. Additionally, 35 plants displayed the round, green seed phenotype, indicating the parental genotype as WW gg. The observed ratio of 67 : 35 is consistent with the anticipated 2 : 1 ratio of genotypes. A high concordance was observed between the projected relative frequencies of various genotypes and the actual outcomes when plants were cultivated from F2 seeds of either round, yellow, or wrinkled yellow phenotypes. Consistent with expectations, the plants cultivated from the wrinkled, green seeds, characterized by the anticipated homozygous recessive genotype ww gg, exclusively yielded wrinkled, green seeds (MacLeod et al., 2005).

Figure 4.10. The F2 progeny of the dihybrid cross for seed shape and seed color. (Source: Matt, Creative Commons License)

The principle of independent assortment, also called Mendel's second law, encompasses Mendel's observation of the independent segregation of two pairs of alleles (Wang, 2013).

The principle of independent assortment is a fundamental concept in genetics that describes the random distribution of alleles for different traits during the formation of gametes. The independent segregation of alleles within a pair is not influenced by the segregation of other pairs during the formation of gametes (Hwang et al., 2010).

Did You Know?

Every other human on Earth and you share 99% of your DNA, making us more similar than distinct.

4.8 DIHYBRID TESTCROSSES

Mendel performed a testcross using F1 generation individuals with heterozygous genotypes for both genes (Ww Gg) to test the independent assortment theory. In a testcross, one parental genotype usually has numerous homozygous recessive alleles, like ww and gg. Figure 4.11 shows that double heterozygotes produce four gametes—WG, Wg, wG, and wg—with equal frequency. ww gg plants generate just wg gametes. Thus, the offspring will have four phenotypes: round yellow, round green, wrinkled yellow, and wrinkled green, in a 1:1:1:1 ratio (Marshall, 2008). This phenotypic ratio represents the double heterozygote's gamete ratio. The lack of dominant alleles from the ww gg parent provides clear findings. Mendel observed 55 round-shaped, yellow-colored offspring, 51 round-shaped, green-colored children, 49 wrinkled-shaped, yellow-colored offspring, and 53 wrinkled-shaped, green-colored offspring. These findings match the predicted 1:1:1:1. In the reciprocal cross, the female parent was the double heterozygote, and the male parent was the homozygous recessive. This discovery corroborated Mendel's premise that male and female gametes had a roughly equal distribution of genotypes (Moemeka et al., 1998).

4.8.1 The Big Experiment

In addition to his initial hypothesis, Mendel further conducted crosses between varieties exhibiting variations in three distinct traits: seed shape (round or wrinkled, denoted by alleles W and w), seed color (yellow or green, denoted by alleles G and g), and the color of the flower (purple or white, denoted by alleles P and p). The observed phenotype of the F1 seeds, resulting from the crossing of three different traits, exhibited a round shape and a yellow color. Furthermore, the subsequent growth of plants derived from these seeds displayed purple flowers (Fairbanks & Rytting, 2001). If we consider the dihybrid cross as an analogy and assume that the alleles of all three genes undergo independent assortment, it can be hypothesized that self-fertilization of the F1 flowers would lead to combinations of phenotypes that can be represented by the successive terms in the multiplication of [(3 4)D + (1 4)R]3. This multiplication yields a ratio of 27 : 9 : 9 : 9 : 3 : 3 : 3 : 1. The multiplication for Mendel's cross is executed in Figure 4.12. The dominant form of all three traits is observed in the most frequent phenotype, accounting for 27 out of 64 occurrences. The second most frequent phenotype, accounting for 9 out of 64 occurrences, exhibits the dominant form of two traits. The dominant form of only one trait is present in the next most frequent phenotype, accounting for 3 out of 64 occurrences. Lastly, the triple recessive phenotype is the least frequent, occurring only once out of 64 occurrences. It can be observed that when one of the traits is considered independently, while the other two are ignored, the ratio of phenotypes is 3:1. Similarly when any two of the traits are considered together, the ratio of phenotypes is 9:3:3:1. This implies that the overall three-gene segregation encompasses all potential independent segregations involving one or two genes (Venema, 2009).

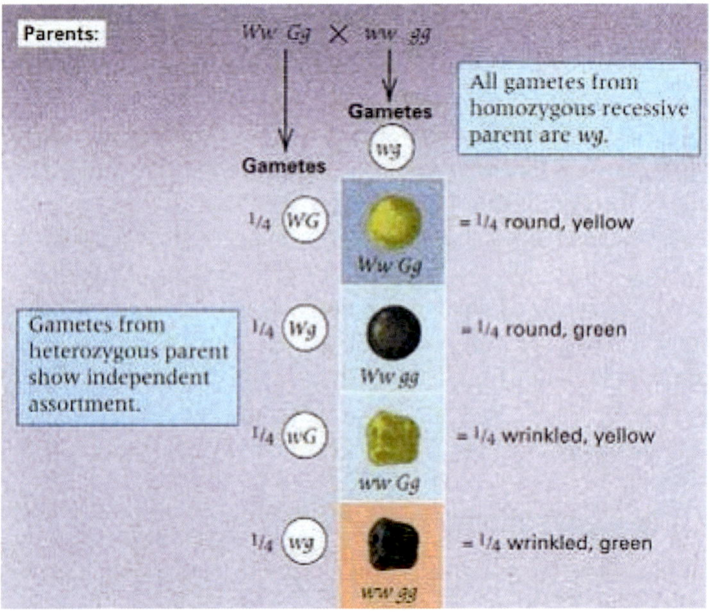

Figure 4.11. Genotypes and phenotypes resulting from a testcross of a Ww Gg double heterozygote (Source: Matt, Creative Commons License)

Figure 4.12's numbers match the independent assortment concept. Mendel was still dissatisfied. The scientists concluded that F2 offspring should have 27 genotypes. Thus, he self-fertilized all 639 plants to determine their genotypes for the three attributes. This experiment was the most trying, according to

Mendel. Figure 4.13 shows the experiment results. The genotypes PP, Pp, and pp show vertical segregation of the alleles W and w from G and g in the three Punnett squares. Under independent assortment, the red numerical number in each cellular unit represents the genotype-specific plant count. The black number represents the genotype-specific plant count. The data's great agreement strongly supported Mendel's main conclusion (Kloos et al., 2004).

"Pea hybrids form germinal and pollen cells that in their composition correspond in equal numbers to all the constant forms resulting from the combination of traits united through fertilization."

Though complicated, Mendel's approach incorporated segregation and independent assortment (Scheinberg, 1956). The author wants to emphasize that hybrid plants' gametes contain an equitable distribution of all possible combinations of alleles acquired from the initial true-breeding progenitors, whose crossbreeding produced the hybrids. The F1 child of WW gg and ww GG genotypes is Ww Gg. These F1 individuals generate equally WG, Wg, wG, and wg gametes. The 1:1 ratio of W:w and G:g gametes show segregation, but the equal frequencies of WG, Wg, wG, and wg show independent assortment (Kline & Sanders, 1984).

$(^3/_4\,W\text{—} + {}^1/_4\,ww) \times (^3/_4\,G\text{—} + {}^1/_4\,gg) \times (^3/_4\,P\text{—} + {}^1/_4\,pp)$

	Genotype	Phenotype	Observed number	Expected number
27/64	*W— G— P—*	Round, yellow, purple	269	270
9/64	*W— G— pp*	Round, yellow, white	98	90
9/64	*W— gg P—*	Round, green, purple	86	90
9/64	*ww G— P—*	Wrinkled, yellow, purple	88	90
3/64	*W— gg pp*	Round, green, white	27	30
3/64	*ww G— pp*	Wrinkled, yellow, white	34	30
3/64	*ww gg P—*	Wrinkled, green, purple	30	30
1/64	*ww gg pp*	Wrinkled, green, white	7	10

- For any one gene, the ratio of phenotypes is 48 : 16 = 3 : 1
- For any pair of genes, the ratio of phenotypes is 36 : 12 : 12 : 4 = 9 : 3 : 3 : 1

Figure 4.12. With independent assortment, the expected ratio of phenotypes in a trihybrid cross is obtained by multiplying the three independent 3 : 1 ratio of the dominant and recessive phenotypes. (Source: Matt, Creative Commons License)

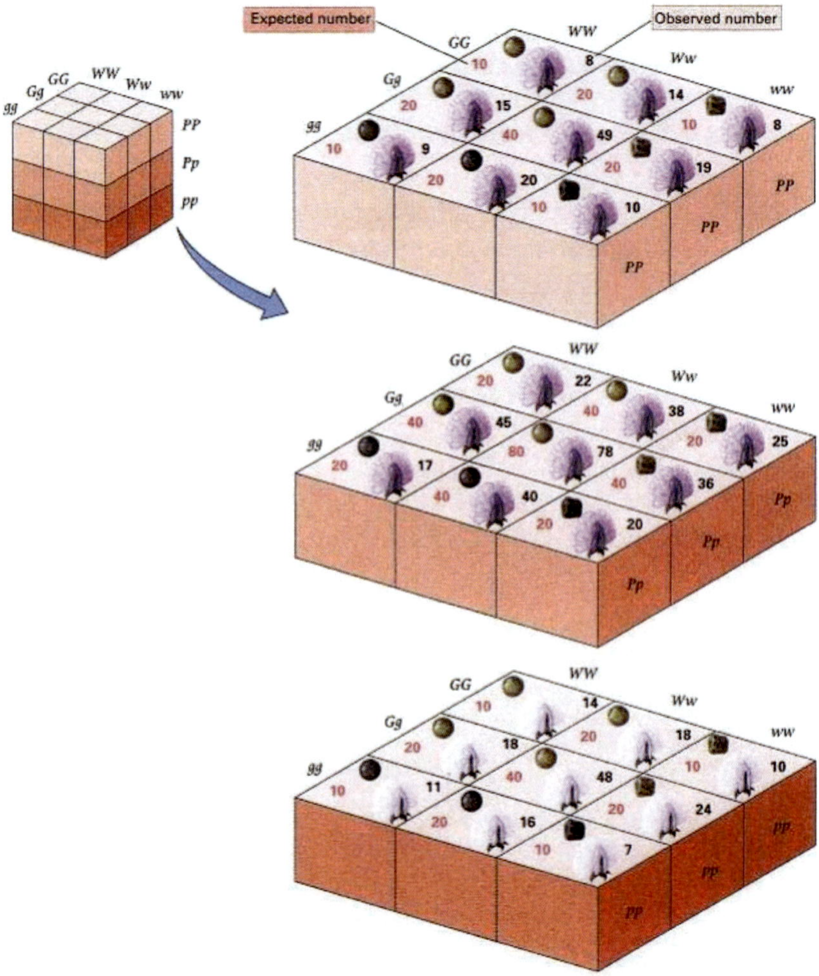

Figure 4.13. Results of Mendel's analysis of the genotypes formed in the F2 generation of a trihybrid cross with the allelic pairs W, w and G, g and P, p. (Source: Sean, Creative Commons License)

4.9 MENDELIAN INHERITANCE & PROBABILITY

Understanding the principles of probability is fundamental to comprehending the transmission of hereditary traits. Initially, it is important to note that the collective outcome of multiple discrete fertilization events determines the distribution of various offspring phenotypes resulting from a genetic cross. Moreover, it should be noted that in every instance of fertilization, the specific amalgamation of dominant and recessive alleles is determined randomly and susceptible to stochastic fluctuations (Sellers et al., 1990).

In examining genetic crosses, the likelihood of an occurrence can be regarded as analogous to the ratio of occurrences anticipated to be observed in a multitude of iterative experiments. Similarly, the ratio of occurrences of an event expected in multiple repeated trials is equal to the probability of its occurrence in a single trial. In the F2 generation of the pea hybridization experiment, Mendel observed 5474 round and 1850 wrinkled seeds (Table 4.1) (Chen et al., 2014).

In this particular instance, the ratio of seeds with wrinkles was calculated to be 1/3.96, or approximately 1/4, by dividing the number of wrinkled seeds (1850) by the sum of the wrinkled and non-wrinkled seeds (1850 + 5474). Hence, it can be inferred that 1/4 can be considered as the estimated ratio of wrinkled seeds that can be anticipated within a substantial population of offspring resulting from this particular genetic cross (Marschall et al., 2013). Alternatively, we can consider the value of 1/4 as the probability associated with selecting a random seed, indicating the likelihood of it being wrinkled. Assessing the likelihood of a genetic occurrence typically necessitates comprehension of the inheritance mechanism and familiarity with the specific cross. To assess the likelihood of obtaining a round seed from a specific cross, it is imperative to know the presence of two alleles, namely 'W' and 'w', with 'W' exhibiting dominance over 'w'. Additionally, the specific cross under consideration must be considered, as the probability of round seeds is contingent upon the nature of the cross (Duncan et al., 2016).

In numerous genetic crosses, the potential outcomes of fertilization exhibit equal likelihood. Let us consider a scenario with n potential outcomes, all equally probable. Among these outcomes, m corresponds to a specific outcome of interest. In this case, the probability of the outcome of interest can be calculated as m divided by n (Slavov et al., 2004). In probability theory, it is customary to refer to a desired result as an event. To illustrate, let us examine the offspring of self-pollination in an Aa plant. Four potential genotypes (outcomes) are equally probable: AA, Aa, aA, and aa. Among the four potential outcomes, it is observed that two of them exhibit heterozygosity. Consequently, the likelihood of an individual being heterozygous can be calculated as 2 out of 4, which can be expressed as a fraction of ½ (Douglas et al., 2002).

4.9.1 Mutually Exclusive Events: The Addition Rule

Occasionally, a desired result can be articulated concerning multiple potential outcomes. As an illustration, a seed exhibiting the phenotypic trait of being "round" can possess one of two possible genotypes: WW or Ww. It is not possible for a seed with a round shape to possess both genotypes simultaneously. In genotypes such as WW or Ww, it is important to note that only one event can manifest in a given organism. Furthermore, the occurrence of one event in an organism prevents the occurrence of other events within the same organism. In this particular instance, the genotype WW in a plant prevents the manifestation of the genotype Ww in the same plant and vice versa. Mutually exclusive events are those that exclude one another. When two events are mutually exclusive, their probabilities are combined using the addition rule (Dzikowski et al., 2006).

Addition Rule: *The Addition Rule states that the probability of either event A or B occurring when the two events are mutually exclusive equals the sum of their probabilities.*

The addition rule is written in symbols, where Prob is used to mean probability.

$$Prob(A+B) = Prob(A) + Prob(B)$$

The addition rule can be applied to determine the proportion of round seeds expected from the cross Ww × Ww. The round-seed phenotype results from the expression of either of two genotypes, WW and Ww, which are mutually exclusive. In any particular progeny organism, the probability of genotype WW is 1/4, and that of Ww is ½ (Uren et al., 2008). Hence the overall probability of either WW or Ww is

$$Prob(WW\ or\ Ww) = Prob(WW) + Prob(Ww) = \frac{1}{4} + \frac{1}{2} = \frac{3}{4}$$

The probability of an individual seed being round is 3/4, which implies that the expected proportion of round seeds among a large number of progeny is also ¾ (Riaz et al., 2017).

4.9.2 Independent Events: The Multiplication Rule

Independent occurrences do not alter the chance of other events. Seed form and color are independent in Mendel's experiments. Thus, the F2 generation should have 9/16 round yellow seeds, 3/16 round green seeds, 3/16 wrinkled yellow seeds, and 1/16 wrinkled green seeds. Since qualities are independent, assessing them separately yields the proportions. Seed shape research predicts 75% round and 25% wrinkled seeds in the F2 generation (Hartwell et al., 1974). Based on seed color, the F2 generation will have 75% yellow and 25% green seeds. Because characteristics are inherited independently, 75% of round seeds will likewise be yellow. Thus, 9/16 seeds are spherical and yellow. One-fourth of the three-fourths of the seeds will be green. Thus, three-fourths by one-fourth equals three-sixteenths, the percentage of spherical green seeds. Similar methods can estimate the remaining phenotypic class proportions. The fundamental principle asserts that multiplying their probabilities yields their combined probability when occurrences are independent (Honeycutt & Pierce, 2007).

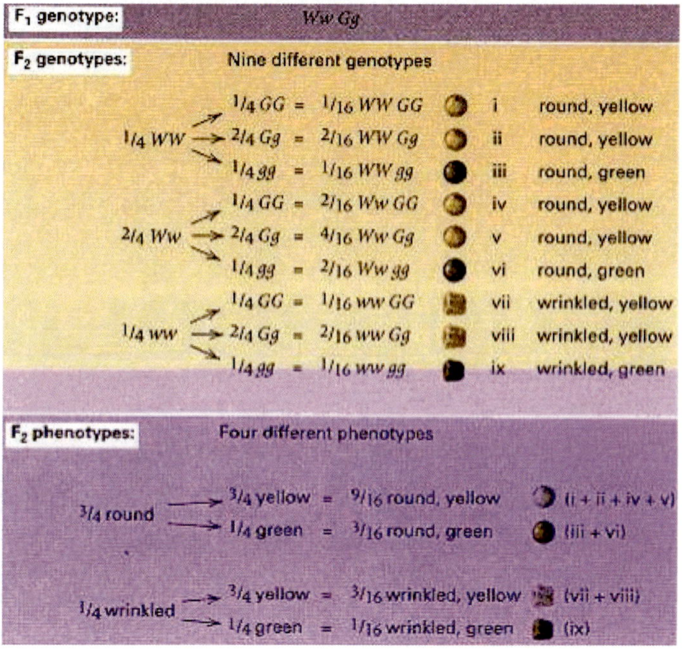

Figure 4.14. This study uses addition and multiplication techniques to calculate the odds of the nine genotypes and four phenotypes in the F2 offspring. (Source: Sean, Creative Commons License).

The genotypes of subsequent progeny resulting from a cross are considered independent events, indicating that earlier offspring genotypes do not affect the relative proportions of genotypes observed in later offspring. The phenomenon of successive offspring exhibiting independence challenges the prevailing notion that within each human family, the ratio of girls to boys tends to equalize at around 1:1. Consequently, the common belief that if a family already consists of, for instance, four girls, the likelihood of having a boy in the next pregnancy increases is contradicted. However, this assertion lacks theoretical support and is further contradicted by empirical evidence concerning the sex ratios observed within human sibships. The term "sibship" denotes a collective of progeny originating from a common set of parents. The data suggests that the probability of a human family having a girl on the next birth is not influenced by the gender composition of the family's previous children, whether they are all boys or all girls. The statistical rationale behind this phenomenon is that while sex ratios tend to equalize when averaged across a substantial number of sibling groups, they are not necessarily required to achieve equilibrium within individual sibling groups. Therefore, within families comprising five children, the occurrence of sibships consisting of five boys is equivalent to the occurrence of sibships consisting of five girls, resulting in an overall sex ratio of 1:1. Nevertheless, both of these sibling groups exhibit atypical sex distributions (Kowles & Kowles, 2001).

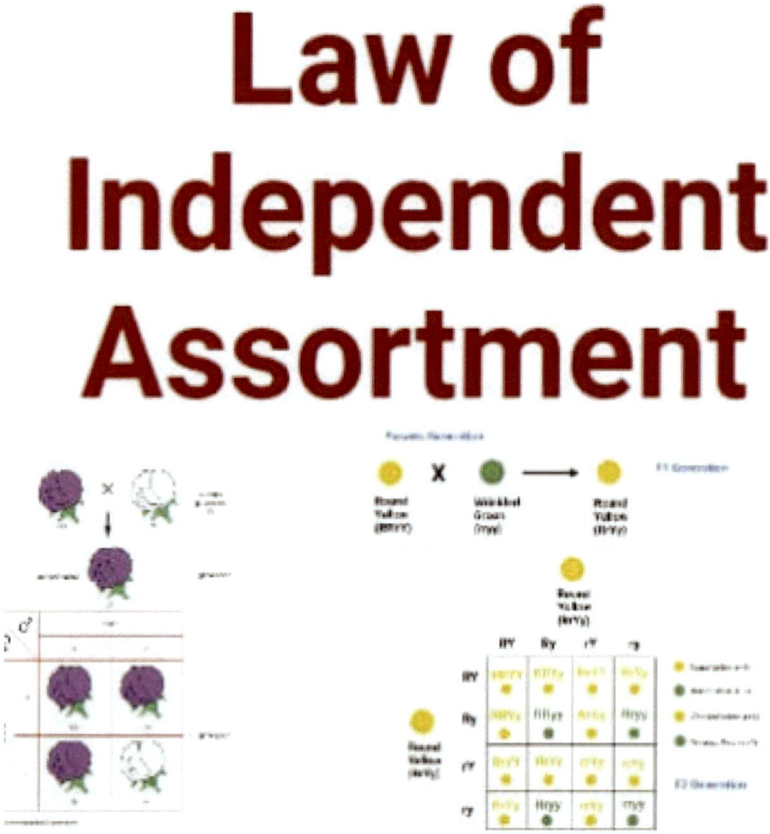

Figure 4.15. Genetics involves two important types of independence. Independent allele segregation is the first kind. (Source: Ridley, Creative Commons License)

When events exhibit independence, such as independent traits or successive offspring resulting from a cross, the probabilities are combined using the multiplication rule (Hasibuan et al., 2017).

Multiplication Rule*: The probability of the joint occurrence of two independent events, denoted as A and B, can be determined by multiplying their probabilities.*

In symbols, the multiplication rule is

$$Prob(A\,and\,B) = Prob(A).Prob(B)$$

The multiplication rule is a mathematical principle that can be applied to address inquiries similar to the following example (Shannon, 1993): Given a cross between two individuals with the genotype Aa, the objective is to determine the probability of both offspring exhibiting the dominant phenotype. The probability of an offspring exhibiting the dominant phenotype is 3/4 due to the mating of Aa × Aa. According to the multiplication rule in probability theory, the likelihood of both offspring exhibiting the dominant phenotype can be calculated by multiplying the individual probabilities of each offspring having the dominant phenotype. In this case, the probability of each offspring having the dominant phenotype is ¾ (Jacob & Monod, 1961). Therefore, the probability of both offspring having the dominant phenotype is determined by multiplying 3/4 by 3/4, resulting in a probability of 9/16. The present inquiry exemplifies a common genetic query that can be resolved by applying the addition and multiplication principles simultaneously. Given a cross between two individuals with the genotypes Aa and Aa, we are interested in determining the likelihood of producing two offspring, one displaying a dominant phenotype and the other displaying a recessive phenotype. Sibships consisting of one dominant phenotype and one recessive phenotype can arise through two distinct mechanisms: with the dominant phenotype being born first or with the dominant phenotype being born second. It is important to note that these two outcomes are mutually exclusive. The probability of the first case can be expressed as the product of 3/4 and 1/4, while the probability of the second case can be expressed as the product of 1/4 and 3/4. Since the events are mutually exclusive, their probabilities can be combined through addition. Consequently, the answer is thus (Amorim & Pinto, 2018).

$$\left(\frac{3}{4}x\frac{1}{4}\right)+\left(\frac{1}{4}x\frac{3}{4}\right)=2\left(\frac{3}{4}\right)\left(\frac{1}{4}\right)$$

Using multiplication and addition rules in genetic calculations is highly effective in determining the probabilities of genetic occurrences. The application of rules for estimating the expected proportions of the nine different genotypes among the F2 progeny resulting from self-pollination of a Ww Gg dihybrid is illustrated in Figure 4.14.

In genetics, the concept of independence is not limited solely to the offspring resulting from mating but also extends to the genes that segregate according to the principle of independent assortment (see Figure 4.15). Independence refers to the condition where the multiplication rule can be applied to ascertain the probability of different progeny types resulting from a cross involving independent assortment among multiple pairs of alleles. The principle mentioned above serves as the theoretical foundation for the anticipated offspring phenotypes resulting from a trihybrid cross, as depicted in Figure 4.12.

The multiplication rule can determine the probability of a particular genotype within the offspring resulting from a genetic cross. In the scenario where a quadruple heterozygote with the genotype Aa Bb Cc Dd undergoes self-fertilization, the likelihood of producing a quadruple heterozygote offspring with the genotype Aa Bb Cc Dd is determined by multiplying the probabilities of inheriting each allele independently. Assuming an independent assortment of all four pairs of alleles, the probability can be calculated as (1/2) (1/2) (1/2) (1/2) = (1/2)4, which simplifies to 1/16 (Darlington, 1969).

Remember

Despite having somewhat different neck lengths, humans and giraffes both have seven neck vertebrae. This is as a result of animals maintaining a genetic blueprint.

4.10 SEGREGATION IN HUMAN PEDIGREES

To determine the genetics of a characteristic, organisms must be mated and produce a large number of offspring for phenotypic categorization. This segregation analysis approach is rarely applicable to humans or large domestic animals. However, phenotypic manifestations that show allele segregation across several generations of networked individuals can sometimes reveal a trait's inheritance mechanism. A pedigree—a genealogical chart showing each person's traits—is the standard method. The pedigree analysis uses genetic probability (Jeffreys et al., 1986).

Figure 4.16 shows most human pedigree symbols. Circles represent females and square guys in this picture. When gender is unknown, a diamond is used. Shaded symbols indicate the phenotype of interest. Heterozygous carriers of recessive alleles are portrayed using half-filled symbols. A horizontal line connects the emblems of a female and a male's progeny, indicating reproduction. Siblings—regardless of gender—are sorted chronologically from left to right (Ott, 1977).

Figure 4.17 shows a pedigree of Huntington's disease, a dominant allele-related genetic disorder. The pedigree's numerical identifiers help identify people. Roman numbers indicate succeeding cohorts. Each generation assigns ascending number values from left to right. The female I-1 and male I-2 start the genealogical record (Mullaart et al., 1995). Huntington's disease is a neurodegenerative ailment that causes neuron function to deteriorate, usually around middle life gradually. It causes severe physical and cognitive disability and death. The pedigree shows that afflicted parents pass on the characteristic to their kids, affecting both genders. Half of the kids of afflicted parents also have the characteristic. Simple Mendelian dominance has these traits. Huntington's disease's dominant allele, HD, is rare. The afflicted family members have the heterozygous genotype HD hd, whereas the unaffected have the homozygous normal genotype hd hd. Figure 4.18 shows a homozygous recessive allele-related pedigree pattern (Zaidi et al., 2019).

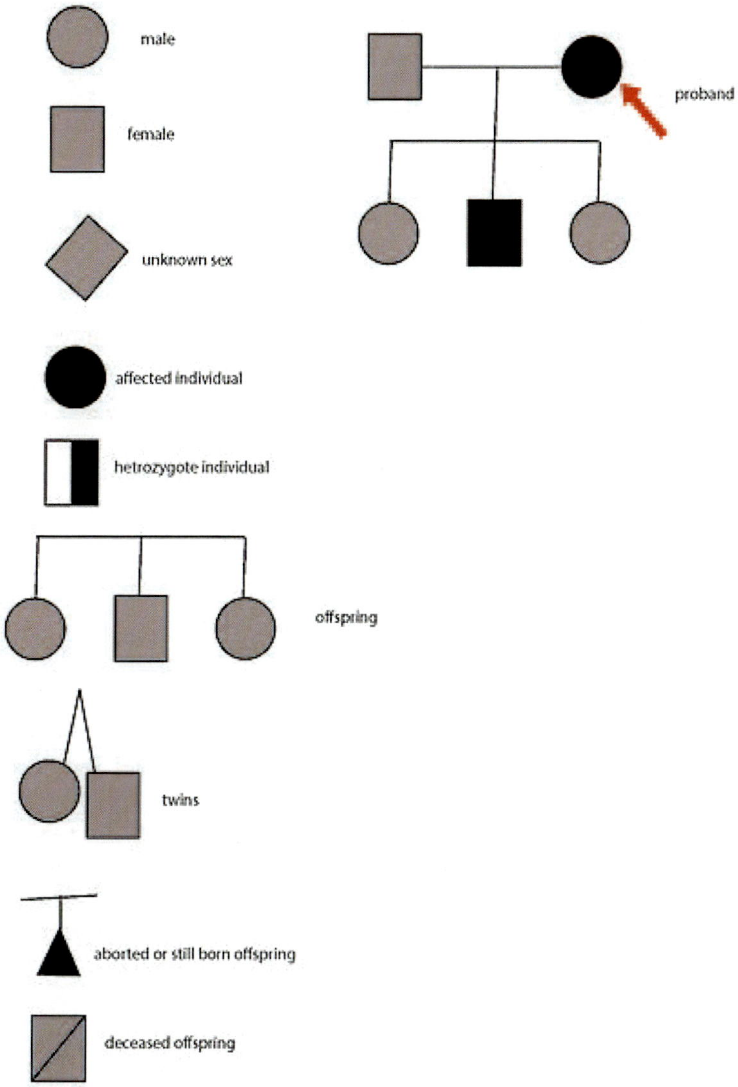

Figure 4.16. Conventional symbols used in depicting human pedigrees (Source: Https://en.wikipedia.org/wiki/File:Pedigreechart.jpg, Creative Commons License)

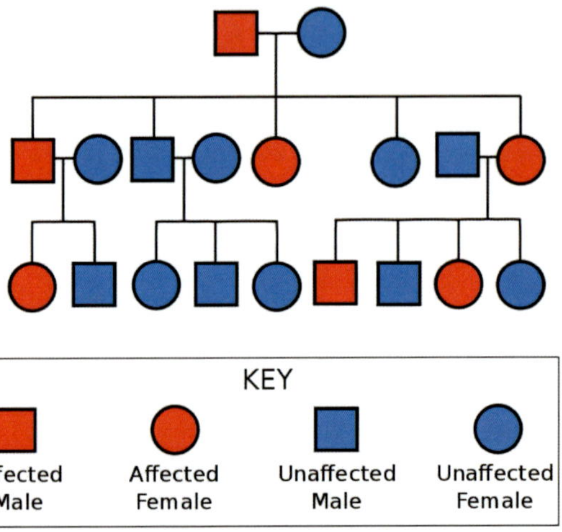

Figure 4.17. Pedigree of a human family showing the inheritance of the dominant gene for Huntington's disease. (Source: Https://www.wikilectures.eu/w/Huntington%27s_Disease#/media/File:Autosomal_Dominant_Pedigree_Chart2.svg, Creative Commons License)

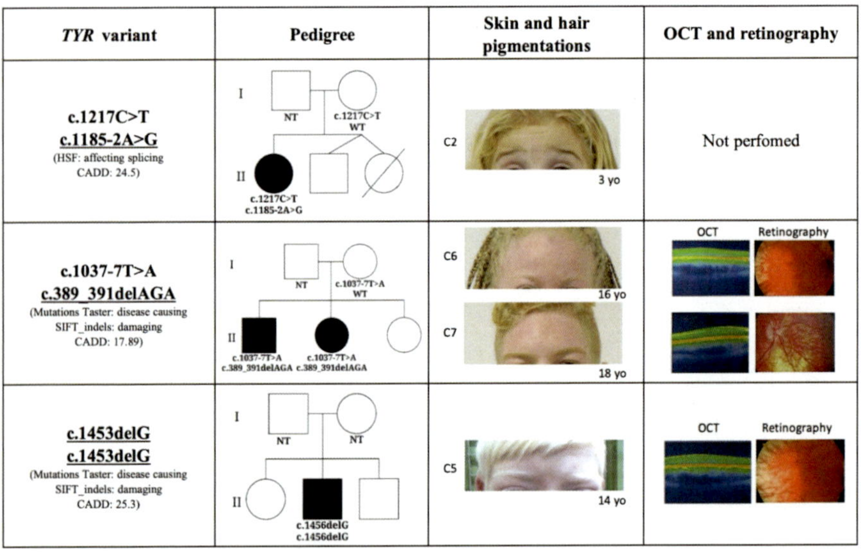

Figure 4.18. Pedigree of albinism. Afflicted individuals (filled symbols) frequently have unaffected parents when inheritance is recessive. (Source: Schidlowski L, Liebert F, Iankilevich PG, Rebellato PRO, Rocha RA, Almeida NAP, Jain A, Wu Y, Itan Y, Rosati R and Prando C (2020) Non-syndromic Oculocutaneous Albinism: Novel Genetic Variants and Clinical Follow Up of a Brazilian Pediatric Cohort. Front, Creative Commons License)

Albinism is the lack of pigmentation in skin, hair, and eyes. The disorder can afflict both genders without requiring affected parents (Guo & Thompson, 1992). Carriers have a heterozygous genotype for the recessive gene and do not have the characteristic. Aa x Aa carriers have a 1/4 chance of passing on the trait. The provided pedigree shows that recessive characteristics, especially uncommon ones, often have consanguineous parents. A dual-line indicates a consanguineous union between first cousins (Zhao, 1994).

Fun fact:

Cows develop close relationships with other cows, and genetics can have an impact on these relationships.

Consanguineous matings help detect rare recessive alleles. Homozygosity for a rare recessive allele is more likely when inherited from a common ancestor than from unrelated parents. An uncommon allele can be passed on to many offspring, creating many carriers in succeeding generations. In a first-cousin mating, 25% of recessive allele carriers will become homozygous (Deng et al., 2002).

4.11 GENETIC ANALYSIS

A geneticist's claim that a single gene with alleles P and p determines pea plant blossom color does not mean this gene is the only factor. The recessive p mutation, which changes blossom color from purple to white when homozygous, was found to identify this gene. Purple flower coloring requires numerous genes, including gene P (Elston & Stewart, 1971). Enzymes that produce the purple pigment anthocyanin are in this category. Geneticists studying flower color seldom stop at identifying the P gene, which causes purple color. A genetic study of flower color should find and isolate at least one mutation in each gene that causes purple coloring. Further examination of these mutant variants' phenotypes would reveal each gene's role in this trait (Hill & Mulder, 2010).

4.11.1 The Complementation Test In Gene Identification

A geneticist would start a flower color genetic investigation by isolating many unique mutants with white blooms. Radiation and some substances can increase mutation rates.

Mutant screens detect and isolate mutants with consistent phenotypes. Some isolated mutants will contain mutations in known genes. A genetic analysis of pea blossom color may show one or more new mutations that have changed the wildtype 'P' allele into a defective 'p' allele, preventing purple pigment production. The recently discovered 'p' alleles are faulty 'P' variants that prevent purple pigment production. Mutant screening can find uncharacterized gene mutations. Like P p mutations, newly found genes may have numerous mutations (Engebrecht & Silverman, 1984).

In a mutant screen to determine flower color, all novel mutations were detected in white-flowering plants. Most new mutant alleles encode a non-functional protein necessary for purple pigment production. The homozygous recessive genotype cannot produce the wildtype protein needed for pigmentation, making the mutant alleles recessive. The blossoms will be white. The heterozygous genotype, with one mutant allele and one wildtype allele, produces purple flowers. The wildtype allele's functioning protein compensates for the mutant allele's faulty protein (Coelho et al., 2008).

Mutations in many genes cause genotypes and white blooms. Two genotypes with white flowers may be homozygous recessive for the same or distinct genes. A mutant screen allows the geneticist to identify allelic mutation pairs and separate them from non-allelic ones. Three white-flowering plant variants in Figure 4.19 illustrate the issue. Homozygosity for a recessive mutation causes white blooms of all kinds. Conversely, various types may have different gene mutations, resulting in alleles. Each mutation might occur in a different gene, preventing them from becoming alleles (Atabekov et al., 1999).

The progeny phenotype from regulated mating across kinds resolves the mutation allelism issue. Figure 4.19 shows two cross-experiment outcomes. It shows purple flowers on the F1 offspring, whereas the other shows white flowers on the mutant offspring. Complementation occurs when the children have purple blooms and the parental plants have mutations in different genes. Complementation suggests that purple blooms require two genes. Mutant strain 1 has a homozygous genotype, pp, suggesting the recessive 'p' allele. The second parent is homozygous for a cc gene mutant allele (Soussy et al., 1993).

Thus, mutant strain 1 must be represented as pp CC and mutant strain 2 as PP cc. The cross produced a heterozygous F1 genotype for Pp and Cc, suggesting one dominant and one recessive gene for each trait. F1 children with the purple flower phenotype have recessive alleles 'p' and 'c'. Figure 4.19 shows another cross result. Mutant F1 progeny have white blooms in this case. Noncomplementation describes it. Since neither mutant strain can provide the genetic function missing in the other, the absence of complementation shows that both parental strains have a gene mutation. Mutant strain 1 has genotype pp. Mutant strain 3 has homozygosity for a unique mutation, suggesting a 'p' mutation recurrence (Ferguson & Horvitz, 1985).

White flowers in the F1 generation suggest a homozygous recessive genotype (pp). However, mutant strain 3 is homozygous for the pp genotype and the only origin of the second p allele. The mutant strain 3 allele is labeled p * to indicate that it was independently mutated. Figure 4.19 shows a complementation cross. This approach determines the allelic connection between recessive mutations in two strains.

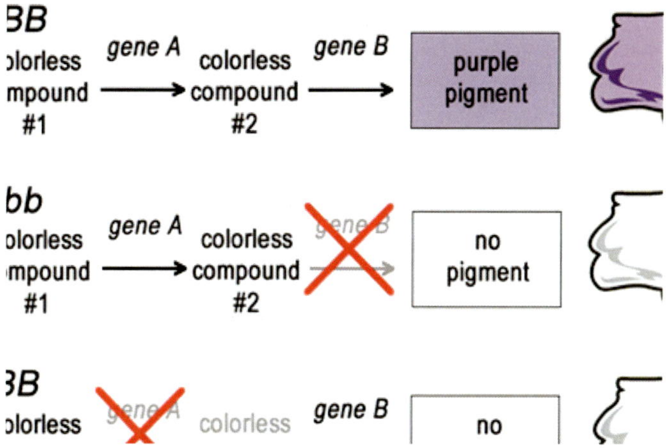

Figure 4.19. The complementation test reveals whether two recessive mutations are alleles of the same gene. (Source: Https://bio.libretexts.org/Courses/City_College_of_San_Francisco/Introduction_to_Genetics/04%3A_Mutation_and_Variation/4.05%3A_Complementation_tests_and_Allelism, Creative Commons License)

The complementation test is an important genetic experiment that determines allelism. Imagine a mutant screen to find new pea mutations that cause white flowers. The experiment uses a genetically stable purple flower strain. We use x-ray-irradiated pollen to fertilize ovules and get seeds. After self-fertilization, the F2 plants are grown from F1 seeds. White flower-colored F1 seeds may have a new mutation. The recessive white phenotype produces purple blooms (Heiss et al., 2008).

However, F1 children will be heterozygous for the unique white mutation; therefore, self-fertilization will produce F2 progeny with a 3:1 purple-to-white bloom ratio. Mutations that cause a particular phenotypic are rare, even when initiated by radiation; therefore, only a small percentage of self-fertilized plants will have white flowers. Let's imagine we've obtained four unique mutations along with the p and c mutations from Figure 4.19's complementation test (Service, 2004).

How will these four new mutations be named? No gene count may be assumed. All four examples might repeat "p" or "c." Conversely, the four examples may indicate unique gene mutations that determine floral coloration.

The new mutations are xl, x2, x3, and x4. The prefix "x" indicates that these mutations were caused by X-ray irradiation, not a gene. The white blossoms of the homozygous recessive F2 seeds revealed each mutation's recessive characteristic.

The complementation test classifies "x" mutations into groups (Witzel-Schlömp et al., 1997).

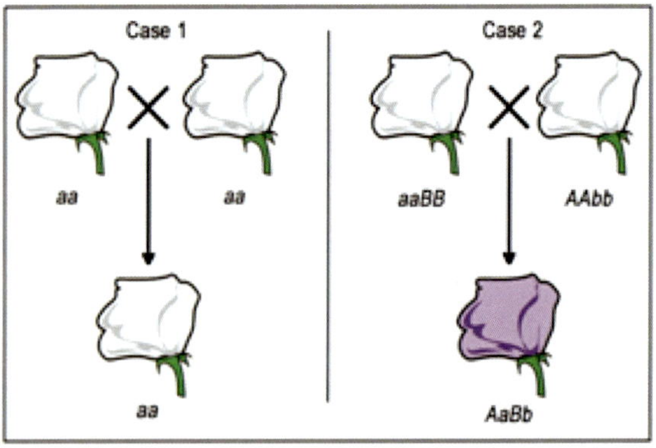

Figure 4.20. Results of complementation tests. (Source: Https://bio.libretexts.org/Courses/City_College_of_San_Francisco/Introduction_to_Genetics/04%3A_Mutation_and_Variation/4.05%3A_Complementation_tests_and_Allelism, Creative Commons License)

A complementation test's results in a triangular arrangement of (+) and (-) symbols. A "+" mark in the junction of the imaginary lines representing the male and female parents indicates F1 offspring with the wildtype phenotype, purple flowers. The crosses with white-flowering F1 offspring are marked with a hyphen (-). + signals show complementation between mutant alleles in parental individuals. The reciprocal of each cross produces F1 offspring with identical genotypes and traits, making the lower section of the triangle unnecessary. Diagonal elements are unnecessary since mating two animals with the same mutation (e.g., xl xl × xl xl) would generate homozygous recessive children for the mutation and hence have the mutant phenotype. (Perou et al., 1996).

***The Principle of Complementation**: In the scenario where two recessive mutations represent alleles of a single gene, the resulting phenotype of an organism harboring both mutations is classified as mutant. Conversely, if these mutations correspond to alleles of distinct genes, the phenotype of an organism carrying both mutations is categorized as wildtype or nonmutant.*

When analyzing complementation data, such as those presented in Figure 4.20, the principle is applied in the opposite direction. Analysis of the phenotypic characteristics exhibited by the first-generation offspring resulting from every conceivable cross provides insight into identifying mutations allelic variants of the identical gene (Shyu & Hu, 2008).

A complementation test showing a mutant phenotype from two recessive mutations shows that they are alleles of the same gene. If these mutations produce a wildtype phenotype, they are alleles of different genes.

Alleles can be arranged circularly for data analysis. If mutations lack complementation, link each pair directly. The lines must link alleles with mutations based on complementation. In a complementation test, no complementation means the mutations are alleles. Mutation x3 is an allele of 'p', proving that 'x3' and 'p' are separate mutant alleles of the gene 'P'. 'x2', 'x4', and 'c' are gene 'C' mutant alleles. All mutations complement 'x1'. This discovery shows that flower coloration is affected by a genetic component other than the 'P' and 'C' genes (Hitabatuma et al., 2021).

In an analysis, each group of non-complementing mutations is called a complementation group. As we have seen, each complementation group defines a gene.

A gene is defined experimentally as a set of mutations that comprise one complementation group. Any pair of mutations in such a group fail to complement one another, resulting in an organism with an observable mutant phenotype.

Thus, mutations in three genes cause white blooms. Alleles 'p' and 'x3' represent gene 'P', while 'c', 'x2', and 'x4' represent gene 'C'. 'xl' is a third gene unique from 'P' and 'C'. All genes are complementation groups (Kerppola, 2009).

To identify valid alleles in a genomic study, mutations might be renamed. Due to the pre-existing nomenclature of the 'p' allele before the mutation screen to acquire additional flower-color mutations, the newly discovered allele of 'p', 'x3', must be renamed to accurately reflect its allelic relationship with p. Renaming the 'x3' mutation as 'p3' with a subscript indicates that it evolved independently from 'p'. To acknowledge their allelic link with the original c mutation and distinguish their origins, the 'x2', and 'x4' mutations should be called 'c2' and 'c4', respectively. An allele of a recently identified gene, the 'x1' mutation, can be named non-specifically. For example, one could call the mutation "albus" (from the Latin word for white) and the 'x1' allele "alb." The wildtype dominant allele of the alb gene, which is necessary for purple coloration, can be called Alb or alb+. Geneticists divide new mutations into complementation groups and label them based on allelic correlations to identify genes and alleles. As genetics advances, alleles are renamed (Remy & Michnick, 2004).

Did You Know?

Genetic testing has revealed that all individuals with blue eyes share a common ancestor who lived between 6,000 and 10,000 years ago.

4.11.2 Why Does Complementation Test Work?

According to a Chinese saying, accurately naming entities is the first step to wisdom. Genetics illustrates this. The prudent act of assigning appropriate vocabulary to the p, c, and alb mutations, specifying their allelic relationship to the three genes, establishes a terminology that effectively communicates the resultant phenotype for each potential genotype to flower color. Purple flowers require one copy of each wild-type P, C, and Alb allele. Two gene P mutant alleles will produce white blooms. This investigation found pp, p3p3, and pp3. Any genotype with two C mutant alleles has white flowers. This investigation found genotypes cc, c2 c2, c4 c4, cc2, and cc4. Any genotype with two Alb mutant alleles has white flowers (Hawley & Gilliland, 2006).

The metabolic pathway synthesizing flowers' purple pigment is responsible for mutations in all three genes during flower-color mutant screening. The biochemical route explains the complementation test's effectiveness. Anthocyanin converts a colorless precursor into two colorless intermediate molecules, 'X' and 'Y'. The graphic shows a metabolic change from one substance to another with each arrow. The gene's wildtype allele encodes an enzyme needed for each step. Encoding the last pathway enzyme is the 'P' allele. Enzymatically, 'Y' becomes anthocyanin. This enzyme helps convert intermediate chemical 'Y' into anthocyanin. Reports say this stage slows route advancement. The flower stays white because purple pigment production is blocked, even if the cellular environment has a higher 'Y' content (Hamza et al., 2015).

The mutant alleles along the pathway's horizontal axis encode dormant enzymes. Homozygous genotypes cannot produce active enzymes. The mutant 'alb' allele, which encodes an inactive version, causes the initial process step to lack enzyme activity. Thus, alb alb genotypes lack this enzyme. The homozygous mutant's first pathway blockage prevents precursor conversion into intermediate X. Y cannot be produced without X, preventing anthocyanin formation. The homozygous alb alb genotype produces white flowers. The alb allele is recessive because, in the heterozygous genotype Alb alb, the wildtype 'Alb' allele encodes a functioning enzyme that facilitates the first stage of the route, preventing any blockage (Hopkins & Nightingale, 2006).

Mutant 'C' gene alleles stop the process at step two. The homozygous genotypes cc, c2 c2, and c4 c4 produce an inactive enzyme, as do the heterozygous genotypes cc2, cc4, and c2 c.4. In the three genotypes, each mutant allele encodes a separate, non-functional enzyme variation that blocks the process at the second stage, resulting in white blooms. The 'c', 'c2', and 'c4' alleles share the same complementation group because they encode enzymatically inactive versions of the same enzyme and cannot complement each other. P gene mutations are similar. The wildtype P allele encodes the enzyme that completes anthocyanin production. Any genotype pp, pp3, or p3p3 without a functioning enzyme blocks the process at its ultimate stage, resulting in white blooms. Due to their P gene mutations, p and p3 belong to the same complementation group (Adams et al., 1991).

4.11.3 Multiple Alleles

The C and P genes exemplify the concept of multiple alleles, wherein more than two alternative forms of a specific gene exist. The presence of the wildtype variant of each gene should be acknowledged as an allele, resulting in a total of two alleles for the Alb gene (Alb and alb), four alleles for the C gene (C, c, c2, and c4), and three alleles for the P gene (P, p, and p3) (Latter, 1972).

When conducting a complementation test and observing that two distinct mutations are alleles of the same gene, it is uncertain whether the mutant alleles possess identical nucleotide sequences in the DNA. At the molecular level, a gene can be defined as a linear arrangement of nucleotides that encodes the precise order of amino acids within a protein. Every nucleotide consists of a base, namely A (adenine), T (thymine), G (guanine), or C (cytosine). Consequently, a gene comprising n nucleotides can undergo mutation at any position, allowing for substitution with any of the remaining three nucleotides. The total count of potential single-nucleotide variations within a gene of length n equals 3 multiplied by n. If any of these DNA sequences are present within the population, they are classified as alleles. For an instance where n equals 5000, it is possible to have a total of 15,000 alleles, excluding any scenarios involving multiple nucleotide substitutions. Most potential alleles may not be concurrently present; however, a considerable number can be found within a given population. The following rules govern the number of alleles (Oren et al., 2019).

- A gamete possesses a single allele for each gene.
- Each organism or cell has the potential to possess a maximum of two distinct alleles.
- A population of organisms can potentially encompass a diverse array of alleles.

Numerous genes exhibit the presence of multiple alleles. An individual's blood group, which can be classified as A, B, O, or AB, is determined by three distinct alleles: I A, I B , and I O. The specific combination of alleles present in an individual's genotype determines their blood group. In this particular case, it is worth noting that there exist two slightly distinct variants of the IA allele, thereby allowing for the distinction of four alleles (Briles et al., 1950).

In contemporary genetics, multiple alleles are observed in two primary contexts. In genetic analysis, a situation may arise where a mutant screen produces multiple mutant alleles for many genes. In the early 1980s, researchers conducted mutant screens in Drosophila intending to identify novel recessive mutations that impeded embryonic development, ultimately resulting in the demise of homozygous recessive embryos. The identification of approximately 18,000 mutations through the screens conducted in this study ultimately led to the recognition and subsequent awarding of the 1995 Nobel Prize to Christiane Nüsslein-Volhard and Eric Wieschaus. It is worth noting that they shared this prestigious prize with Edward B. Lewis, who had already made significant contributions to Drosophila development genetics (Morgenthaler & Thilly, 2007).

In the realm of genetic research, geneticists frequently come across the presence of multiple alleles when investigating natural populations of organisms. In most populations, such as the human population, it is common for each gene to possess multiple alleles exhibiting slight variations in their nucleotide sequence. Despite exhibiting variations in one or more nucleotides within the DNA sequence, most of these alleles retain the capacity to perform the typical gene function and do not manifest any discernible alterations in phenotype (Kritzik et al., 1998).

It is not uncommon for a gene to possess multiple alleles within human populations. Genes in DNA profiling, such as those employed in forensic investigations, typically exhibit multiple alleles within the population. Each individual is limited to possessing a maximum of two alleles for each of these genes, although it is common for the population to exhibit 20 or more alleles. Therefore, it is improbable for two unrelated individuals to possess identical genotypes, particularly when considering multiple genes with numerous alleles. In studies conducted on affected children worldwide, it has been observed that a considerable number of defective alleles, exceeding 200, have been identified for each gene in the case of inherited recessive diseases such as cystic fibrosis and phenylketonuria. Numerous alternative forms accompany the conventional manifestation of each gene. In most genes within various populations, the "normal" or "wildtype" allele is not represented by a singular nucleotide sequence but rather by a collection of distinct nucleotide sequences. Each of these sequences can perform the typical functions associated with the gene.

Multiple alleles of a gene can be attributed to random occurrences and can be seen as a reflection of the evolutionary history of mutations within a population. These mutations are then spread among different population subgroups through migration and interbreeding. In certain instances, there exist biological mechanisms that exhibit a preference for the preservation of a substantial quantity of alleles (Han et al., 2017). This phenomenon can be observed in certain flowering plants, where genes responsible for regulating self-sterility exhibit significant diversity in allelic variations. The phenomenon of self-sterility is observed in various species of red clover that exhibit natural growth in numerous pasture environments. The presence of self-sterility genes inhibits the process of self-fertilization by requiring a pollen grain to possess a distinct self-sterility allele that differs from both alleles found within the flower it lands on to undergo pollen tube growth and fertilization.

To clarify, a pollen grain harboring an allele already present within a flower cannot exert its function on a said flower. Due to self-sterility alleles within a plant, all pollen grains produced by said plant must possess one of these alleles. Consequently, the pollen cannot function on the same plant from which it originated, thereby preventing self-fertilization. In such circumstances, a novel allele in any given plant confers a selective advantage, as the pollen carrying the new allele can fertilize all flowers except those located on the same plant. Throughout evolutionary processes, red clover populations have accumulated numerous alleles of the self-sterility gene. Many of these alleles have been isolated and subsequently subjected to DNA sequencing, enabling their characterization. Numerous alleles exhibit variations at multiple nucleotide sites, indicating a considerable antiquity of the alleles within the population (Whiting, 1943).

4.12 COMPLICATIONS IN THE CONCEPT OF DOMINANCE

Mendel found dominant-recessive patterns for all characteristics. This fortunate event helped him make important findings. Deviations from dominance occur often. In the traditional attribute of round vs. wrinkled pea seeds, round seeds are not dominant. As WW and Ww have identical seed morphologies, the round phenotype dominates. However, each gene can affect several traits. It is usual for a pair of alleles to dominate one attribute but not another. Compared to spherical seeds, wrinkled seeds lack a functioning form of starch-branching enzyme I (SBEI). This enzyme helps produce amylopectin, a branching starch. Heterozygous Ww seeds had 50% lower SBEI levels than WW seeds. Homozygous ww seeds lack SBEI. WW-homozygous peas have large, spherical starch granules. These seeds maintain moisture and shrink evenly during ripening, eliminating wrinkling. The starch grains of ww-homozygous seeds lack amylopectin. Thus, starch granules are irregular. These seeds wrinkle due to excessive water loss and unequal shrinking during maturity (Murea et al., 2012).

In Ww heterozygotes, the w allele affects starch granule shape. Heterozygous seeds have intermediate starch granules. The seeds have enough amylopectin to shrink without creases. Thus, supremacy paradoxically occurs. Round seeds are dominant, whereas wrinkled seeds are recessive. Only two phenotypes exist. Round seeds outnumber wrinkled seeds in phenotypic expression. Only two phenotypes exist. Starch grain morphology distinguishes the three genotypes under the microscope. The WW genotype has big, spherical starch granules, while Ww has large, irregular grains. The ww genotype has tiny, irregular starch granules. The Ww genotype has an intermediate SBEI enzyme level between the WW and ww genotypes. The round, wrinkled pea example shows that "dominance" cannot be assigned to one pair of genes, regardless of phenotypic consequences. When a gene influences numerous features, a pair of alleles may dominate certain qualities but not others. Overarching principle (Dekker & Sibai, 1998):

The phenotype consists of many physical and biochemical attributes, and dominance may be observed for some of these attributes and not others. Thus dominance is a property of a pair of alleles concerning a particular attribute of phenotype.

4.12.1 Amorphs, Hypomorphs, & Other Types of Mutations

A wild-type allele can undergo mutation at several DNA nucleotide locations, creating many gene alleles. Certain alleles may affect phenotype more than others in the setting of numerous alleles. One mutant allele can completely inactivate the enzyme, whereas another can only partially reduce enzyme activity. Geneticists sometimes classify mutations by effect.

Amorphs, like Mendel's wrinkled mutation, produce inactive 1,4-alpha-glucan branching (SBEI) enzymes. An amorphic mutation might result from a gene loss or an amino acid alteration that inactivates the enzyme. Hypomorphs are genetic changes that reduce enzyme output without eliminating it. Amino acid changes that impair enzymatic activity or enzyme synthesis sometimes cause hypomorphic mutations. "Hyper" in "hypermorph" denotes a high enzyme level, frequently caused by a mutation that affects gene control and produces too much gene product (Heitzler et al., 1996).

Most mutations affect the gene's protein output as amorphs, hypomorphs, or hypermorphs. These mutations affect the wild-type, non-mutated allele's enzyme activity. However, additional mutations can arise. Neomorphs are genetic mutations that change gene function. Neomorphism can activate a gene in an inactive cell or tissue type. Neomorphs can activate a gene during development while the wild-type gene is dormant. Neomorphic mutations in Drosophila's eyeless gene can trigger wild-type gene expression in tissues that don't develop eyes. Mutations can cause complex eye components, including pigmentation, to develop in unusual anatomical locations. The wildtype eyeless gene expresses in legs, mouthparts, abdomen, and wings. Ectopic expression occurs when a wild-type gene is expressed in an unexpected area (Bushey et al., 2009).

An antimorph is a mutation whose result opposes the gene's typical product. Due to an amino acid sequence change, a mutant interacts with the wild-type protein to produce an inactive complex. Herman J. Muller named mutations in 1931. X-ray radiation causes mutations, according to Muller's studies. X-ray radiation mutations cause DNA sequence disruptions and abnormal gene expression. Muller's language helped define these mutations and later spread to other genetic changes (Gindhart & Kaufman, 1995).

4.12.2 Incomplete Dominance

Incomplete dominance is observed when the phenotypic expression of a heterozygous genotype falls within the spectrum of phenotypes exhibited by the homozygous genotypes. The majority of genes are responsible for encoding enzymes, and it is common for each allele within a genotype to independently contribute to the overall quantity of the enzyme present within the cell or organism. In genetic heterozygosity, the observable characteristics of the organism, known as the phenotype, are situated within the spectrum between the phenotypes exhibited by the corresponding homozygous individuals. The current state of terminology regarding this situation lacks consensus, as various terms, such as incomplete dominance, partial dominance, and semi-dominance, are employed interchangeably (Ren et al., 2019).

An illustrative instance of incomplete dominance pertains to the phenomenon of flower color in the Antirrhinum snapdragon plant. A series of enzymatic reactions facilitate the formation of a red type of anthocyanin pigment in wild-type flowers. The rate of the overall reaction is constrained by a wild-type enzyme, which is encoded by the 'I' allele. Consequently, the quantity of red pigment is contingent upon the quantity of enzyme synthesized by the 'I' allele.

The alternative allele '' encodes for an enzyme that is not active, resulting in flowers with an ivory color in individuals with genotype ii. The reduction in the quantity of the critical enzyme in individuals with Ii heterozygosity leads to a corresponding decrease in the amount of red pigment present in the flowers.

Consequently, the dilution effect manifests itself, causing the flowers to have a pink hue.

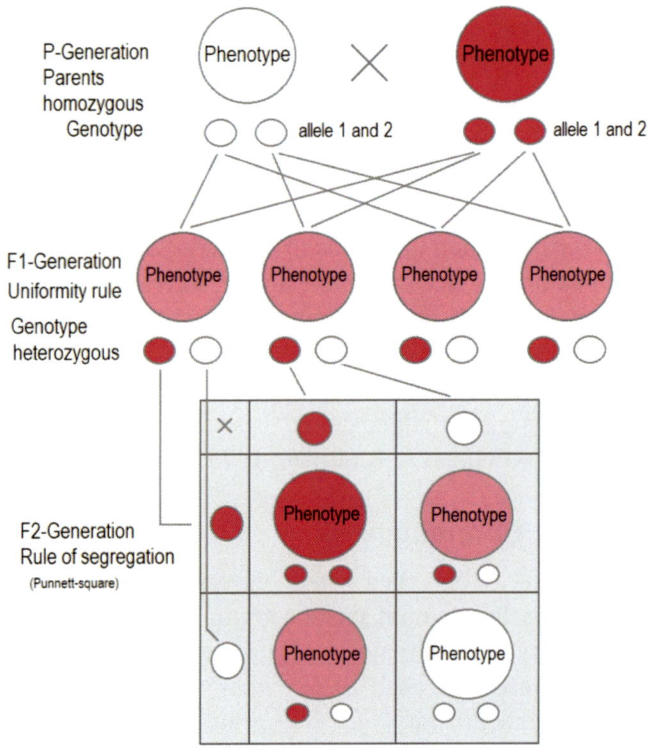

Figure 4.21. Absence of dominance in the inheritance of flower color in snapdragons. (Source: By Sciencia58 - Own work, CC0, https://commons.wikimedia.org/w/index.php?curid=82708902, Creative Commons License)

The direct observation of Mendelian segregation occurs when snapdragons with contrasting flower colors are subjected to crossbreeding. As an illustration, the hybridization of plants derived from a homozygous red-flowered variety and a homozygous ivory-flowered variety yields F1 plants exhibiting flowers of a pink hue. In the second filial generation (F2) derived from the self-fertilization of the first filial generation (F1) hybrids, an experiment yielded 22 plants exhibiting red flowers, 52 plants displaying pink flowers, and 23 plants showcasing ivory flowers. The numerical data exhibits a notable concurrence with the Mendelian ratio of 1 dominant homozygote to 2 heterozygotes to 1 recessive homozygote.

Consistent with the anticipated outcomes based on the principles of Mendelian inheritance, the F2 plants with red flowers exclusively yielded offspring with red flowers, while the F2 plants with ivory flowers exclusively produced progeny with ivory flowers. Conversely, the F2 plants with pink flowers generated offspring with red, pink, and ivory flowers in a ratio of 1/4 red: 1/2 pink: 1/4 ivory (Ghanizadeh & Harrington, 2018).

Quantitative phenotypes often show incomplete dominance. Quantitative qualities are continuous. Height, weight, egg production, plant blossoming, and enzyme content are features. A discrete trait has a dichotomy where variants are mutually exclusive. Examples are round vs. wrinkled seeds and yellow versus green seeds. Heterozygotes usually have a quantitative phenotype within the homozygote range. Incomplete dominance describes it (Klymiuk et al., 2022).

4.12.3 Codominance & the human ABO blood groups

"Codominance" refers to an organism's phenotype (observable attributes) having traits from both homozygous genotypes. Pink snapdragons show that heterozygous phenotypes do not have an intermediate stage between homozygous genotypes. It exhibits qualities from both homozygous genotypes.

Codominance is an example of"possessing the attributes of both homozygous genotypes". The following alleles determine the A, B, AB, and O blood types. Polysaccharides, sugar-based polymers on the outside of erythrocytes, determine blood type. Synthesize polysaccharides A and B. Both are modified by the I A or B allele's enzymatic product from a precursor molecule. Transferase enzymes may connect two kinds of sugar units to the precursor molecule.

Red blood cells with genotype IA IA contain just the A polysaccharide, making them blood type A. Genotype IB IB red blood cells express just the B polysaccharide, resulting in blood type B. AB blood type is caused by heterozygous genotype IA IB erythrocytes with both A and B polysaccharides. Genotype IA IB shows codominance since it has features from both homozygous genotypes. The heterozygous genotype has both A and B carbohydrates in red blood cells (Wilson et al., 1988).

Table 4.3. Genetic control of human ABO blood group (Source: Bruce, Creative Commons License)

Genotype	Antigens present on red blood cells	ABO blood group phenotype	Antibodies present in the blood fluid	Blood types that can be tolerated in transfusion	Blood types that can accept blood for transfusion
I^AI^A	A	Type A	Anti-B	A& O	A & AB

$I^{A}I^{0}$	A	Type A	Anti-B	A & O	A & AB
$I^{B}I^{B}$	B	Type B	Anti-A	B & O	B & AB
$I^{B}I^{0}$	B	Type B	Anti-A	B & O	B & AB
$I^{A}I^{B}$	a &B	Type AB	Neither anti-A nor anti-B	A, B, AB & O	AB only
$I^{0}I^{0}$	Neither A nor B	Type 0	Anti A & Anti B	O only	A, B, AB &O

The ABO blood groups hold significant relevance in medicine due to the frequent necessity of blood transfusions. An essential characteristic of the ABO system is the presence of antibodies against either the A or B polysaccharide in most human blood. Antibodies, also known as immunoglobulins, are protein molecules synthesized by the immune system in response to the presence of an antigenic stimulus. These proteins possess the ability to bind to the corresponding antigen specifically. Antibodies typically exhibit specificity by selectively recognizing and binding to a single antigen. Certain antibodies can bind with antigens, resulting in the formation of substantial molecular aggregates that have the potential to precipitate (Bajpai et al., 2022).

Antibodies play a crucial role in the immune system's response to viral and bacterial pathogens and other cellular entities by facilitating their elimination from the host organism. While it is generally understood that antibodies typically require prior stimulation by the antigen to form, individuals who can produce anti-A and anti-B antibodies can generate them. These antibodies can be generated by antigens resembling polysaccharides A and B, which are found on the outer membranes of numerous prevalent bacterial strains. Nevertheless, the phenomenon known as tolerance serves as a mechanism that inhibits an organism's production of antibodies targeting its antigens. This mechanism guarantees that the presence of A antigen or B antigen stimulates the production of antibodies solely in individuals whose own red blood cells lack A or B, respectively. The outcome:

People of blood type O make both anti-A and anti-B antibodies; those of blood type A make anti-B antibodies; those of blood type B make anti-A antibodies; and those of blood type AB make neither type of antibody.

The antibodies present in the plasma of individuals belonging to each of the ABO blood types are displayed in the fourth column of Table 4.3 (Storry & Olsson, 2009). The ABO blood groups possess clinical significance due to the occurrence of agglutination reactions when individuals who produce antibodies against A or B red-cell antigens receive blood transfusions containing these antigens. This reaction leads to the clumping of the donor's red blood cells. In the context of this reaction, it can be observed that the anti-A antibody can induce agglutination in red blood cells that possess either blood type A or blood type AB. This phenomenon arises due to the A antigen in both blood types. Likewise, anti-B antibody results in the agglutination of red blood cells belonging to individuals with blood type B or AB. The phenomenon of blood cell agglutination can result in the obstruction of numerous blood vessels, leading to a state of shock and potential fatality for the recipient of the transfusion. The phenomenon of incompatibility in the opposite direction, wherein the donated blood carries antibodies targeting the recipient's red blood cells, is generally deemed permissible due to the rapid dilution of the donor's antibodies, thereby preventing the occurrence of agglutination. The last two columns of Table 4.3 display the various categories of blood transfusions that are considered compatible. It is important to acknowledge that individuals with blood type AB can receive blood from individuals of any other ABO type. This unique characteristic of type AB blood is commonly referred to as being a universal recipient. In contrast, individuals with blood type O can donate blood to individuals of any ABO type, thus earning the designation of a universal donor (Nwosu et al., 2022).

4.12.4 Incomplete Penetrance &Variable Expressivity

In certain cases, the expected monohybrid Mendelian ratios, such as 3: 1 (or 1 : 2 : 1 in the presence of an intermediate heterozygote), may not be consistently observed, despite the trait being governed by the influence of a single recessive allele. The consistent ratios observed suggest that organisms possessing identical genotypes also manifest identical phenotypes (Bateman et al., 2018). In contrast, it is generally observed that organisms with the same genotype exhibit similar phenotypes; this is not universally true, especially in natural populations where researchers can control neither mating patterns nor environmental conditions. The phenotypic expression of a specific genotype can exhibit variation due to the influence of other genes on the phenotype or the susceptibility of the biological processes responsible for the phenotype to environmental conditions. The categorization of variable gene expression typically encompasses two main groups (Benson et al., 1998):

- **Variable expressivity** is a phenomenon in which genes are manifested to varying extents across diverse organisms. In the context of human beings, inherited genetic diseases frequently exhibit variability in their phenotypic manifestation across individuals. One patient may exhibit a more pronounced illness, while another with an identical disease may experience a milder manifestation. Variable expressivity refers to the phenomenon wherein a single mutant gene can manifest as a severe disease manifestation in one individual while presenting as a milder form in another. The degrees of expression frequently exhibit a continuum ranging from complete manifestation to near absence of the anticipated phenotypic traits (Oetjens et al., 2019).

- **Incomplete penetrance** is when the anticipated phenotype resulting from a specific genotype is not consistently manifested. An individual with a genetic predisposition to lung cancer may avoid developing the ailment by refraining from tobacco consumption. The absence of gene expression can arise due to environmental factors, as exemplified by the choice to refrain from smoking or due to interactions with other genetic elements. Incomplete penetrance represents an extreme manifestation of variable expressivity, wherein the observable phenotype is so subtle that it cannot be readily detected. The penetrance of a genotype refers to the proportion of organisms exhibiting a phenotype that aligns with their underlying genetic makeup for a specific trait. A genotype that consistently manifests its associated phenotype is said to possess a penetrance of 100 percent (Wright, 2022).

Did You Know?

During the 1970s, researchers achieved the successful integration of the gene responsible for human insulin into bacterial cells, resulting in the synthesis of artificial insulin for the purpose of managing diabetes.

SUMMARY

Inheritance of traits is governed by discrete units known as genes. In organisms of higher complexity, such as plants or animals, genetic material is typically organized in pairs. Each gene pair consists of one member inherited from the maternal parent and another from the paternal parent. Variations in DNA sequence can give rise to distinct forms of a gene. The various manifestations of a gene are referred to as alleles. An organism's genotype is determined by its specific combination of alleles. The phenotype of an organism is comprised of its observable characteristics. In an organism, when the two alleles of a gene pair exhibit identical characteristics (e.g., AA or aa), the genotype is considered homozygous for the 'A' or 'a' allele.

Conversely, if the alleles differ (Aa), the genotype is classified as heterozygous. In cases where the observable characteristics of an organism with two different alleles are identical to those of an organism with two identical alleles, the expressed allele is referred to as dominant. In contrast, the allele that remains unexpressed is referred to as recessive.

In genetics, the offspring from a mating event are called the F1 generation. The F2 generation is produced through mating between individuals from the F1 generation. In a monohybrid cross, such as AA × aa, involving the consideration of a single gene, the genotypic ratio observed in the F2 generation is 1:2:1 for dominant homozygotes (AA), heterozygotes (Aa), and recessive homozygotes (aa), respectively. The observed phenotypic distribution in the F2 generation follows a Mendelian ratio 3:1, with three individuals exhibiting the dominant phenotype and one displaying the recessive phenotype. The Mendelian ratios of genotypes and phenotypes arise from the segregation process during gamete formation, wherein the alleles of each gene pair separate into distinct gametes, followed by the random fusion of gametes during fertilization.

The phenomena of segregation, independent assortment, and random union of gametes adhere to the principles of probability, serving as the foundation for making predictions about the results of genetic crosses. The addition and multiplication rules are two fundamental principles for combining probabilities. The principle of addition applies to mutually exclusive events. It asserts that the probability of either one or the other of two events occurring is equal to the sum of their respective probabilities. The multiplication rule applies to events that are independent of each other. It asserts that the probability of both events occurring simultaneously equals the product of their probabilities. In certain organisms, such as human beings, the execution of controlled crosses is not feasible, necessitating the utilization of pedigree analysis to conduct genetic investigations across multiple generations of a family tree. The pedigree analysis involves the identification of potential genotypes among family members within a pedigree, as well as the assessment of the likelihood that an individual member possesses a specific genotype.

The complementation test serves as the functional definition of a gene. If a mating between two individuals with homozygous recessive mutations produces offspring that do not exhibit the mutant phenotype, these mutations are regarded as alleles of distinct genes. These alleles are commonly referred to as complementary to each other. Conversely, when a cross between two individuals with homozygous recessive mutations produces offspring displaying mutant characteristics, these mutations are regarded as alleles of the same gene. These alleles are commonly referred to as non-complementary. To assess the complementation of a given set of recessive mutations, it is necessary to perform a comprehensive complementation test by crossing the homozygous recessive individuals in all possible pairwise combinations.

Natural populations and mutant screens frequently give rise to multiple alleles. Multiple alleles refer to the existence of more than two alternative forms of a gene. A substantial quantity of alleles can be observed in various genetic phenomena, such as the genes employed in DNA typing and the self-sterility alleles found in certain flowering plants. While a population may possess various alleles, it is important to note that a gamete can only contain a single allele for each gene, and an organism can carry a maximum of two distinct alleles for each gene.

Dihybrid crosses exhibit variation in two distinct genes, as exemplified by mating individuals with genotypes AA BB and aa bb. The phenotypic ratios observed in the dihybrid F2 generation are 9 : 3 : 3 : 1, assuming that both the A and B alleles exhibit dominance and that the genes segregate independently. The 9 : 3 : 3: 1 ratio can undergo modifications through gene interactions, specifically through epistasis. Various forms of epistasis can manifest dihybrid ratios, such as 9:7, 12:3:1, 13:3, 9:4:3, and 9:6:1.

In cases where individuals possess heterozygous genotypes, it is not always possible to observe complete dominance of one allele over the other. In most instances, an individual possessing one copy of a normal allele and one copy of a mutant allele that encodes a faulty gene product will typically exhibit a reduced gene product production compared to an individual with two copies of the normal allele. In cases where the phenotype is influenced by the quantity of wild-type gene product rather than its mere presence, heterozygotes typically exhibit an intermediate phenotype. The phenomenon at hand is commonly referred to as incomplete dominance. Codominance is a genetic phenomenon in which both alleles present in a heterozygous individual are fully expressed, manifesting phenotypic traits characteristic of both homozygous genotypes. The phenomenon of codominance is demonstrated by the presence of the IA and IB alleles in individuals who possess the blood group AB. Codominance is frequently observed in proteins when each alternative allele encodes a distinct amino acid substitution. This is advantageous as it allows for the differentiation of the alternative protein forms through chemical or physical methods. The expression of genes can vary across different organisms, a phenomenon known as variable expressivity. Incomplete penetrance refers to the phenomenon wherein certain organisms do not express a particular genotype.

REFERENCES

1. Adams, M. D., Kelley, J. M., Gocayne, J. D., Dubnick, M., Polymeropoulos, M. H., Xiao, H., ... & Venter, J. C. (1991). Complementary DNA sequencing: expressed sequence tags and human genome project. *Science*, *252*(5013), 1651-1656.
2. Altenberg, L., & Feldman, M. W. (1987). Selection, generalized transmission and the evolution of modifier genes. I. The reduction principle. *Genetics*, *117*(3), 559-572.
3. Amorim, A., & Pinto, N. (2018). Big data in forensic genetics. *Forensic science international: genetics*, *37*, 102-105.
4. Atabekov, J. G., Malyshenko, S. I., Morozov, S. Y., Taliansky, M. E., Solovyev, A. G., Agranovsky, A. A., & Shapka, N. A. (1999). Identification and study of tobacco mosaic virus movement function by complementation tests. *Philosophical Transactions of the Royal Society of London. Series B: Biological Sciences*, *354*(1383), 629-635.
5. Awang-Kanak, F., Masnoddin, M., Matawali, A., Daud, M. A., & Jumat, N. R. (2016). Difficulties experience by science foundation students on basic mendelian genetics topic: A preliminary study. *Transactions on Science and Technology*, *3*(1-2), 283-290.
6. Bajpai, A., Jonnalagadda, V., Sridevi, B., Muthukottiappan, P., & Jesudasan, R. A. (2022). Single-cell variations in the expression of codominant alleles A and B on RBC of AB blood group individuals. *Journal of Genetics*, *101*(2), 36.
7. Bateman, M. S., Collinson, M. N., Bunyan, D. J., Collins, A. L., Duncan, P., Firth, R., ... & Barber, J. C. (2018). Incomplete penetrance, variable expressivity, or dosage insensitivity in four families with directly transmitted unbalanced chromosome abnormalities. *American Journal of Medical Genetics Part A*, *176*(2), 319-329.
8. Belyaev, D. K., Fomicheva, I. I., Taranin, A. V., & Baranov, O. K. (1986). Genetic polymorphism of IgG in mink. II. A genetic analysis of allotypes. *Experimental and Clinical Immunogenetics*, *3*(2), 65-74.
9. Benson, D. W., Sharkey, A., Fatkin, D., Lang, P., Basson, C. T., McDonough, B., ... & Seidman, C. E. (1998). Reduced penetrance, variable expressivity, and genetic heterogeneity of familial atrial septal defects. *Circulation*, *97*(20), 2043-2048.
10. Bhattacharyya, D., Bandyopadhyay, S. K., & Das, P. (2007). Handwritten Signature Watermarking and Extraction Technique Inspired by Principle of Segregation. *International Journal of Security and Its Application*, *1*(1), 35-44.

11. Bison, O., Ramalho, M. A. P., Rezende, G. D. S. P., Aguiar, A. M., & De Resende, M. D. V. (2006). Comparison between open pollinated progenies and hybrids performance in Eucalyptus grandis and *Eucalyptus urophylla*. *Silvae Genetica*, *55*(1-6), 192-196.
12. Bowden, J., Davey Smith, G., & Burgess, S. (2015). Mendelian randomization with invalid instruments: effect estimation and bias detection through Egger regression. *International journal of epidemiology*, *44*(2), 512-525.
13. Briles, W. E., McGibbon, W. H., & Irwin, M. R. (1950). On multiple alleles effecting cellular antigens in the chicken. *Genetics*, *35*(6), 633.
14. Brown, R. L. (1939). The fundamental principles of segregation. *The Institute of Fuel*, *13*, 15-23
15. Buchanan, D. S. (2011). Genetics: Mendelian. *genetics*, *2*, 4.
16. Bushey, D., Tononi, G., & Cirelli, C. (2009). The Drosophila fragile X mental retardation gene regulates sleep need. *Journal of Neuroscience*, *29*(7), 1948-1961.
17. Champagne, A. M. (1973). The segregation academy and the law. *The Journal of Negro Education*, *42*(1), 58-66.
18. Chen, N., Van Hout, C. V., Gottipati, S., & Clark, A. G. (2014). Using Mendelian inheritance to improve high-throughput SNP discovery. *Genetics*, *198*(3), 847-857.
19. Coelho, D., Suormala, T., Stucki, M., Lerner-Ellis, J. P., Rosenblatt, D. S., Newbold, R. F., ... & Fowler, B. (2008). Gene identification for the cblD defect of vitamin B12 metabolism. *New England Journal of Medicine*, *358*(14), 1454-1464.
20. Darlington, C. D. (1969). The genetics of society. *Past & Present*, (43), 3-33.
21. Davey Smith, G., & Ebrahim, S. (2003). 'Mendelian randomization': can genetic epidemiology contribute to understanding environmental determinants of disease? *International journal of epidemiology*, *32*(1), 1-22.
22. Davvaz, B., Nezhad, A. D., & Heidari, M. M. (2013). Inheritance examples of algebraic hyperstructures. *Information Sciences*, *224*, 180-187.
23. Dekker, G. A., & Sibai, B. M. (1998). Etiology and pathogenesis of preeclampsia: current concepts. *American journal of obstetrics and gynecology*, *179*(5), 1359-1375.
24. Deng, H. W., Livshits, G., Yakovenko, K., Xu, F. H., Conway, T., Davies, K. M., ... & Recker, R. R. (2002). Evidence for a major gene for bone mineral density/content in human pedigrees identified via probands with extreme bone mineral density. *Annals of Human Genetics*, *66*(1), 61-74.

25. Douglas, J. A., Skol, A. D., & Boehnke, M. (2002). Probability of detection of genotyping errors and mutations as inheritance inconsistencies in nuclear-family data. *The American Journal of Human Genetics*, *70*(2), 487-495.
26. Dudits, D., Maroy, E., Praznovszky, T., Olah, Z., Gyorgyey, J., & Cella, R. (1987). Transfer of resistance traits from carrot into tobacco by asymmetric somatic hybridization: regeneration of fertile plants. *Proceedings of the National Academy of Sciences*, *84*(23), 8434-8438.
27. Duncan, R. G., Castro-Faix, M., & Choi, J. (2016). Informing a learning progression in genetics: Which should be taught first, Mendelian inheritance or the central dogma of molecular biology?. *International Journal of Science and Mathematics Education*, *14*, 445-472.
28. Dzikowski, R., Frank, M., & Deitsch, K. (2006). Mutually exclusive expression of virulence genes by malaria parasites is regulated independently of antigen production. *PLoS Pathogens*, *2*(3), e22.
29. Edlin, G. J. (1987). Inappropriate use of genetic terminology in medical research: a public health issue. *Perspectives in Biology and Medicine*, *31*(1), 47-56.
30. Elston, R. C., & Stewart, J. (1971). A general model for the genetic analysis of pedigree data. *Human heredity*, *21*(6), 523-542.
31. Emerson, C., James, S., Littler, K., & Randazzo, F. (2017). Principles for gene drive research. *Science*, *358*(6367), 1135-1136.
32. Engebrecht, J., & Silverman, M. (1984). Identification of genes and gene products necessary for bacterial bioluminescence. *Proceedings of the National Academy of Sciences*, *81*(13), 4154-4158.
33. Fairbanks, D. J., & Rytting, B. (2001). Mendelian controversies: a botanical and historical review. *American Journal of Botany*, *88*(5), 737-752.
34. Fasoula, V. A., & Fasoula, D. A. (2002). Principles underlying genetic improvement for high and stable crop yield potential. *Field Crops Research*, *75*(2-3), 191-209.
35. Feng, Y., Niazadeh, R., & Saberi, A. (2022, July). Near-optimal bayesian online assortment of reusable resources. In *Proceedings of the 23rd ACM Conference on Economics and Computation* (pp. 964-965).
36. Ferguson, E. L., & Horvitz, H. R. (1985). Identification and characterization of 22 genes that affect the vulval cell lineages of the nematode Caenorhabditis elegans. *Genetics*, *110*(1), 17-72.
37. Ford, R. H. (2000). Inheritance of kernel color in corn: explanations & investigations. *The American Biology Teacher*, 10(5), 181-188.

38. Ghanizadeh, H., & Harrington, K. C. (2018). Restricted glyphosate translocation in Lolium multiflorum is controlled by a single incomplete dominant nuclear gene. *New Zealand Journal of Crop and Horticultural Science*, *46*(4), 346-353.
39. Gindhart Jr, J. G., & Kaufman, T. C. (1995). Identification of Polycomb and trithorax group responsive elements in the regulatory region of the Drosophila homeotic gene Sex combs reduced. *Genetics*, *139*(2), 797-814.
40. Grattapaglia, D., & Sederoff, R. (1994). Genetic linkage maps of *Eucalyptus grandis* and *Eucalyptus urophylla* using a pseudo-testcross: mapping strategy and RAPD markers. *Genetics*, *137*(4), 1121-1137.
41. Guo, S. W., & Thompson, E. A. (1992). A Monte Carlo method for combined segregation and linkage analysis. *American Journal of Human Genetics*, *51*(5), 1111.
42. Hain, R., Stabel, P., Czernilofsky, A. P., Steinbiß, H. H., Herrera-Estrella, L., & Schell, J. (1985). Uptake, integration, expression and genetic transmission of a selectable chimaeric gene by plant protoplasts. *Molecular and General Genetics MGG*, *199*, 161-168.
43. Hallowell, N., Arden Jones, A., Eeles, R., Foster, C., Lucassen, A., Moynihan, C., & Watson, M. (2006). Guilt, blame and responsibility: men's understanding of their role in the transmission of BRCA1/2 mutations within their family. *Sociology of Health & Illness*, *28*(7), 969-988.
44. Hamza, A., Tammpere, E., Kofoed, M., Keong, C., Chiang, J., Giaever, G., ... & Hieter, P. (2015). Complementation of yeast genes with human genes as an experimental platform for functional testing of human genetic variants. *Genetics*, *201*(3), 1263-1274.
45. Han, Y., Cameron, J. N., Wang, L., & Beavis, W. D. (2017). The predicted cross value for genetic introgression of multiple alleles. *Genetics*, *205*(4), 1409-1423.
46. Hartwell, L. H., Culotti, J., Pringle, J. R., & Reid, B. J. (1974). Genetic Control of the Cell Division Cycle in Yeast: A model to account for the order of cell cycle events is deduced from the phenotypes of yeast mutants. *Science*, *183*(4120), 46-51.
47. Hasibuan, N. A., Yusmiarti, K., Waruwu, F. T., & Rahim, R. (2017). Expert systems with genetics probability. *International Journal of Research In Science & Engineering*, *3*(4), 5720-5725.
48. Hawley, R. S., & Gilliland, W. D. (2006). Sometimes the result is not the answer: the truths and the lies that come from using the complementation test. *Genetics*, *174*(1), 5-15.

49. Heiss, K., Nie, H., Kumar, S., Daly, T. M., Bergman, L. W., & Matuschewski, K. (2008). Functional characterization of a redundant Plasmodium TRAP family invasin, TRAP-like protein, by aldolase binding and a genetic complementation test. *Eukaryotic cell*, *7*(6), 1062-1070.

50. Heitzler, P., Haenlin, M., Ramain, P., Calleja, M., & Simpson, P. (1996). A genetic analysis of pannier, a gene necessary for viability of dorsal tissues and bristle positioning in Drosophila. *Genetics*, *143*(3), 1271-1286.

51. Hill, W. G., & Mulder, H. A. (2010). Genetic analysis of environmental variation. *Genetics research*, *92*(5-6), 381-395.

52. Hitabatuma, A., Pang, Y. H., Yu, L. H., & Shen, X. F. (2021). A competitive fluorescence assay based on free-complementary DNA for ochratoxin A detection. *Food Chemistry*, *342*, 128303.

53. Honeycutt, B. B., & Pierce, B. A. (2007). Illustrating probability in genetics with hands-on learning: making the math real. *The American Biology Teacher*, *69*(9), 544-551.

54. Hopkins, M. M., & Nightingale, P. (2006). Strategic risk management using complementary assets: Organizational capabilities and the commercialization of human genetic testing in the UK. *Research policy*, *35*(3), 355-374.

55. Hwang, M., Bronnenberg, B. J., & Thomadsen, R. (2010). An empirical analysis of assortment similarities across US supermarkets. *Marketing Science*, *29*(5), 858-879.

56. Jacob, F., & Monod, J. (1961). Genetic regulatory mechanisms in the synthesis of proteins. *Journal of Molecular Biology*, *3*(3), 318-356.

57. James, D. R., & Taeuber, K. E. (1985). Measures of segregation. *Sociological methodology*, *15*, 1-32.

58. Jeffreys, A. J., Wilson, V., Thein, S. L., Weatherall, D. J., & Ponder, B. A. (1986). DNA" fingerprints" and segregation analysis of multiple markers in human pedigrees. *American Journal of Human Genetics*, *39*(1), 11.

59. Ka, K., Ryoo, R., Jang, Y., Park, Y., Jeong, Y., Kang, J., ... & Jeon, S. (2019). Characteristics of fruiting bodies formed upon monohybrid cross of *Lentinula edodes* strains. *Korean Journal of Mycology*, *47*(2), 173-179.

60. Kabat, E. A., Wu, T. T., & Bilofsky, H. O. W. A. R. D. (1980). Evidence indicating independent assortment of framework and complementarity-determining segments of the variable regions of rabbit light chains. Delineation of a possible J minigene. *The Journal of Experimental Medicine*, *152*(1), 72-84.

61. Kalina, J. (2014). Gregor Mendel, his experiments and their statistical evaluation. *Acta Musei Moraviae, Scientiae biologicae*, *99*, 87-99.

62. Kerppola, T. K. (2009). Visualization of molecular interactions using bimolecular fluorescence complementation analysis: characteristics of protein fragment complementation. *Chemical Society Reviews*, *38*(10), 2876-2886.
63. Kessler, L., Collier, A., & Halbert, C. H. (2007). Knowledge about genetics among African Americans. *Journal of Genetic Counseling*, *16*, 191-200.
64. Kline, K., & Sanders, B. G. (1984). Analyses of muscular dystrophy and Con A deficiency traits in testcross progeny of chickens. *Journal of Heredity*, *75*(1), 31-33.
65. Kloos, W. E., George, C. G., & Sorge, L. K. (2004). Inheritance of the flower types of *Gerbera hybrida*. *Journal of the American Society for Horticultural Science*, *129*(6), 802-810.
66. Klymiuk, V., Chawla, H. S., Wiebe, K., Ens, J., Fatiukha, A., Govta, L., ... & Pozniak, C. J. (2022). Discovery of stripe rust resistance with incomplete dominance in wild emmer wheat using bulked segregant analysis sequencing. *Communications Biology*, *5*(1), 826.
67. Kowles, R., & Kowles, R. (2001). Additional Genetic Analysis Using Statistics. *Solving Problems in Genetics*, 58, 447-472.
68. Kritzik, M., Savage, B., Nugent, D. J., Santoso, S., Ruggeri, Z. M., & Kunicki, T. J. (1998). Nucleotide polymorphisms in the α2 gene define multiple alleles that are associated with differences in platelet α2β1 density. *Blood, The Journal of the American Society of Hematology*, *92*(7), 2382-2388.
69. Kumar, B., & Patra, N. K. (2010). Gene frequency-based estimation of natural outcrossing in opium poppy (Papaver somniferum L.). *Molecular breeding*, *26*, 619-626.
70. Laird, N. M., Horvath, S., & Xu, X. (2000). Implementing a unified approach to family based tests of association. *Genetic Epidemiology: The Official Publication of the International Genetic Epidemiology Society*, *19*(S1), S36-S42.
71. Lanie, A. D., Jayaratne, T. E., Sheldon, J. P., Kardia, S. L., Anderson, E. S., Feldbaum, M., & Petty, E. M. (2004). Exploring the public understanding of basic genetic concepts. *Journal of Genetic Counseling*, *13*(4), 305-320.
72. Lanigan, T. M., Kopera, H. C., & Saunders, T. L. (2020). Principles of genetic engineering. *Genes*, *11*(3), 291.
73. Latter, B. D. H. (1972). Selection in finite populations with multiple alleles. III. Genetic divergence with centripetal selection and mutation. *Genetics*, *70*(3), 475-490.
74. Lewis, J., Leach, J., & Wood-Robinson, C. (2000). All in the genes?—

young people's understanding of the nature of genes. *Journal of Biological Education*, *34*(2), 74-79.

75. Li, Y. F., & Selloni, A. (2014). Mosaic texture and double c-axis periodicity of β-NiOOH: insights from first-principles and genetic algorithm calculations. *The Journal of Physical Chemistry Letters*, *5*(22), 3981-3985.
76. Liu, J. S., Wehner, T. C., & Donaghy, S. B. (1997). SASGENE: a SAS computer program for genetic analysis of gene segregation and linkage. *Journal of Heredity*, *88*(3), 253-254.
77. Lorieux, M., Goffinet, B., Perrier, X., De Leon, D. G., & Lanaud, C. (1995). Maximum-likelihood models for mapping genetic markers showing segregation distortion. 1. Backcross populations. *Theoretical and Applied Genetics*, *90*, 73-80.
78. Lumme, J., & Pohjola, L. (1980). Selection against photoperiodic diapause started from monohybrid crosses in *Drosophila littoralis*. *Hereditas*, *92*(2), 377-378.
79. MacLeod, A., Tweedie, A., McLellan, S., Taylor, S., Cooper, A., Sweeney, L., ... & Tait, A. (2005). Allelic segregation and independent assortment in *T. brucei* crosses: proof that the genetic system is Mendelian and involves meiosis. *Molecular and biochemical parasitology*, *143*(1), 12-19.
80. Marschall, T., Hajirasouliha, I., & Schönhuth, A. (2013). MATE-CLEVER: Mendelian-inheritance-aware discovery and genotyping of midsize and long indels. *Bioinformatics*, *29*(24), 3143-3150.
81. Marshall, P. A. (2008). Mapping linked genes in *Drosophila melanogaster* using data from the F2 generation of a dihybrid cross. *The American Biology Teacher*, *70*(9), 554-556.
82. Matthews, T. C., & Craig Jr, G. B. (1989). Isozyme polymorphisms maintained by lethal loci in inbred strains of *Aedes triseriatus*. *Journal of Heredity*, *80*(1), 53-57.
83. Miedaner, T., Hübner, M., Korzun, V., Schmiedchen, B., Bauer, E., Haseneyer, G., ... & Reif, J. C. (2012). Genetic architecture of complex agronomic traits examined in two testcross populations of rye (Secale cereale L.). *BMC Genomics*, *13*, 1-13.
84. Milligan, J. (2018). Subsidizing Segregation. *Virginia Law Review*, *104*(5), 847-932.
85. Moemeka, A. N., Hildebrandt, A. L., Radaskiewicz, P., & King, T. R. (1998). Brief communication. Shorn (shn): a new mutation causing hypotrichosis in the Norway rat. *Journal of Heredity*, *89*(3), 257-260.
86. Morgenthaler, S., & Thilly, W. G. (2007). A strategy to discover genes

that carry multi-allelic or mono-allelic risk for common diseases: a cohort allelic sums test (CAST). *Mutation Research/Fundamental and Molecular Mechanisms of Mutagenesis*, *615*(1-2), 28-56.

87. Mullaart, E., Verwest, A. M., Børglum, A. D., Uitterlinden, A. G., Temeerman, J. G., Kruse, T. A., & Vijg, J. (1995). Two-dimensional DNA typing of human pedigrees: spot pattern characterization and segregation. *Genomics*, *29*(3), 641-646.
88. Muller, H. J. (1914). A new mode of segregation in Gregory's tetraploid primulas. *The American Naturalist*, *48*(572), 508-512.
89. Murea, M., Ma, L., & Freedman, B. I. (2012). Genetic and environmental factors associated with type 2 diabetes and diabetic vascular complications. *The review of diabetic studies: RDS*, *9*(1), 6.
90. Nelson, C. J., Baetcke, K. P., Frith, C. H., Kodell, R. L., & Schieferstein, G. (1982). The influence of sex, dose, time, and cross on neoplasia in mice given benzidine dihydrochloride. *Toxicology and Applied Pharmacology*, *64*(2), 171-186.
91. Nwosu, N. C., Onwuka, O. M., & Bob-Manuel, I. F. (2022). Tongue rolling and hand clasping among various ABO blood groups in a University community in Eastern Nigeria. *EUREKA: Life Sciences*, (5), 11-17.
92. Oetjens, M. T., Kelly, M. A., Sturm, A. C., Martin, C. L., & Ledbetter, D. H. (2019). Quantifying the polygenic contribution to variable expressivity in eleven rare genetic disorders. *Nature Communications*, *10*(1), 4897.
93. Oltenacu, E. B. (1983). Teaching the principles of genetics with the aid of the herediscope. *Journal of Biological Education*, *17*(1), 59-61.
94. Oren, E., Tzuri, G., Vexler, L., Dafna, A., Meir, A., Faigenboim, A., ... & Gur, A. (2019). The multi-allelic APRR2 gene is associated with fruit pigment accumulation in melon and watermelon. *Journal of Experimental Botany*, *70*(15), 3781-3794.
95. Ott, J. (1977). Counting methods (EM algorithm) in human pedigree analysis: linkage and segregation analysis. *Annals of Human Genetics*, *40*(4), 443-454.
96. Otten, L. A. B. M., De Greye, H., Hernalsteens, J. P., Van Montagu, M., Schieder, O., Straub, J., & Schell, J. (1981). Mendelian transmission of genes introduced into plants by the Ti plasmids of Agrobacterium tumefaciens. *Molecular and General Genetics MGG*, *183*, 209-213.
97. Peay, K. G., Russo, S. E., McGuire, K. L., Lim, Z., Chan, J. P., Tan, S., & Davies, S. J. (2015). Lack of host specificity leads to independent assortment of dipterocarps and ectomycorrhizal fungi across a soil fertility gradient. *Ecology Letters*, *18*(8), 807-816.

98. Perou, C. M., Moore, K. J., Nagle, D. L., Misumi, D. J., Woolf, E. A., McGrail, S. H., ... & Kaplan, J. (1996). Identification of the murine beige gene by YAC complementation and positional cloning. *Nature Genetics*, *13*(3), 303-308.
99. Piaget, J. (2013). *Principles of Genetic Epistemology: Selected Works*. Routledge. (Vol. 7, pp. 87-88).
100. Rabinowitz, D., & Laird, N. (2000). A unified approach to adjusting association tests for population admixture with arbitrary pedigree structure and arbitrary missing marker information. *Human Heredity*, *50*(4), 211-223.
101. Ray, D. L. (2008). Demonstrating independent assortment of chromosomes, square dance style. *The American Biology Teacher*, *70*(7), 394-394.
102. Reardon, S. F., & Firebaugh, G. (2002). Measures of multigroup segregation. *Sociological Methodology*, *32*(1), 33-67.
103. Remy, I., & Michnick, S. W. (2004). A cDNA library functional screening strategy based on fluorescent protein complementation assays to identify novel components of signaling pathways. *Methods*, *32*(4), 381-388.
104. Ren, L., Yan, X., Cao, L., Li, J., Zhang, X., Gao, X., ... & Liu, S. (2019). Combined effects of dosage compensation and incomplete dominance on gene expression in triploid cyprinids. *DNA Research*, *26*(6), 485-494.
105. Riaz, N., Blecua, P., Lim, R. S., Shen, R., Higginson, D. S., Weinhold, N., ... & Reis-Filho, J. S. (2017). Pan-cancer analysis of bi-allelic alterations in homologous recombination DNA repair genes. *Nature Communications*, *8*(1), 857.
106. Ruch, D. G. (1998). A cookie model for the development of the concept of independent assortment. *The American Biology Teacher*, *60*(9), 696-698.
107. Sachs, E. S., Benedict, J. H., Stelly, D. M., Taylor, J. F., Altman, D. W., Berberich, S. A., & Davis, S. K. (1998). Expression and segregation of genes encoding CryIA insecticidal proteins in cotton. *Crop Science*, *38*(1), 1-11.
108. Samerski, S. (2006). The unleashing of genetic terminology: how genetic counselling mobilizes for risk management. *New Genetics and Society*, *25*(2), 197-208.
109. Sanderson, E., Davey Smith, G., Windmeijer, F., & Bowden, J. (2019). An examination of multivariable Mendelian randomization in the single-sample and two-sample summary data settings. *International journal of epidemiology*, *48*(3), 713-727.
110. Sasaaman, C., & Garthwalte, R. L. (1985). Independent assortment of four loci in the Isopod, *Porcellio dilatatus*. *Journal of Heredity*, *76*(6), 470-471.

111. Scheinberg, S. L. (1956). Genetic studies of cellular antigens in the chicken. *Genetics*, *41*(6), 834.
112. Schön, C. C., Utz, H. F., Groh, S., Truberg, B., Openshaw, S., & Melchinger, A. E. (2004). Quantitative trait locus mapping based on resampling in a vast maize testcross experiment and its relevance to quantitative genetics for complex traits. *Genetics*, *167*(1), 485-498.
113. Sellers, T. A., Bailey-Wilson, J. E., Elston, R. C., Wilson, A. F., Elston, G. Z., Ooi, W. L., & Rothschild, H. (1990). Evidence for mendelian inheritance in the pathogenesis of lung cancer. *JNCI: Journal of the National Cancer Institute*, *82*(15), 1272-1279.
114. Service, P. M. (2004). How good are quantitative complementation tests?. *Science of Aging Knowledge Environment*, *2004*(12), pe13-pe13.
115. Shannon, C. E. (1993). *An Algebra for Theoretical Genetics Ph. D. Dissertation, Massachusetts Institute of Technology, 1940.* Wiley-IEEE Press. (Vol. 3, pp. 891-920).
116. Shyu, Y. J., & Hu, C. D. (2008). Fluorescence complementation: an emerging tool for biological research. *Trends in biotechnology*, *26*(11), 622-630.
117. Slavov, G. T., Howe, G. T., Yakovlev, I., Edwards, K. J., Krutovskii, K. V., Tuskan, G. A., ... & Adams, W. T. (2004). Highly variable SSR markers in Douglas-fir: Mendelian inheritance and map locations. *Theoretical and Applied Genetics*, *108*, 873-880.
118. Soussy, C. J., Wolfson, J. S., Ng, E. Y., & Hooper, D. C. (1993). Limitations of plasmid complementation test for determination of quinolone resistance due to changes in the gyrase A protein and identification of conditional quinolone resistance locus. *Antimicrobial Agents and Chemotherapy*, *37*(12), 2588-2592.
119. Stewart, J. H. (1982). Difficulties experienced by high school students when learning basic Mendelian genetics. *The American Biology Teacher*, *44*(2), 80-89.
120. Storry, J. R., & Olsson, M. L. (2009). The ABO blood group system revisited: a review and update. *Immunohematology*, *25*(2), 48-59.
121. Teem, J. L., Ambali, A., Glover, B., Ouedraogo, J., Makinde, D., & Roberts, A. (2019). Problem formulation for gene drive mosquitoes designed to reduce malaria transmission in Africa: results from four regional consultations 2016–2018. *Malaria Journal*, *18*, 1-13.
122. Terao, K., Hamano, M., Cho, F., & Honjo, S. (1988). Human-type ABO blood groups as genetic markers for the management of a squirrel monkey (Saimiri sciureus) breeding colony. *Primates*, *29*, 287-292.

123. Uren, A. G., Kool, J., Matentzoglu, K., de Ridder, J., Mattison, J., van Uitert, M., ... & Berns, A. (2008). Large-scale mutagenesis in p19ARF- and p53-deficient mice identifies cancer genes and their collaborative networks. *Cell*, *133*(4), 727-741.
124. Venema, D. R. (2009). Laboratory exercises to examine recombination & aneuploidy in Drosophila. *The American Biology Teacher*, *71*(6), 336-343.
125. Verbanck, M., Chen, C. Y., Neale, B., & Do, R. (2018). Detection of widespread horizontal pleiotropy in causal relationships inferred from Mendelian randomization between complex traits and diseases. *Nature Genetics*, *50*(5), 693-698.
126. Verhaegen, D., Plomion, C., Gion, J. M., Poitel, M., Costa, P., & Kremer, A. (1997). Quantitative trait dissection analysis in Eucalyptus using RAPD markers: 1. Detection of QTL in interspecific hybrid progeny, stability of QTL expression across different ages. *Theoretical and Applied Genetics*, *95*, 597-608.
127. Wang, R. (2013). Assortment management under the generalized attraction model with a capacity constraint. *Journal of Revenue and Pricing Management*, *12*, 254-270.
128. Werner, C. R., Gaynor, R. C., Gorjanc, G., Hickey, J. M., Kox, T., Abbadi, A., ... & Stahl, A. (2020). How population structure impacts genomic selection accuracy in cross-validation: implications for practical breeding. *Frontiers in plant science*, *11*, 592977.
129. Whiting, P. W. (1943). Multiple alleles in complementary sex determination of Habrobracon. *Genetics*, *28*(5), 365.
130. Wilson, A. F., Elston, R. C., Siervogel, R. M., & Tran, L. D. (1988). Linkage of a gene regulating dopamine-beta-hydroxylase activity and the ABO blood group locus. *American Journal of Human Genetics*, *42*(1), 160.
131. Witzel-Schlömp, K., Späth, P. J., Hobart, M. J., Fernie, B. A., Rittner, C., Kaufmann, T., & Schneider, P. M. (1997). The human complement C9 gene: identification of two mutations causing deficiency and revision of the gene structure. *Journal of Immunology (Baltimore, Md.: 1950)*, *158*(10), 5043-5049.
132. Wright, C. F. (2022). Incomplete penetrance and variable expressivity: from clinical studies to population cohorts. *Frontiers in Genetics*, *13*, 920390.
133. Wu, C. I. (1991). Inferences of species phylogeny in relation to segregation of ancient polymorphisms. *Genetics*, *127*(2), 429-435.

134. Xue, Y., Ankala, A., Wilcox, W. R., & Hegde, M. R. (2015). Solving the molecular diagnostic testing conundrum for Mendelian disorders in the era of next-generation sequencing: single-gene, gene panel, or exome/genome sequencing. *Genetics in Medicine*, *17*(6), 444-451.
135. Yang, M., & Kim, J. W. (2018). Principles of genetic counseling in the era of next-generation sequencing. *Annals of Laboratory Medicine*, *38*(4), 291.
136. Yang, X., Ge, R., Yang, Y., Shen, H., Li, Y., & Tseng, C. C. (2009, September). Interactive computer program for learning genetic principles of segregation and independent assortment through meiosis. In *2009 Annual International Conference of the IEEE Engineering in Medicine and Biology Society*. IEEE. (Vol. 2, pp. 5842-5846).
137. Zaidi, A. A., Wilton, P. R., Su, M. S. W., Paul, I. M., Arbeithuber, B., Anthony, K., ... & Makova, K. D. (2019). Bottleneck and selection in the germline and maternal age influence transmission of mitochondrial DNA in human pedigrees. *Proceedings of the National Academy of Sciences*, *116*(50), 25172-25178.
138. Zhang, H., Chen, W., & Sun, K. (2017). Mendelism: new insights from Gregor Mendel's lectures in Brno. *Genetics*, *207*(1), 1-8.
139. Zhao, L. P. (1994). Segregation analysis of human pedigrees using estimating equations. *Biometrika*, *81*(1), 197-209.

CHAPTER 5

THE GENETIC CODE & TRANSLATION

Contents

5.1 INTRODUCTION

Within the expansive fabric of life on our planet, an elaborate molecular communication system known as the genetic code intricately regulates the fundamental nature of being. The genetic code, which is present in the DNA of all living organisms, functions as a fundamental plan that coordinates the production of proteins, which are essential components of life. The mechanism through which this code is decrypted and converted into operational proteins is a remarkable feat of biological engineering, elucidating the mysteries surrounding the emergence of life's variety and intricacy from a comparatively basic set of principles. This chapter delves into the captivating realm of the genetic code and translation, revealing the fascinating interaction between genotype and phenotype, the relationship between protein structure and function, the universal nature and redundancy of the genetic code, the complexities of the translation process, the significance of RNA-RNA interactions, and the application of these concepts in understanding genetic disorders such as Lesch-Nyhan Syndrome (Koonin & Novozhilov, 2009).

The fundamental essence of all living organisms resides in a central storage unit of genetic data, which is contained within the molecules of DNA (deoxyribonucleic acid). The data presented in the form of nucleotide sequences contain the encoded instructions necessary for constructing and sustaining the complex mechanisms that comprise the fundamental essence of life (Ambrogelly et al., 2007).

The chapter delves into the fundamental concept of the intriguing correlation between genotype and phenotype. The term "genotype" pertains to an organism's precise genetic constitution, denoted by the arrangement of nucleotides in its DNA sequence. In contrast, the concept of phenotype encompasses the range of observable characteristics and traits displayed by an organism, encompassing both physical appearances and physiological functions. The comprehension of the interaction between phenotype and genotype is crucial for unraveling the source of an organism's traits and their inheritability (Neumann, 2012).

The Lesch-Nyhan Syndrome, an uncommon hereditary condition, exemplifies the intricate correlation between genotype and phenotype. The etiology of this syndrome can be attributed to a genetic mutation occurring in the HPRT1 gene, resulting in a deficiency of the hypoxanthine-guanine phosphoribosyltransferase (HPRT) enzyme. The manifestation of major neurological symptoms and behavioral disruptions, including self-injurious behaviors, occurs due to the biochemical imbalance in purine metabolism. Examining this disorder yields significant insights into the significance of particular genetic mutations and their influence on an individual's phenotype, thereby providing a glimpse into the intricacies of genetic disorders (Bossi & Roth, 1980).

Proteins, which serve as the primary functional components within cells, are intricately associated with the genetic code through the biological process known as translation. Proteins, which consist of lengthy sequences of amino acids, perform a diverse range of tasks, including facilitating biochemical reactions and offering structural reinforcement (Rodin et al., 2011).

The genetic code functions via a trinucleotide-based system, where groups of three nucleotides, referred to as codons, are associated with distinct amino acids or act as initiation or termination signals for the process of protein synthesis. The genetic code consists of a total of 64 codons, which are formed by the combination of four nucleotide bases: adenine, cytosine, guanine, and thymine. Among the total of 64 codons, 61 are responsible for encoding amino acids, whereas the remaining three serve as stop codons, indicating the termination of protein synthesis. The present codon system exemplifies the remarkable efficiency and accuracy of the genetic code in converting a wide range of nucleotide sequences into a limited set of amino acids (Yarian et al., 2002).

One notable characteristic of the genetic code is its nearly universal presence among all extant organisms. The genetic code exhibits conservation across various organisms, ranging from bacteria to animals and plants, thereby highlighting the shared evolutionary origins of life on our planet. Moreover, the genetic code exhibits a certain level of depravity, whereby multiple codons have the ability to encode for the identical amino acid. The presence of redundancy in protein synthesis provides a mechanism for enhancing robustness, serving as a protective measure against mutations that could potentially have negative consequences on the observable characteristics of an organism (Novozhilov et al., 2007).

The process of translation is a crucial component of cellular biology, whereby the genetic information contained within the DNA molecule is translated into messenger RNA (mRNA) and then transformed into proteins. The process of protein synthesis occurs within the ribosomes, which are sophisticated molecular structures that play a crucial role in coordinating the incorporation of amino acids to form fully functional proteins. The translation process consists of three primary phases: initiation, elongation, and termination. A distinct set of factors and regulatory elements controls each stage (Hoffmann, 1974).

The genetic code serves as the fundamental basis for the translation process. Still, it is vital to note that various RNA-RNA interactions play a significant role in refining this process and regulating the expression of genes. RNA-binding proteins, microRNAs, and tiny regulatory RNAs are prominent constituents of this regulatory network. Through the modulation of translation rate and efficiency, these interactions facilitate cellular responses to environmental stimuli and exert precise regulation over the expression of particular genes. Consequently, they play a pivotal role in the organism's adaptive capacity to its surroundings (Wong et al., 2016).

In summary, the chapter pertaining to "The Genetic Code and Translation" explores the intricate molecular mechanisms that serve as the foundation for the remarkable phenomena observed in living organisms. This chapter elucidates the complexities of the genetic code, its relationship with the structure and function of proteins, and the remarkable phenomenon of translation. It explores the biological language governing all living organisms' fundamental design. In addition, an examination of the genotype-phenotype correlation, as demonstrated by genetic disorders such as Lesch-Nyhan Syndrome, underscores the significant influence of genetic data on the intricacy and variety of living organisms. The examination of these fundamental biological processes enhances our comprehension of the interconnectedness of living organisms and provides invaluable perspectives into the marvels of the natural realm (Vetsigian et al., 2006).

5.1.1 History

The historical trajectory of the genetic code is a compelling narrative characterized by rigorous scientific investigation, encompassing numerous years of pioneering breakthroughs and cooperative endeavors among esteemed scholars in the disciplines of genetics, biochemistry, and molecular biology. The comprehension of life's fundamental processes has been significantly influenced by the elucidation of the crucial molecular language that regulates the conversion of genetic information into proteins. Let us undertake a scholarly exploration of the significant milestones in the process of unraveling the genetic code throughout history (Shandell et al., 2021).

The origins of the endeavor to unravel the genetic code can be attributed to the groundbreaking contributions of Gregor Mendel, widely regarded as the progenitor of contemporary genetics. During the mid-19th century, Gregor Mendel conducted a series of experiments involving pea plants, which established the fundamental principles underlying the concept of heredity. These groundbreaking investigations catalyzed subsequent generations of scientists, motivating them to delve into the genetic underpinnings that contribute to the remarkable diversity observed in living organisms (Knight et al., 1999).

In the 20th century, there was a notable convergence between genetics and biochemistry, which laid the foundation for significant breakthroughs and transformative advancements. During the 1930s, Friedrich Miescher, a biochemist, made a significant discovery by identifying DNA as a distinct substance present within the cell nucleus. This seminal finding served as the foundation for subsequent investigations into the nature and properties of nucleic acids.

During the 1950s, the utilization of X-ray crystallography by Rosalind Franklin and Maurice Wilkins initiated the invention of the double-helical configuration of DNA (Fournier et al., 2010).

The investigation into the genetic code experienced a surge in progress during the initial years of the 1960s, as researchers specializing in genetics and biochemistry directed their focus towards unraveling the mechanisms by which DNA's genetic data is converted into proteins. The primary focus of this study revolved around the inquiry: What is the mechanism by which the four nucleotides (adenine, cytosine, guanine, and thymine) present in DNA are translated into the twenty amino acids that comprise proteins (Wolf & Koonin, 2007)?

A significant advancement occurred in 1961 when Marshall Nirenberg and Johann Matthaei made a crucial breakthrough in their research. The researchers performed experiments utilizing synthetic RNA composed solely of uracil (poly-U) and determined that this specific RNA sequence encoded the amino acid phenylalanine. This discovery represents the initial instance in which scientists successfully decoded a particular codon within the genetic code, establishing a foundation for subsequent investigations (Kato, 2019).

Expanding upon the advancements above, Nirenberg, in collaboration with Philip Leder and Har Gobind Khorana, initiated a sequence of experiments during the mid-1960s to unravel additional codons and their respective amino acids. By employing an assortment of in vitro translation systems and synthetic RNA molecules, the researchers successfully elucidated a greater number of codons, thereby establishing the universal nature of the genetic code across diverse organisms (Osawa et al., 1992). The Nobel Prize in Physiology was awarded in 1968 to Nirenberg, Khorana, and Robert Holley for their significant contributions to unraveling the genetic code. The acknowledgment served to emphasize the importance of their findings in furthering the comprehension of molecular biology (Sun & Caetano-Anollés, 2008).

As the investigation into the genetic code progressed, it became evident that the genetic code exhibits degeneracy. The identification was made that there exists a phenomenon wherein multiple codons can encode for a single amino acid, thereby facilitating redundancy and enhancing the resilience of the process of protein synthesis. This observation served to enhance scientists' comprehension of the efficacy and adaptability of the genetic code (Wiltschi & Budisa, 2007).

During the late 1960s and early 1970s, significant progress was made in decoding the genetic code, leading to a comprehensive understanding of its structure. Consequently, researchers redirected their efforts toward investigating the mechanisms underlying translation and the crucial involvement of transfer RNA (tRNA) in facilitating this process.

The identification of tRNA molecules, which possess the capacity to recognize codons and convey the corresponding amino acids selectively, established a pivotal connection between the genetic code and the process of protein synthesis (Di Giulio, 1989).

The historical narrative surrounding the genetic code serves as a testament to the formidable impact of scientific cooperation and unwavering determination. The elucidation of the genetic code, starting with Mendel's experiments on pea plants and culminating in the seminal contributions of Nirenberg, Khorana, and other scientists, has had a profound impact on our comprehension of genetics. This breakthrough has not only transformed our understanding of the field but has also laid the groundwork for significant progress in biotechnology, medicine, and molecular biology.

As researchers persist in deciphering the complexities of the genetic code, this captivating endeavor persists as a constantly evolving narrative that contributes to our understanding of the intricate molecular intricacies of life (Freeland & Hurst, 1998).

5.2 LESCH-NYHAN SYNDROME & THE RELATION BETWEEN PHENOTYPE & GENOTYPE

In the year 1962, an investigation was conducted by Dr. William Nyhan and Michael Lesch on a male child who was afflicted with a perplexing array of symptoms indicative of a severe illness. The male child exhibited hematuria, elevated levels of uric acid in his bloodstream, and unmanageable muscular contractions affecting his upper and lower extremities. The individual exhibited cognitive impairments and engaged in self-injurious behaviors by biting his fingers and lips. Upon thoroughly examining the male individual, Nyhan and Lesch determined that he was suffering from an unspecified ailment. Subsequently, additional individuals presenting comparable symptoms were documented, leading to the designation of the condition as the Lesch-Nyhan syndrome (Fu et al., 2014).

One of the initial manifestations of Lesch-Nyhan syndrome is the emergence of orange-colored particulate matter, resembling "sand," which is composed of uric acid crystals, observed in diapers within a few weeks following the infant's delivery. Within a span of approximately one year, the child initiates the manifestation of twisting movements in the extremities, specifically the hands and feet, alongside the occurrence of involuntary spasms. Approximately 50%

of the children experience seizures. Following a period of approximately two to three years, certain individuals within the child population demonstrate a tendency towards compulsive self-inflicted harm (Genao & Torres, 2014).

Lesch-Nyhan syndrome predominantly manifests in males and is passed on as an X-linked recessive disorder. The disease is attributed to the presence of a defective gene on the sole X chromosome of affected males. In the year 1967, a group of scientists affiliated with the National Institutes of Health conducted a study that concluded that the disease in question arises due to a flawed version of the gene responsible for encoding the enzyme hypoxanthine-guanine phosphoribosyl transferase (HGPRT).

In the cellular environment, the degradation of DNA and RNA results in the release of purines. These purines are salvaged by the enzyme HGPRT, which utilizes them to synthesize new nucleotides for RNA and DNA (Sampat et al., 2011). Individuals diagnosed with Lesch-Nyhan syndrome exhibit a genetic mutation within the HGPRT gene, resulting in an alteration of the enzyme's amino acid sequence, thereby causing a loss of functionality. As a consequence, the recycling process of purines is impeded, leading to their accumulation and subsequent conversion into uric acid. The manifestation of disease symptoms is attributed to elevated concentrations of uric acid (Ceballos-Picot et al., 2015).

The Lesch-Nyhan syndrome serves as an exemplification of the connection between genotype and phenotype, wherein a genetic mutation impacts the production of a protein, subsequently giving rise to the manifestation of disease symptoms. The preceding chapters have elucidated the process by which genetic information is encoded within DNA and subsequently transferred from DNA to RNA. This chapter focuses on the analysis of translation. This fundamental biological process involves converting the nucleotide sequence present in mRNA into a protein's corresponding amino acid sequenceng (Nyhan et al., 1970).

5.3 THE MOLECULAR RELATION BETWEEN GENOTYPE & PHENOTYPE

Archibald Garrod was the individual who initially proposed the presence of a correlation between genotype and proteins. In 1908, Garrod put forth the accurate proposition that genes can encode enzymes. Regrettably, his theory failed to impact his peers during that time significantly (Andresen et al., 1997).

Did You Know?

In the past, non-coding regions of DNA were commonly regarded as non-functional and insignificant. However, recent scientific investigations have revealed that these regions actually serve crucial regulatory functions in the process of gene expression.

5.3.1 The One Gene, One Enzyme Hypothesis

Beadle and Tatum employed the filamentous fungus Neurospora as a model organism to investigate the biochemical consequences of genetic mutations. The utilization of Neurospora as a model organism is advantageous due to its ease of cultivation in laboratory settings and the haploid nature of its primary vegetative component. The haploid state facilitates the convenient observation of recessive mutations, as depicted in Figure 5.1 (Horowitz & Leupold, 1951).

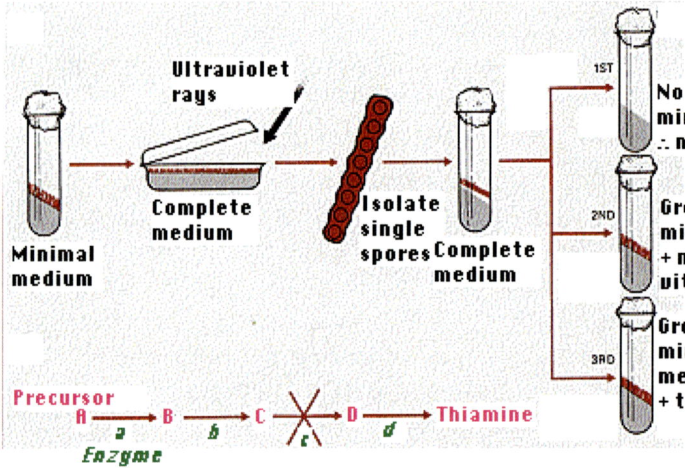

Figure 5.1. Beadle and Tatum employed Neurospora, a fungus with a complicated life cycle, to study gene-protein relationships. (Source: Https://bio.libretexts.org/Bookshelves/Introductory_and_General_Biology/Biology_%28Kimball%29/06%3A_Gene_Expression/6.01%3A_One_Gene_-_One_Enzyme_Theory, Creative Commons License)

Wild-type Neurospora exhibits growth on a minimal medium comprising solely of inorganic salts, nitrogen, a carbon source such as sucrose, and the essential vitamin biotin. The fungus has the capacity to biosynthesize all necessary biological molecules from fundamental compounds. Nevertheless, it is possible for mutations to occur that impede fungal growth by compromising the fungus's capacity to produce one or more vital biological compounds (Hickman & Cairns, 2003).

The auxotrophs, which are mutants lacking essential nutrients, exhibit an inability to thrive in a minimal medium. However, their growth can be sustained in a medium supplemented with the specific substance they are no more capable of synthesizing. The initial step undertaken by Beadle and Tatum involved subjecting spores of Neurospora to irradiation in order to induce mutations. Following irradiation, individual spores were transferred into separate culture tubes containing a complete medium, a nutrient-rich medium that provides all the necessary biological substances for growth. Subsequently, the spores were transferred from each culture to tubes that were filled with a minimal medium. Fungi harboring auxotrophic mutations exhibited growth on a nutrient-rich medium yet failed to increase on a nutritionally limited medium. This observation facilitated the identification of cultures by Beadle and Tatum that possessed such mutations (Rashevsky, 1962). Following identifying an auxotrophic mutation within a specific culture, Beadle and Tatum embarked on a quest to ascertain the precise impact of this mutation. The spores of each mutant strain were transferred from a medium containing all necessary nutrients to a set of tubes. Each tube contained a minimal medium supplemented with a specific essential biological molecule, such as an amino acid.

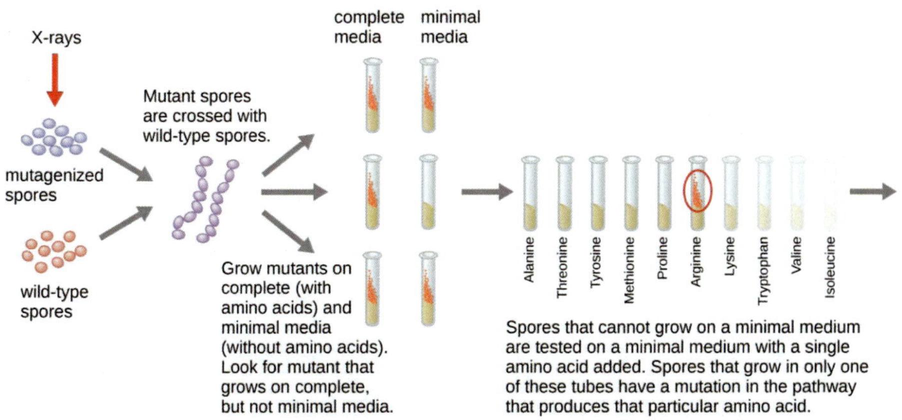

Figure 5.2. Beadle and Tatum found that enzymes influence each process step. (Source: Https://bio.libretexts.org/Courses/Portland_Community_College/Cascade_Microbiology/22%3A_Appendix_B_-_Molecular_Genetics_Review/22.1%3A_Using_Microbiology_to_Discover_the_Secrets_of_Life, Creative Commons License)

Upon observing the growth of spores within a tube, Beadle and Tatum successfully discerned the substance introduced as the biological molecule whose synthesis had been impacted by the mutation. An instance of an auxotrophic mutant, which exhibits growth solely on a minimal medium supplemented with arginine, can be attributed to a mutation that interferes with the biosynthesis of arginine. The utilization of this procedure by scientists facilitated the genetic dissection of complex biochemical pathways involving multiple steps (Yanofsky, 2005). A series of auxotrophic mutants, whose growth was dependent on the presence of arginine, were initially isolated (Smith, 1994). Subsequently, the mutants were subjected to assessment regarding their growth potential on a minimal medium that was supplemented with three specific compounds, namely ornithine, citrulline, and arginine. Based on the obtained results, the mutants were categorized into three distinct groups (as presented in Table 5.1) according to their ability to exhibit growth in the presence of specific substances. The mutants belonging to Group I exhibited growth when cultivated on a minimal medium that was supplemented with ornithine, citrulline, or arginine. The mutants belonging to Group II exhibited growth on a minimal medium that was supplemented with either arginine or citrulline. However, these mutants did not display any growth when the medium was solely supplemented with ornithine (Robinson, 1974).

Table 5.1. Arginine auxotrophic mutant growth on minimum medium with nutrients (Source: Marshal, Creative Commons License)

Mutant Strain Number	Ornithine	Citrulline	Arginine
Group I	+	+	+
Group II	-	+	+
Group III	-	-	+

Ultimately, the group III mutants exclusively exhibited growth when the medium was supplemented with arginine. Srb and Horowitz suggested that the biochemical pathway responsible for the synthesis of the amino acid arginine consists of a minimum of three sequential enzymatic steps (Stark, 1977):

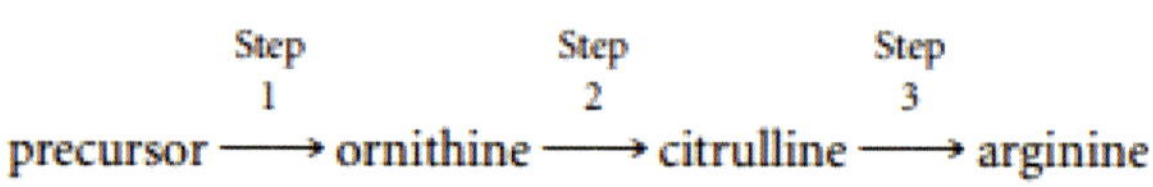

The researchers reached a conclusion that the mutations observed in group I had an impact on the initial step of this pathway. In contrast, the mutations in group II affected the second step, and the mutations in group III affected the third step. However, it remains unclear how the researchers determined the accuracy of the sequence of compounds in the biochemical pathway. It is worth noting that in the event of a mutation obstructing step 1, the introduction of either ornithine or citrulline facilitates growth, as these substances can still undergo conversion into arginine. Similarly, in the event that step 2 encounters an obstruction, the introduction of citrulline facilitates growth, whereas the introduction of ornithine does not yield any discernible impact. In the event that step 3 encounters an obstruction, the proliferation of spores will solely occur upon the introduction of arginine into the medium. The fundamental tenet posits that an auxotrophic mutant is incapable of biosynthesizing any molecule beyond the point at which a mutation hinders the metabolic pathway. By applying this line of reasoning in conjunction with the data presented in Table 5.1, it becomes evident that the inclusion of arginine in the growth medium facilitates the growth of all three categories of mutants. Hence, the biochemical processes influenced by all the mutant variations occur prior to the step leading to the production of arginine. The inclusion of citrulline in the growth medium enables the proliferation of group I and group II mutants, while group III mutants are unable to grow under these conditions. Consequently, it can be inferred that group III mutations impact a biochemical process occurring subsequent to citrulline synthesis but preceding arginine synthesis (Wang et al., 2004).

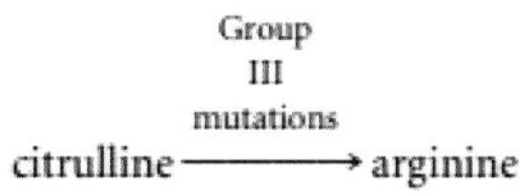

Group I mutants can grow with the addition of ornithine, but group II and group III mutants cannot; this suggests that the mutations in these two groups influence stages beyond the generation of ornithine. We know that group II mutations interfere with citrulline formation. Hence we know that group II mutations prevent ornithine from being converted into citrulline (Lacks & Hotchkiss, 1960).

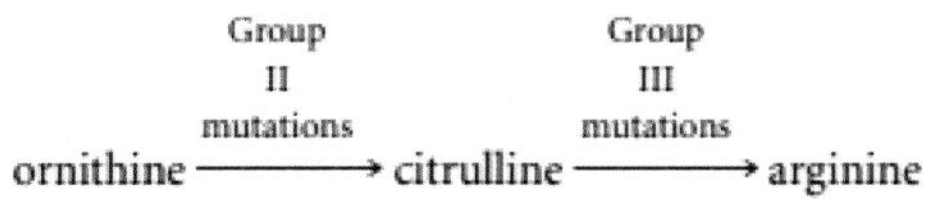

Given that group I mutations impact a specific stage preceding the synthesis of ornithine, it can be inferred that these mutations disrupt the process of converting a precursor substance into ornithine. The biochemical pathway leading to the synthesis of ornithine, citrulline, and arginine can now be delineated (Burn et al., 2002).

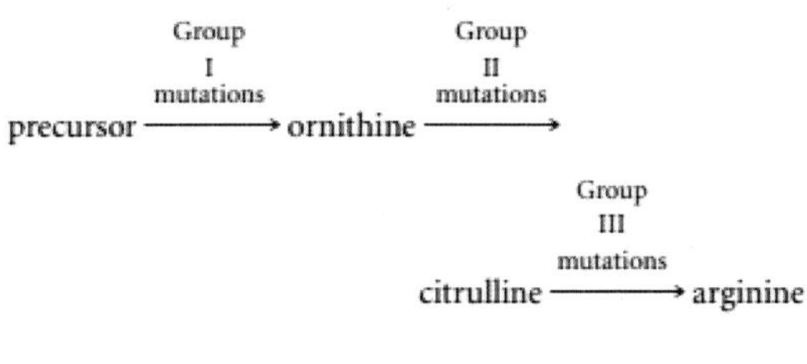

It is essential to acknowledge that this particular procedure does not inherently identify all stages within a given pathway. Instead, it solely identifies the steps that yield the compounds under examination. By employing mutations and employing this particular mode of reasoning, Beadle and Tatum successfully discerned the genes responsible for regulating multiple biosynthetic pathways within the organism Neurospora. It has been determined that individual enzymes regulate each step within a given pathway, as illustrated in Figure 5.2, which specifically pertains to the arginine pathway. The findings from genetic crosses and mapping studies have provided evidence that mutations impacting a specific step within a pathway consistently correspond to the same chromosomal location. Beadle and Tatum postulated that mutations impacting a specific biochemical pathway were associated with a single genetic locus responsible for encoding a specific enzyme (Awai et al., 2006).

5.3.2 The Structure & Function of Proteins

Proteins play a pivotal role in all biological processes (Figure 5.3). Numerous proteins exhibit enzymatic properties, serving as pivotal catalysts that facilitate the chemical reactions within cellular systems. Additionally, certain proteins fulfill structural roles by offering a framework and reinforcement for various biological structures such as membranes, filaments, bone, and hair.

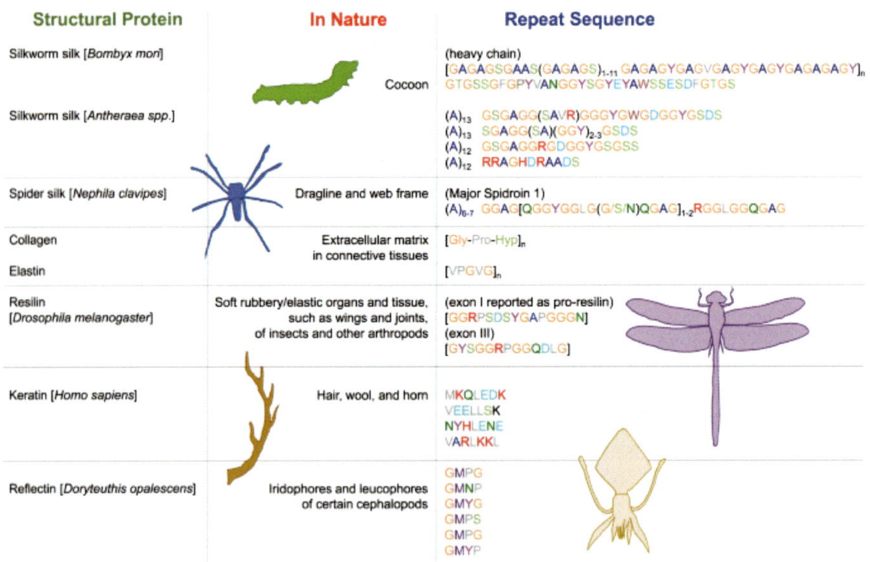

Figure 5.3. Proteins are essential to all cellular activities and serve a wide variety of biological purposes. (Source: Numata, K. How to define and study structural proteins as biopolymer materials. Polym J 52, 1043–1056 (2020). https://doi.org/10.1038/s41428-020-0362-5, Creative Commons License)

Certain proteins play a crucial role in facilitating the transportation of substances, while others serve as regulators, facilitators of communication, or defenders within biological systems (Johnson & Barford, 1993).

Proteins are comprised of amino acids that are connected in a linear fashion. Proteins consist of a set of 20 frequently occurring amino acids, as illustrated in Figure 5.4, which displays their respective three-letter and one-letter abbreviations. Modified forms of the common amino acids are occasionally observed in proteins. The 20 amino acids that are commonly found share a similar structural framework, with the only distinguishing factor being the variations in the structures of their R (radical) groups (Ellis & Jones, 1998).

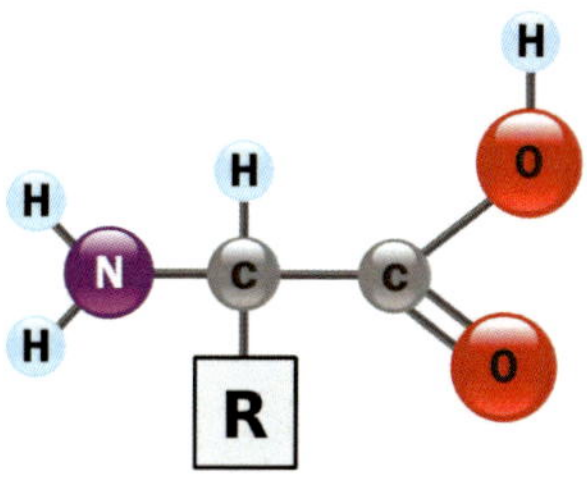

Figure 5.4. Typical amino acids share structural similarities. There are four types of groups that can bind to an amino acid's core carbon atom (Source: By This W3C-unspecified vector image was created with Inkscape . - Own work, Public Domain, https://commons.wikimedia.org/w/index.php?curid=2551977, Creative Commons License)

The linkage between amino acids in proteins occurs through peptide bonds (as depicted in Figure 5.4), resulting in the formation of polypeptide chains (Schumann et al., 1990). A protein is composed of one or multiple polypeptide chains. Similar to nucleic acids, polypeptides exhibit polarity, characterized by the presence of a free amino group (NH2) at one end and a free carboxyl group (CO2H) at the other end. Certain proteins are composed of a limited number of amino acids, while others possess a significantly larger quantity, potentially reaching into the thousands. Similar to nucleic acids, proteins exhibit multiple levels of molecular organization in their structure. The fundamental configuration of a protein is determined by the linear arrangement of its constituent amino acids (Figure 5.6a). The folding and twisting of a polypeptide chain into a secondary structure, as depicted in Figure 5.6b, is facilitated by the interactions among adjacent amino acids. The beta-pleated sheet and the alpha helix are two frequently observed secondary structures in proteins. The secondary structures of proteins engage in interactions and subsequently undergo folding processes to give rise to a tertiary structure (as depicted in Figure 5.6c). The tertiary structure represents the complete, three-dimensional conformation of the protein (Helmreich & Hofmann, 1996).

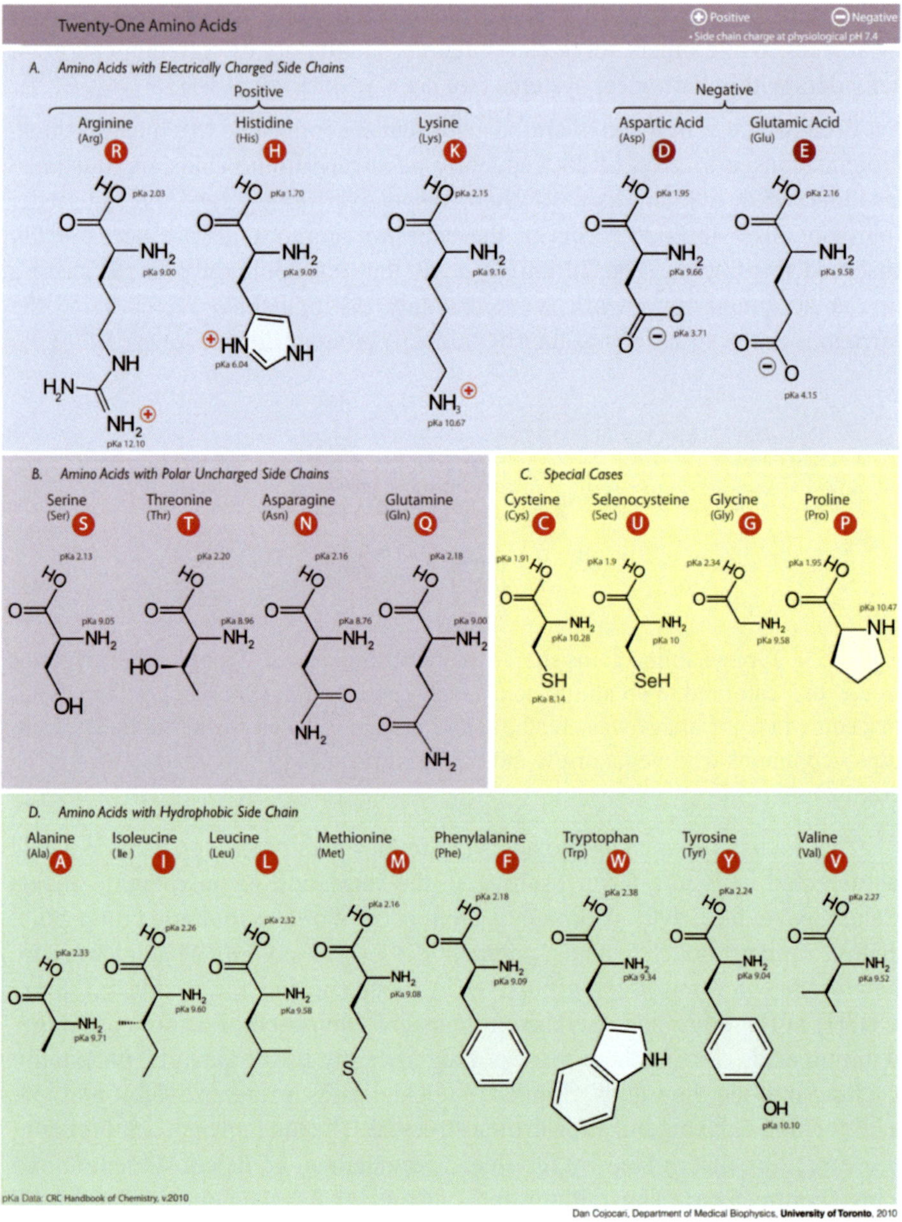

Figure 5.5. Peptide bonds connect amino acids together. (Source: By Dan Cojocari - Template:Own, Adobe Illustrator, CC BY-SA 3.0, https://commons.wikimedia.org/w/index.php?curid=8604352, Creative Commons License)

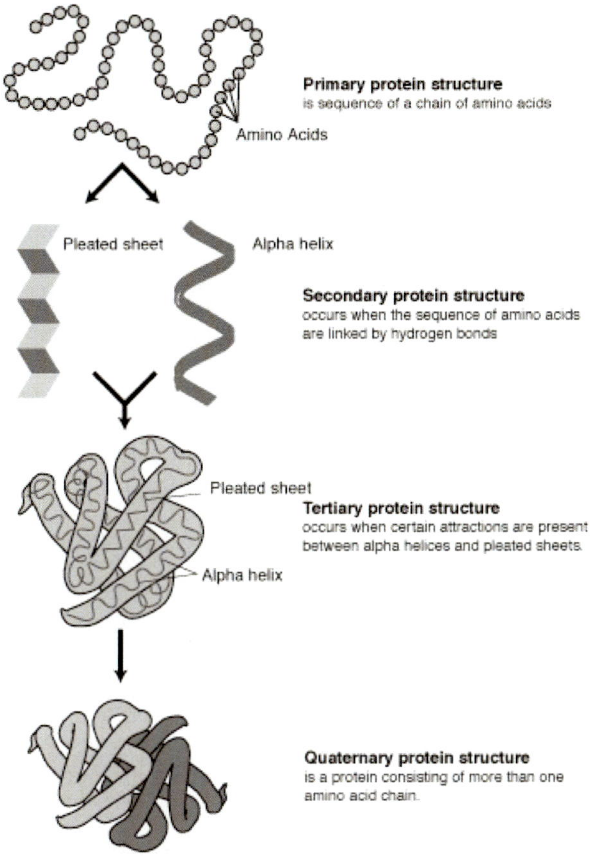

Figure 5.6. Proteins are hierarchically structured. (Source: By NHGRI - Courtesy: National Human Genome Research Institute, Public Domain, https://commons.wikimedia.org/w/index.php?curid=1318488, Creative Commons License)

Did You Know?

In recent years, significant progress has been made in the field of genetic engineering, enabling researchers to manipulate the genetic material of various organisms. This breakthrough holds promise for addressing a range of issues, including the treatment of diseases and the mitigation of environmental challenges.

5.4 THE GENETIC CODE

In the year 1953, the structure of DNA was successfully explained by Watson and Crick. Nevertheless, the mechanism by which the DNA base sequence dictated the arrangement of amino acids in proteins, known as the genetic code, was not readily apparent and continued to elude scientists for an additional decade. One of the initial inquiries regarding the genetic code pertaining to the requisite number of nucleotides for the specification of a singular amino acid (Clary & Wolstenholme, 1985).

The fundamental component of the genetic code, responsible for encoding a single amino acid, is referred to as a codon. It was acknowledged by numerous early researchers that codons necessitate a minimum of three nucleotides. The mRNA molecule consists of nucleotide positions that can be filled with any of the four bases: adenine (A), guanine (G), cytosine (C), or uracil (U). If a codon were composed of a single nucleotide, the limited number of nucleotide options (A, G, C, and U) would be insufficient to encode the 20 distinct amino acids that are typically present in proteins.

If codons were composed of two nucleotides, such as GU or AC, the total number of possible codons would be 16 (4 x 4). However, this number would still be insufficient to encode all 20 amino acids. The number of possible codons can be determined by considering that each codon consists of three nucleotides. With four possible nucleotides (A, C, G, and T), the total number of codons can be calculated as 4 x 4 x 4, resulting in 64 possible codons. This number is more than sufficient to encode the 20 different amino acids.

Hence, the utilization of a triplet code, which necessitates the inclusion of three nucleotides per codon, emerges as the optimal approach for encoding the complete set of 20 amino acids. In 1961, Francis Crick and his colleagues conducted experiments involving mutations in bacteriophage, which provided confirmation that the genetic code operates as a triplet code (Sánchez et al., 2006).

5.4.1 Breaking The Genetic Code

Once it had been definitively proven that the genetic code is composed of codons comprising three nucleotides, the subsequent objective was to ascertain the specific grouping of three nucleotides that correspond to each amino acid. The accomplishment of this objective necessitated the creation of a cell-free system dedicated to protein synthesis. This system would enable the investigation of the translation process pertaining to a specific mRNA molecule (Kwon et al., 2003).

From a logical standpoint, the most straightforward approach to deciphering the code would involve ascertaining the fundamental sequence of a specific RNA fragment, subsequently introducing it into a protein-synthesizing system devoid of cells, and facilitating the synthesis of a protein under the guidance of said RNA fragment (Hornos et al., 1999).

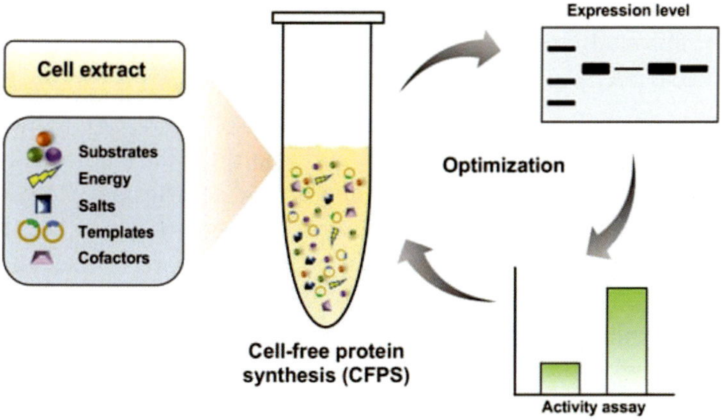

Figure 5.7. A cell-free protein synthesis mechanism was essential for deciphering the genetic code. (Source: Meng Y, Yang M, Liu W, Li J. Cell-Free Expression of a Therapeutic Protein Serratiopeptidase. Molecules. 2023; 28(7):3132. https://doi.org/10.3390/molecules28073132, Creative Commons License)

The amino acid sequence of the recently synthesized protein can subsequently be ascertained, and its sequence can be juxtaposed with that of the RNA. Regrettably, during that period, the determination of the nucleotide sequence of an RNA fragment was not feasible. Consequently, alternative approaches were required to decipher the genetic code. The initial indications regarding the genetic code were discovered in 1961 through the research conducted by Marshall Nirenberg and Johann Heinrich Matthaei. The researchers generated artificial ribonucleic acids (RNAs) through the utilization of polynucleotide phosphorylase, an enzyme. In contrast to RNA polymerase, polynucleotide phosphorylase does not necessitate a template for its activity. It possesses the capability to non-specifically connect RNA nucleotides that are accessible in a random manner (Nirenberg, 2004).

The initial synthetic messenger RNAs (mRNAs) employed by Nirenberg and Matthaei were composed of homopolymers, which are RNA molecules comprised solely of a single nucleotide type. As an illustration, the introduction of polynucleotide phosphorylase to a solution comprising uracil nucleotides resulted in the production of RNA molecules exclusively composed of uracil nucleotides, consequently comprising solely UUU codons. The poly(U) RNAs were subsequently introduced into 20 separate tubes, each of which contained a cell-free protein synthesis system along with a mixture of 20 distinct amino acids, with one of them being radioactively labeled. The process of translation occurred in all twenty tubes. However, the presence of radioactive protein was observed solely in one of the tubes, specifically the tube that contained phenylalanine labeled with a radioactive marker (refer to Figure 5.8). The findings of this study indicate that the codon UUU is responsible for encoding the amino acid phenylalanine.

The outcomes of comparable experiments involving poly(C) and poly(A) RNA have indicated that the codon CCC corresponds to the amino acid proline, while the codon AAA corresponds to the amino acid lysine. However, due to technical limitations, the results obtained from poly(G) were inconclusive and could not be reliably interpreted. In order to acquire further knowledge regarding extra codons, Nirenberg and his colleagues conducted an experiment involving the synthesis of RNA molecules that incorporated two or three distinct nucleotide bases. The RNAs synthesized by polynucleotide phosphorylase exhibit random incorporation of nucleotides, resulting in the presence of mixed bases (Lenstra, 2014).

Consequently, these RNAs are referred to as random copolymers. In the context of this experiment, it was observed that the combination of adenine and cytosine nucleotides in the presence of polynucleotide phosphorylase resulted in the generation of RNA molecules exhibiting a total of eight distinct codons. These codons were specifically identified as AAA, AAC, ACC, ACA, CAA, CCA, CAC, and CCC. The utilization of cell-free protein-synthesizing systems resulted in the production of proteins comprising six distinct amino acids, namely asparagine, glutamine, histidine, lysine, proline, and threonine, through the involvement of poly (AC) RNAs (Tejedor & Valcárcel, 2010).

The distribution of amino acids within proteins was contingent upon the relative abundance of the two nucleotides employed during the synthesis of artificial mRNA. Furthermore, the likelihood of encountering a specific codon could be theoretically determined by evaluating the ratios of the constituent bases. If a ratio of 4:1 of C to A was employed in the synthesis of RNA, the likelihood of C appearing at any specific position within a codon is determined to be, while the probability of A being present within the codon is calculated to be 1/5 (Hornos & Hornos, 1993).

When bases are incorporated randomly, the probability of observing a codon with two Cs and one A can be calculated as follows: 4/5 x 4/5 x 1/5 = 16/125= 0.13, or 13%. Similarly, the probability of observing a codon with two As and one C (AAC, ACA, or CAA) can be calculated as 1/5 x 1/5 x 4/5 = 4/125= 0.032 or about 3%. Hence, it can be inferred that the prevalence of an amino acid encoded by two cytosines (C) and one adenine (A) would be higher compared to an amino acid encoded by two adenines (A) and one cytosine (C).

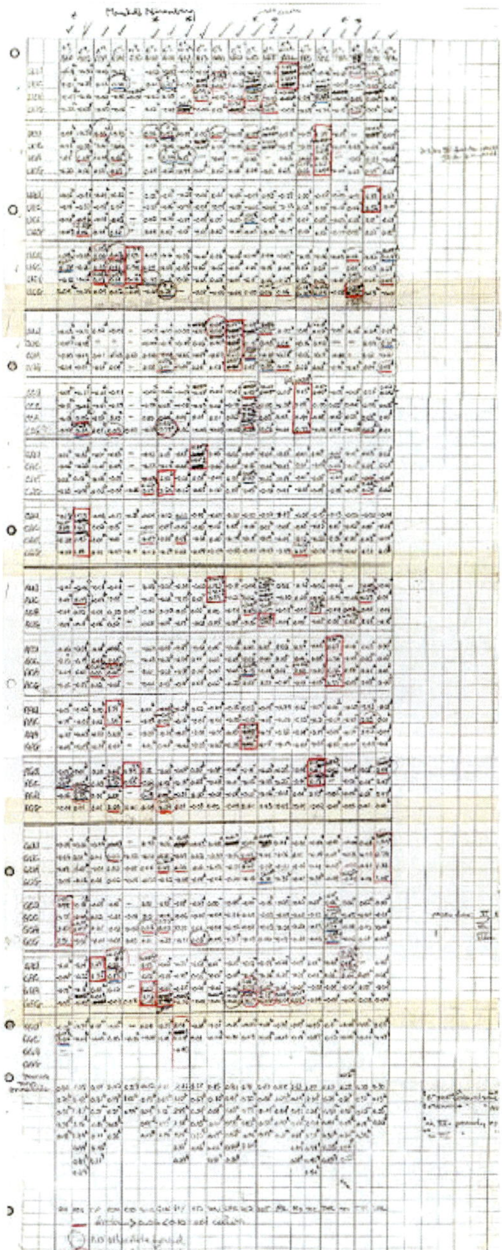

Figure 5.8. Nirenberg and Matthaei devised a homopolymer-specific amino acid identification approach. (Source: By Marshall Nirenberg - http://history.nih.gov/exhibits/nirenberg/popup_htm/05_chart_lg.htm, Public Domain, https://commons.wikimedia.org/w/index.php?curid=12854694, Creative Commons License)

The base composition of codons was derived by comparing the percentages of amino acids in proteins produced by random copolymers with the expected theoretical frequencies for the codons. The experiments conducted did not yield any significant findings regarding the codon base sequence. It was observed that histidine was encoded by a codon consisting of two Cs and one A. However, the specific identity of this codon, whether it was ACC, CAC, or CCA, remained undetermined (Antoneli & Forger, 2011).

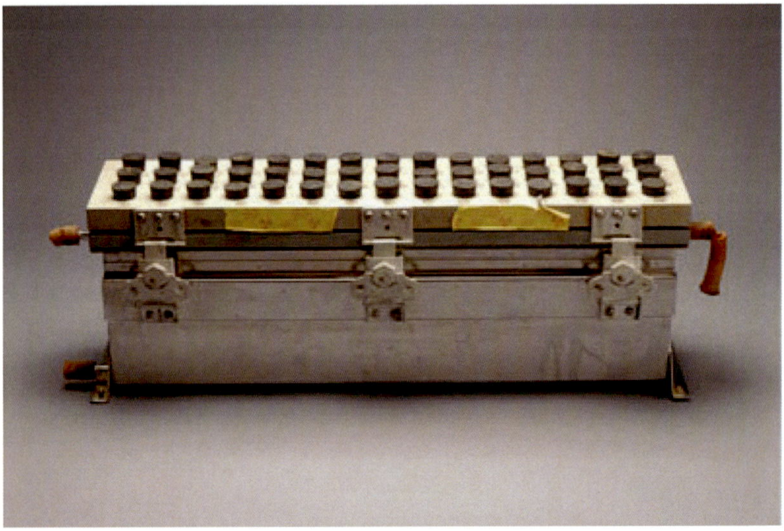

Figure 5.9. Nirenberg and Matthaei's random copolymers revealed the genetic code. (Source: By NIH - http://history.nih.gov/exhibits/nirenberg/HS5_cracked.htm, Public Domain, https://commons.wikimedia.org/w/index.php?curid=8283613, Creative Commons License)

In 1964, Nirenberg and Philip Leder devised an alternative method utilizing ribosome-bound transfer RNAs (tRNAs) to address the constraints associated with random copolymers. It was discovered that a concise mRNA sequence, even comprising only a single codon, can bind with a ribosome (Mat et al., 2010). The codon on the shorter messenger RNA molecule would subsequently form base pairs with the corresponding anticodon located on a tRNA molecule, which transports the amino acid stated by the codon (refer to Figure 5.10). The mRNA that was associated with ribosomes was combined with transfer RNAs (tRNAs) and amino acids and subsequently subjected to filtration using a nitrocellulose filter. The transfer RNAs (tRNAs) that formed pairs with the mRNA molecules bound to the ribosome were observed to adhere to the filter, while tRNAs that were not bound remained able to pass through. One notable benefit of this system lies in its compatibility with concise synthetic mRNA molecules, which can be readily synthesized with a predetermined sequence. Nirenberg and Leder synthesized more than 50 short messenger RNAs (mRNAs) that contained codons with known sequences. These synthesized mRNAs were subsequently introduced individually into a mixture consisting of ribosomes and transfer RNAs (tRNAs). Subsequently, the researchers proceeded to isolate the tRNAs bound to the ribosomes and ascertained the specific amino acids associated with these bound tRNAs. As an illustration, the presence of a tRNA molecule carrying valine was observed in synthetic RNA sequences containing the codon GUU, while no such tRNA association was observed in RNA sequences containing the codons UGU and UUG. By employing this methodology, Nirenberg and his colleagues successfully ascertained the specific amino acids that are encoded by over 50 codons (Martin, 1984).

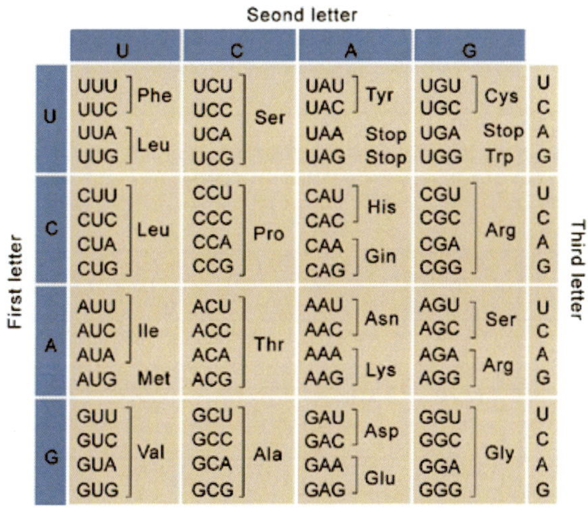

Figure 5.10. Nirenberg and Leder used ribosome-bound tRNAs to deliver genetic code information. (Source: By NIH - http://history.nih.gov/exhibits/nirenberg/HS5_cracked.htm, Public Domain, https://commons.wikimedia.org/w/index.php?curid=33555245, Creative Commons License)

An additional method was employed to provide supplementary information regarding the genetic code. Gobind Khorana and his coworkers employed chemical methodologies to artificially produce RNA molecules comprising established repetitive sequences (Greenough, 2023). The researchers postulated that if an mRNA sequence consisted of alternating uracil and guanine nucleotides (UGUG UGUG), it would be interpreted during the process of translation as two alternating codons, UGU GUG UGU GUG. Consequently, this would result in the synthesis of a protein consisting of two alternating amino acids. Upon introduction of the synthetic mRNA into a cell-free protein synthesis system, Khorana and his colleagues observed the generation of a protein composed of cysteine and valine residues arranged in an alternating pattern. The aforementioned technique was unable to ascertain the specific assignment of cysteine to either of the two codons, UGU or GUG. However, when utilized in conjunction with other methodologies, it played a pivotal role in deciphering the genetic code (Li et al., 2013). The comprehensive comprehension of the genetic code was achieved in 1968. In the subsequent section, an analysis will be conducted on several attributes of the code, which holds significant importance in contemporary biology. Francis Crick has drawn a parallel between its significance and that of the periodic table of elements in the field of chemistry (Boch et al., 2009).

UUU F	UUC F	UCU S	UCC S	CUU L	CUC L	CCU P	CCC P
UUA L	UUG L	UCA S	UCG S	CUA L	CUG L	CCA P	CCG P
UAU Y	UAC Y	UGU C	UGC C	CAU H	CAC H	CGU R	CGC R
UAA stop	UAG stop	UGA stop	UGG W	CAA Q	CAG Q	CGA R	CGG R
AUU I	AUC I	ACU T	ACC T	GUU V	GUC V	GCU A	GCC A
AUA I	AUG M	ACA T	ACG T	GUA V	GUG V	GCA A	GCG A
AAU N	AAC N	AGU S	AGC S	GAU D	GAC D	GGU G	GGC G
AAA K	AAG K	AGA R	AGG R	GAA E	GAG E	GGA G	GGG G

Figure 5.11. 64 codons specify amino acids in the genetic code. (Source: Négadi T. The Genetic Codes: Mathematical Formulae and an Inverse Symmetry-Information Relationship. Information. 2017; 8(1):6. https://doi.org/10.3390/info8010006, Creative Commons License)

5.4.2 The Degeneracy of the Code

A single amino acid is represented by a sequence of three consecutive nucleotides in messenger RNA. Each nucleotide within this sequence can be one of four possible bases: adenine (A), guanine (G), cytosine (C), or uracil (U). Consequently, there are a total of 64 potential codons, as calculated by raising 4 to the power of 3 (4^3), as illustrated in Figure 5.11. Three of these codons are classified as stop codons, which serve the purpose of indicating the termination of the translation process. Hence, a total of 61 codons, referred to as sense codons, are responsible for encoding amino acids. The degeneracy of the genetic code is attributed to the presence of 61 sense codons, which surpasses the 20 distinct amino acids commonly observed in proteins (Subramaniam et al., 2013).

Consequently, the code possesses an excess of information required for amino acid specification. The aforementioned phrase does not imply any moral corruption in the genetic code. Rather, the term "degenerate" was adopted by Francis Crick from the field of quantum physics, where it denotes various physical states that possess equivalent significance (Lehmann & Libchaber, 2008).

5.4.2.1 Iso-accepting tRNAs

tRNAs function as adapter molecules that selectively bind specific amino acids and transport them to ribosomes. Within ribosomes, these amino acids are subsequently joined together to form polypeptide chains. Every individual type of tRNA is specifically associated with a singular type of amino acid. The cellular composition of the majority of organisms typically includes a range of approximately 30 to 50 distinct transfer RNAs (tRNAs), despite the fact that proteins are comprised of only 20 unique amino acids. Hence, certain amino acids are transported by multiple transfer RNA molecules. Isoaccepting tRNAs are a group of transfer RNA molecules that share the ability to bind to and transport the same amino acid, despite having distinct anticodons. Certain codons that have the same meaning encode for distinct isoacceptor tRNAs (Araya et al., 1975).

Remember

The utilization of DNA analysis on ancient remains has yielded valuable insights pertaining to the migratory patterns and interconnections among civilizations of antiquity.

5.4.2.2 Wobble

Several codons that have the same meaning exhibit variation only in the third position (refer to Figure 5.11). As an illustration, the amino acid alanine is represented by the codons GCU, GCC, GCA, and GCG, all of which possess a common GC initiation. When the codon present on the messenger RNA (mRNA) molecule and the anticodon found on the transfer RNA (tRNA) molecule come together (as depicted in Figure 5.12), the initial (5') base of the codon aligns with the third base (3') of the anticodon, adhering strictly to the base pairing rules established by Watson and Crick: adenine (A) pairs with uracil (U), while cytosine (C) pairs with guanine (G). Subsequently, the intermediate bases of the codon and anticodon pair adhere rigorously to the principles established by Watson and Crick. Once hydrogen bonding has occurred between these pairs, the third bases form a weak pairing characterized by potential flexibility or wobble. The wobble hypothesis, formulated by Francis Crick in 1966, posited the possibility of nonstandard base pairings at the third position of a codon. As an illustration, within the anticodon, a guanine (G) nucleotide has the potential to form a base pair with either a cytosine (C) or a uracil (U) nucleotide in the third position of the codon, as indicated in Table 5.2 (Alkatib et al., 2012).

Table 5.2 The wobble rules indicate which bases at the third position (3' end) of the mRNA codon can couple with nucleotides at the first (5' end) of the tRNA anticodon. (Source: By Fdardel - Own work, CC BY-SA 3.0, https://commons.wikimedia.org/w/index.php?curid=2636730, Creative Commons License)

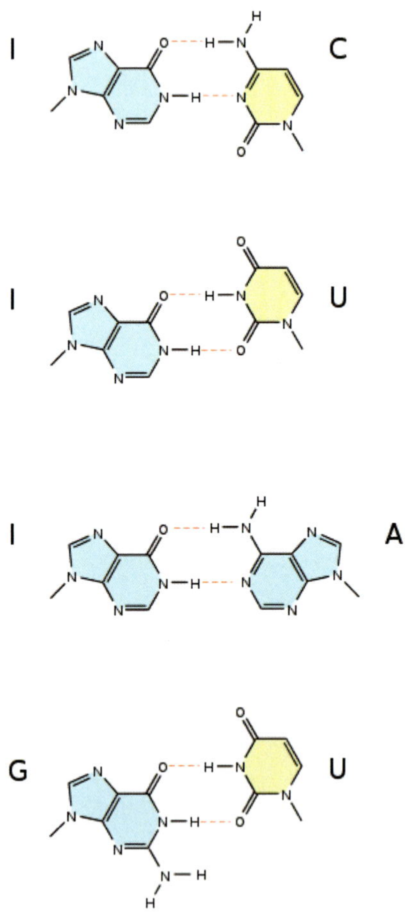

It is crucial to bear in mind that wobble facilitates the ability of certain transfer RNAs (tRNAs) to form base pairs with multiple codons on a messenger RNA (mRNA) molecule. Consequently, this phenomenon enables approximately 30 to 50 distinct tRNAs to establish pairing with the 61 sense codons. Certain codons exhibit synonymous properties due to a phenomenon known as wobble (Rozov et al., 2016).

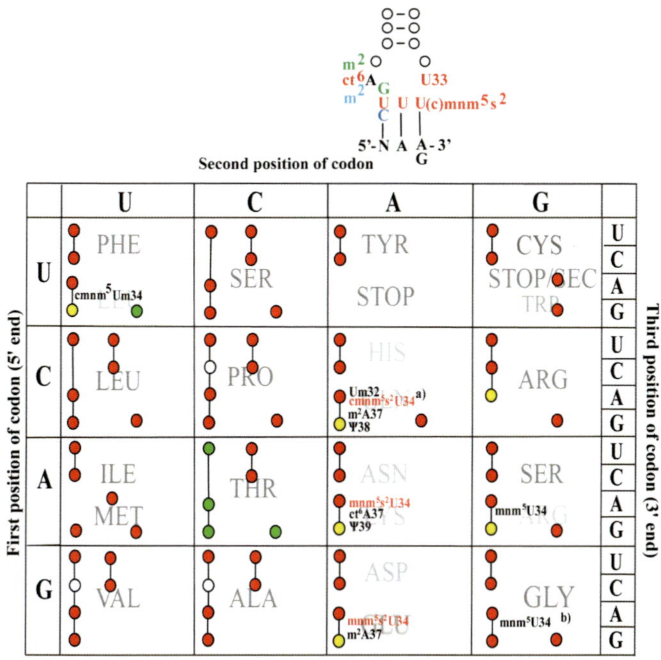

Figure 5.12. The phenomenon of wobble is observed in the interaction between a codon present on messenger RNA (mRNA) and its corresponding anticodon found on transfer RNA (tRNA) (Source: Nilsson K, Jäger G, Björk GR (2017) An unmodified wobble uridine in tRNAs specific for Glutamine, Lysine, and Glutamic acid from Salmonella enterica Serovar Typhimurium results in nonviability—Due to increased missense errors? PLoS ONE 12(4): e0175092. h, Creative Commons License)

5.4.3 The Reading Frame & Initiation Codons

The initial investigations into the genetic code revealed that it predominantly exhibits a nonoverlapping characteristic. A characteristic of an overlapping code is the inclusion of a single nucleotide in multiple codons, as depicted in Figure 5.13. Typically, however, every nucleotide sequence of a messenger RNA (mRNA) encodes a single amino acid. Viruses often exhibit the presence of overlapping codes, wherein a single sequence of mRNA may encode two distinct proteins. In the context of nucleotide sequences, it is important to note that there exist three distinct sets of codons, each representing a different way in which the sequence can be interpreted when read in groups of three (Falkner & Moss, 1988). The term "reading frame" refers to the various ways in which a sequence can be interpreted. For any given sequence of nucleotides, there are three possible reading frames. The three reading frames exhibit distinct collections of codons, thereby resulting in the specification of proteins possessing entirely dissimilar sequences of amino acids. Therefore, it is imperative for the translational apparatus to employ the appropriate reading frame accurately. Establishing the correct reading frame is a fundamental process in molecular biology. The establishment of the reading frame is determined by the initiation codon, which represents the initial codon within the mRNA sequence responsible for encoding an amino acid (Slightom et al., 1986).

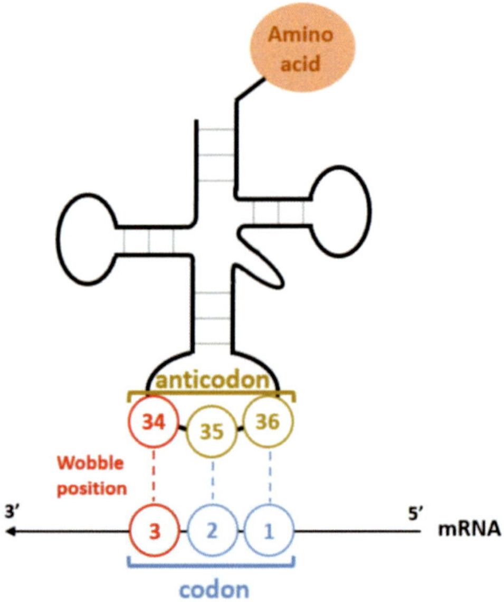

Figure 5.13. Different parts of the genetic code do not overlap. (Source: Fages-Lartaud M, Hohmann-Marriott MF. Overview of tRNA Modifications in Chloroplasts. Microorganisms. 2022; 10(2):226. https://doi.org/10.3390/microorganisms10020226, Creative Commons License)

Following the initiation codon, the subsequent codons are interpreted as consecutive sets of three nucleotides. There is a lack of omission of bases between the codons, resulting in the absence of punctuation marks for codon separation. The initiation codon is predominantly represented by the nucleotide sequence AUG, although there are infrequent instances where GUG and UUG are utilized (Sieber et al., 2018).

Did You Know?

Certain organisms, such as salamanders, exhibit the remarkable capability of regenerating body parts, which can be attributed to distinct genetic mechanisms.

5.4.4 Termination Codes

There are three codons, namely UAA, UAG, and UGA, which do not possess the ability to encode amino acids. The termination of protein synthesis is indicated by specific nucleotide sequences known as stop codons, termination codons, or nonsense codons. There is an absence of transfer RNA (tRNA) molecules that possess anticodons capable of forming complementary base pairs with termination codons (Swart et al., 2016).

5.4.5 The Universality of the Code

The universality of the genetic code, which implies that each codon consistently encodes the identical amino acid across all organisms, has been widely accepted for an extended period of time. It has been established that the genetic code exhibits a high degree of universality, although a limited number of exceptions have been identified. The majority of these exceptions pertain to termination codons, although there exist a limited number of instances where one sense codon is substituted for another (Effros et al., 2002).

5.5 THE PROCESS OF TRANSLATION

Having acquired knowledge of the genetic code, we can now commence the examination of the process through which amino acids are synthesized into proteins. Given the greater amount of knowledge available on the subject, our primary focus will be on bacterial translation. Eukaryotic translation exhibits a high degree of similarity in various aspects, albeit with notable distinctions that will be elucidated as we progress through the successive stages of translation. The translation occurs within ribosomes, which can be conceptualized as mobile apparatuses responsible for protein synthesis. In recent years, significant advancements have been made in elucidating the intricate structure of the ribosome through the implementation of various methodologies (Imamoto, 1973). These advancements have significantly enhanced our comprehension of the translational process. The translation involves the attachment of a ribosome to the 5' end of an mRNA strand, followed by its movement in the direction of the 3' end. As the ribosome moves, it translates the codons present on the mRNA strand. This process is illustrated in Figure 5.14. Protein synthesis commences at the amino terminus of the polypeptide chain, and the elongation of the protein occurs through the successive incorporation of additional amino acids at the carboxyl terminus. The process of protein synthesis can be conveniently categorized into four distinct stages. Firstly, there is the initial stage involving the binding of amino acids to the transfer RNAs (tRNAs). Secondly, the initiation stage occurs, during which the necessary components for translation are assembled at the ribosome. Following this, the elongation stage takes place, wherein amino acids are sequentially added to the growing polypeptide chain. Lastly, the termination stage is reached, where protein synthesis concludes at the termination codon, and the translation components are released from the ribosome (Kimball & Jefferson, 2006).

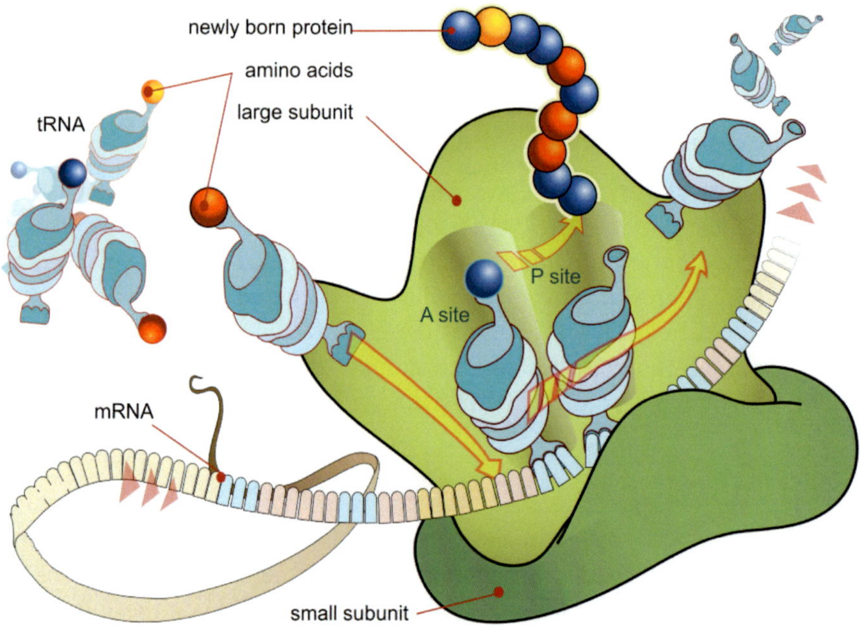

Figure 5.14. The translation of an mRNA molecule takes place on a ribosome. (Source: By LadyofHats - Own work using:nih.govmiami.edumolecularassembler.comufl.eduphschool.comuic.eduyoutube.comyoutube.com, Public Domain, https://commons.wikimedia.org/w/index.php?curid=4889777, Creative Commons License)

5.5.1 The Binding of Amino Acids

The initial step in the process of translation involves the specific pairing of transfer RNA (tRNA) molecules with their corresponding amino acids. The tRNA molecule, upon binding with its corresponding amino acid, facilitates the transportation of said amino acid to the ribosome. Subsequently, the anticodon of the tRNA molecule establishes base pairing with the codon present on the mRNA molecule. This procedure facilitates the linkage of amino acids according to the sequence specified by the messenger RNA. In order to achieve accurate translation, it is imperative to ensure the accurate pairing of transfer RNA molecules with their corresponding amino acids. As previously stated, a cellular entity generally harbors 30 to 50 distinct transfer RNAs, collectively serving as carriers for the 20 diverse amino acids. Every transfer RNA molecule exhibits specificity towards a distinct type of amino acid. The presence of the sequence CCA characterizes the 3' end of all tRNAs (Jefferson & Kimball, 2003).

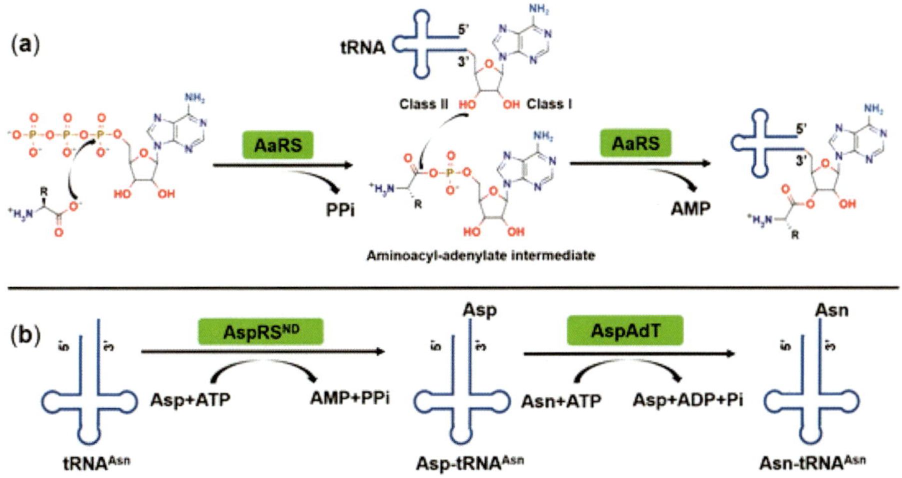

Figure 5.15. An amino acid binds at the 3' end of a transfer RNA (tRNA). (Source: Pang L, Weeks SD, Van Aerschot A. Aminoacyl-tRNA Synthetases as Valuable Targets for Antimicrobial Drug Discovery. International Journal of Molecular Sciences. 2021; 22(4):1750. https://doi.org/10.3390/ijms22041750, Creative Commons License)

A crucial determinant of specificity in the interaction between an amino acid and its corresponding transfer RNA molecule is facilitated by a group of enzymes known as aminoacyl-tRNA synthetases. A cellular entity possesses a total of 20 distinct aminoacyl-tRNA synthetases, with each synthetase specifically responsible for the attachment of one of the 20 amino acids. Every synthetase possesses the ability to identify a specific amino acid, along with all the transfer RNAs that are capable of accepting that particular amino acid. The ability of a synthetase to identify the correct amino acid is contingent upon the variations in size, charge, and R groups exhibited by the amino acids. The transfer RNAs, on the other hand, exhibit a high degree of similarity in their tertiary structure. How does a synthetase differentiate between different transfer RNAs? The ability of a synthetase to identify tRNAs is contingent upon the distinct nucleotide sequences exhibited by the tRNAs. The identification of crucial nucleotides in recognition processes has been accomplished by manipulating various nucleotides within a specific transfer RNA (tRNA) molecule, followed by the assessment of whether the modified tRNA is still acknowledged by its corresponding synthetase enzyme (Mader et al., 1995).

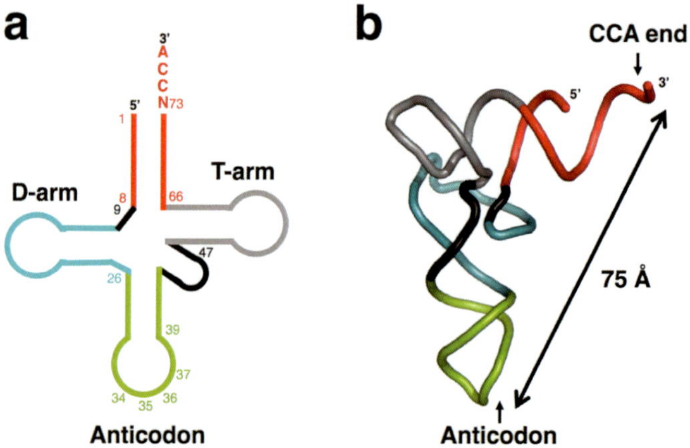

Figure 5.16. The correct aminoacyl-tRNA synthetase recognizes certain sites on tRNA molecules. (Source: Tamura K. Origins and Early Evolution of the tRNA Molecule. Life. 2015; 5(4):1687-1699. https://doi.org/10.3390/life5041687, Creative Commons License)

The process of tRNA charging, which involves the binding of a tRNA molecule to its corresponding amino acid, necessitates the expenditure of energy. This energy is derived from the hydrolysis of adenosine triphosphate (ATP).

$$amino\,acid + tRNA + ATP \rightarrow aminoacyl - tRNA + AMP + PPii$$

The hydrolysis of adenosine triphosphate (ATP) results in the removal of two phosphate groups, leading to the formation of adenosine monophosphate (AMP) and pyrophosphate (PPi). Additionally, this process yields the amino-acylated tRNA, which refers to the transfer RNA molecule covalently linked to its corresponding amino acid. The reaction occurs in two distinct stages, as depicted in Figure 5.17. In order to determine the amino-acylated tRNA that is formed, it is customary to denote the three-letter abbreviation for the respective amino acid preceding the tRNA. For instance, the amino acid alanine (Ala) binds to its corresponding tRNA, resulting in the formation of its aminoacyl-tRNA (Ala-tRNAAla). The incidence of errors in tRNA charging is infrequent, with a prevalence ranging from approximately 1 in 10,000 to 1 in 100,000 reactions (Proud, 2004).

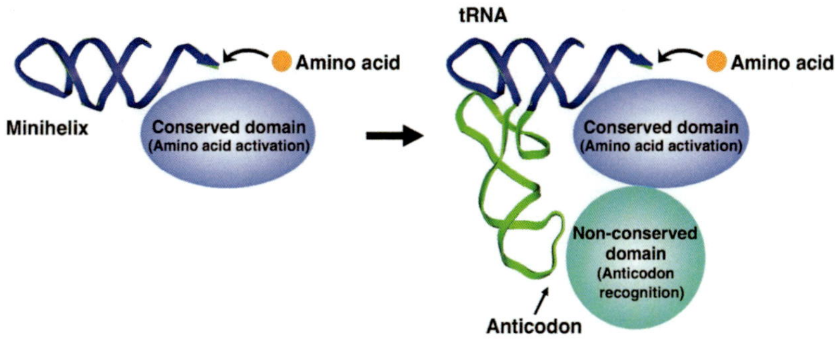

Figure 5.17. In a two-step procedure, an amino acid is linked to its tRNA. (Source: Tamura K. Origins and Early Evolution of the tRNA Molecule. Life. 2015; 5(4):1687-1699. https://doi.org/10.3390/life5041687, Creative Commons License)

Did You Know?

Genetic factors can influence mate selection, leading to preferences for certain physical traits and health indicators.

5.5.2 The Initiation of Translation

Initiation represents the subsequent phase within the protein synthesis process. During the process of initiation, the various constituents essential for the synthesis of proteins come together, including (1) messenger RNA (mRNA); (2) the smaller and larger subunits comprising the ribosome; (3) a group of three proteins known as initiation factors; (4) initiator transfer RNA carrying N-formyl methionine (fMet-tRNAfMet); and (5) guanosine triphosphate (Holcik et al., 2000). The process of initiation consists of three primary stages. Initially, messenger RNA (mRNA) forms a binding interaction with the smaller subunit of the ribosome. Additionally, the initiator tRNA establishes a binding interaction with the mRNA molecule using complementary base pairing between the codon sequence on the mRNA and the anticodon sequence on the tRNA. Furthermore, the initiation complex is joined by the larger ribosome. Let us examine each of these steps in greater detail (Kozak, 2002).

A functional ribosome comprises two distinct subunits, namely the small 30S subunit and the large 50S subunit, as observed in bacterial cells. When the process of translation is not actively occurring, the two subunits maintain a state of dynamic equilibrium characterized by their continuous association and dissociation (see Figure 5.18). The binding of an mRNA molecule to the small ribosome subunit is only possible when the subunits are in a dissociated state. During the initiation phase, the binding of initiation factor 3 (IF-3) to the small subunit of the ribosome effectively inhibits the binding of the large subunit. Experiments have successfully identified specific key sequences on the mRNA molecule that are essential for the binding of ribosomes (Sussman et al., 1996).

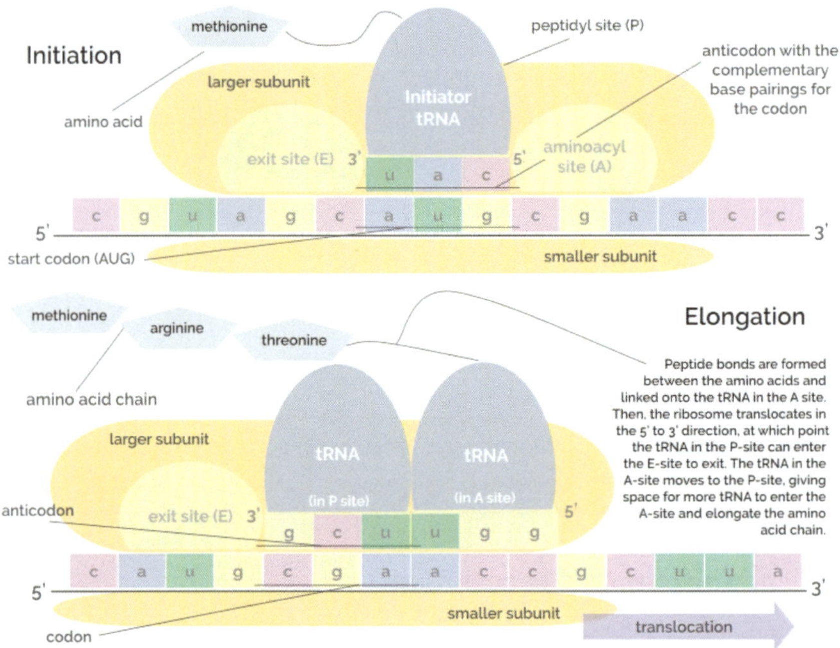

Figure 5.18. Multiple starting points and GTP are needed for translation to begin in bacterial cells. (Source: By Jordan Nguyen - Own work, CC BY-SA 4.0, https://commons.wikimedia.org/w/index.php?curid=79164021, Creative Commons License)

Following the attachment of the ribosome to the mRNA in these experimental procedures, the introduction of ribonuclease occurs, resulting in the degradation of all mRNA molecules except for the specific region that is encompassed by the ribosome. The residual mRNA can be isolated from the ribosome and subjected to further analysis (Pause et al., 1994). The range of nucleotides encompassed by the ribosome during the initiation process spans from 30 to 40 and encompasses the AUG initiation codon. The Shine-Dalgarno consensus sequence (Figure 5.19) is located within the ribosome-binding site and exhibits complementarity to a nucleotide sequence positioned at the 3' terminus of 16S rRNA, which is a constituent of the ribosome's small subunit (Jackson et al., 2010).

Subsequently, the initiator fMet-tRNAfMet molecule becomes associated with the initiation codon. The initiation of this step necessitates the presence of initiation factor 2, which associates with guanosine triphosphate to form a complex. Initiation factor 1 is an additional factor that facilitates the separation of the large and small ribosomal subunits. Currently, the initiation complex comprises several components, namely: (1) the small subunit of the ribosome; (2) the messenger RNA (3) the initiator transfer RNA along with its associated amino acid; (4) a single molecule of guanosine triphosphate and (5) initiation factors IF-3, IF-2, and IF-1 (Christian & Spremulli, 2009). The assemblage of these constituents is commonly referred to as the 30S initiation complex. During the concluding phase of initiation, the dissociation of IF-3 from the small subunit occurs, thereby facilitating the incorporation of the large subunit of the ribosome into the initiation complex. The GTP molecule, which is supplied by IF-2, undergoes hydrolysis to form guanosine diphosphate (GDP), resulting in the departure of both IF-1 and IF-2 (refer to Figure 5.18d). The 70S initiation complex is formed when the large subunit becomes part of the initiation complex (Touriol et al., 2003).

Within bacterial cells, the 16S rRNA sequences found in the small subunit of the ribosome engage in binding interactions with the Shine-Dalgarno sequence present in mRNA. This specific binding event serves to accurately position the ribosome directly over the start codon. There is no equivalent consensus sequence found in eukaryotic mRNA. In eukaryotic mRNA, the cap located at the 5' end assumes a pivotal function in the initiation of translation (Macejak & Sarnow, 1991). The diminutive subunit of the eukaryotic ribosome, in collaboration with initiation factors, identifies the cap and forms a bond with it; subsequently, the diminutive subunit proceeds to traverse (scan) the mRNA until it locates the initial AUG codon. The facilitation of start codon identification is enabled by the existence of a consensus sequence, commonly referred to as the Kozak sequence, which encompasses the start codon (Hyatt et al., 2010):

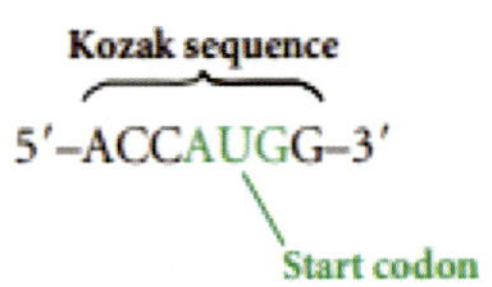

An additional noteworthy distinction lies in the fact that eukaryotic initiation necessitates a greater number of initiation factors. Several factors are responsible for maintaining the separation of ribosomal subunits, similar to the role played by IF-3 in bacterial cells. The presence of the 5' cap on mRNA is acknowledged by other cellular components, facilitating the binding of the small subunit of the ribosome at this site. Furthermore, certain individuals exhibit RNA helicase activity, which serves the purpose of unraveling secondary structures that might be present in the 5' untranslated region of mRNA. This action facilitates the movement of the small subunit along the mRNA molecule until it reaches the initiation codon (Osterman et al., 2013).

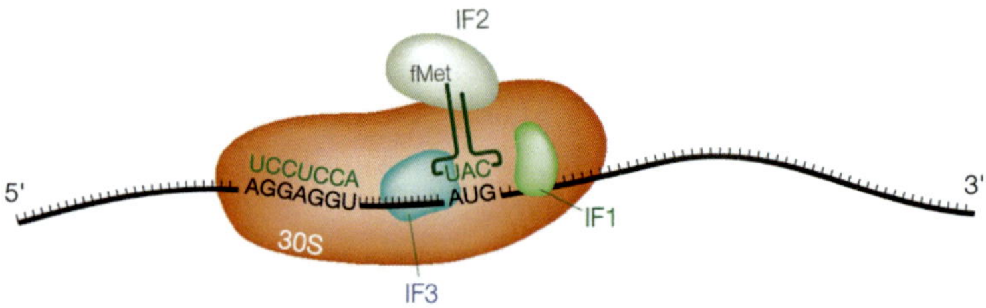

Figure 5.19. The binding of the small subunit of the ribosome necessitates the presence of Shine-Dalgarno consensus sequences within mRNA. (Source: Https://commons.wikimedia.org/wiki/File:TranslationInitiationBacteria.png, Creative Commons License)

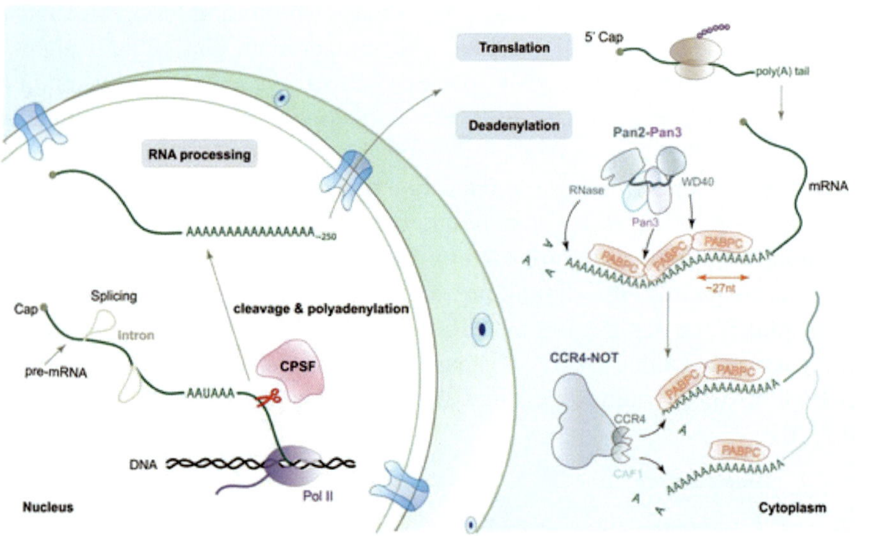

Figure 5.20. Eukaryotic mRNA's 3' poly(A) tail initiates translation. (Source: Liu J, Lu X, Zhang S, Yuan L, Sun Y. Molecular Insights into mRNA Polyadenylation and Deadenylation. International Journal of Molecular Sciences. 2022; 23(19):10985. https://doi.org/10.3390/ijms231910985, Creative Commons License)

Additional initiation factors facilitate the recruitment of the initiator tRNA and methionine (Met-tRNAfMet) to the initiation complex. The involvement of the poly(A) tail located at the 3' terminus of eukaryotic mRNA in the initiation of translation is also significant. Proteins that form associations with the poly(A) tail engage in interactions with proteins that specifically attach to the 5' cap, thereby facilitating the augmentation of the binding process between the small subunit of the ribosome and the 5' terminus of the mRNA. The observed interaction between the 5' cap and the 3' tail implies that there is a bending of the mRNA in a reverse direction during the initiation of translation, resulting in the formation of a circular structure (Arai et al., 1997).

5.5.3 Elongation

The subsequent phase in the process of protein synthesis involves elongation, wherein amino acids are interconnected to form a polypeptide chain. The process of elongation necessitates the presence of four key components: (1) the aforementioned 70S complex, (2) transfer RNAs (tRNAs) that have been loaded with their respective amino acids, (3) a set of elongation factors including EF-Ts, EF-Tu, and EF-G, and (4) guanosine triphosphate (GTP). A ribosome possesses three distinct sites that are capable of being occupied by transfer RNAs (tRNAs). These sites include the aminoacyl site, also known as the A site, the peptidyl site, also known as the P site, and the exit site, also known as the E site. The P site is promptly occupied by the initiator tRNA, as it is the sole site that can accommodate the fMet-tRNAfMet. However, all other tRNAs initially enter the A site. Following the initiation process, the ribosome becomes associated with the mRNA molecule. At this stage, the fMet-tRNAfMet is accurately positioned above the AUG start codon within the P site of the ribosome, while the neighboring A site remains vacant (Shalgi et al., 2013).

The process of elongation can be divided into three distinct stages. The initial stage involves the transportation of a tRNA molecule that has been charged with its corresponding amino acid to the A site. The existence of elongation factor Tu (EF-Tu), elongation factor Ts (EF-Ts), and guanosine triphosphate is necessary for this process. The EF-Tu protein initially associates with guanosine triphosphate before subsequently interacting with a transfer RNA molecule that carries a specific amino acid, resulting in the formation of a trimeric complex. The trimeric complex infiltrates the A site of the ribosome, where the anticodon on the transfer RNA aligns with the codon on the messenger RNA (mRNA). Once the charged transfer RNA molecule has successfully occupied the A site, guanosine triphosphate (GTP) undergoes hydrolysis to guanosine diphosphate (GDP), resulting in the detachment of the EF-Tu-GDP complex. The factor EF-Ts facilitates the regeneration of EF-Tu–GDP to EF-Tu–GTP. Within eukaryotic cells, a comparable series of reactions facilitates the transportation of the charged transfer RNA to the A site (Rajkovic & Ibba, 2017). The subsequent stage of elongation involves the establishment of a peptide bond connecting the amino acids that are bound to transfer RNAs located in the P and A sites. The peptide bond formation results in the liberation of the amino acid in the P site from its corresponding transfer RNA. The process by which peptide bonds are formed in the ribosome is commonly known as peptidyl transferase. For an extended period, it was widely believed that this particular activity is executed by one of the proteins located in the larger subunit of the ribosome (Spencer et al., 2012).

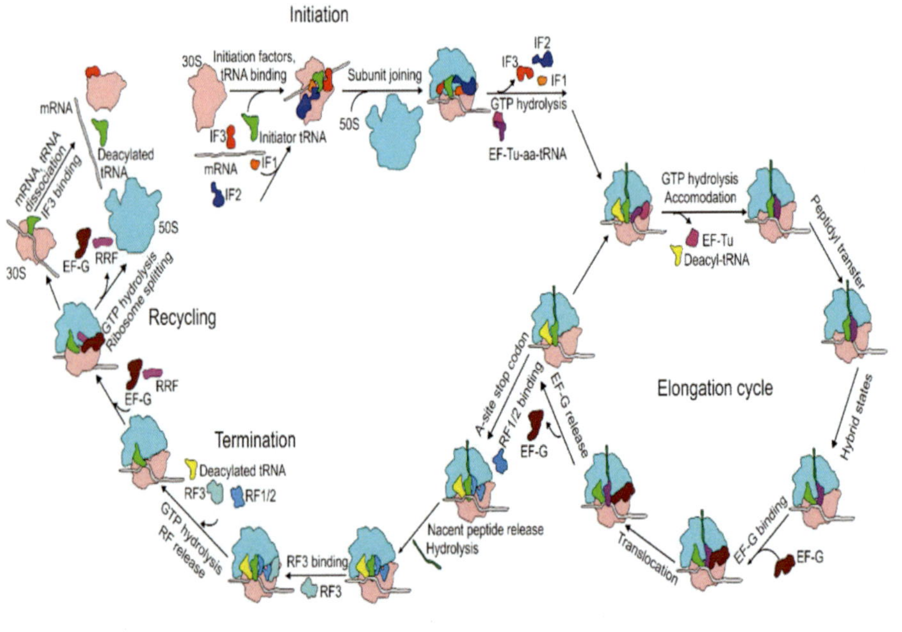

Figure 5.21. The elongation of translation comprises three steps (Source: Xu B, Liu L and Song G (2022) Functions and Regulation of Translation Elongation Factors. Front. Mol. Biosci. 8:816398. doi: 10.3389/fmolb.2021.816398, Creative Commons License)

The subsequent stage in the process of elongation involves translocation, which refers to the directional movement of the ribosome along the mRNA molecule in the 5' to 3' direction. The subsequent step involves the alignment of the ribosome with the following codon, which necessitates the involvement of elongation factor G (EF-G) and the hydrolysis of guanosine triphosphate (GTP) to guanosine diphosphate (GDP). Due to the presence of codon-anticodon pairing, the tRNAs located in the P and A sites remain bound to the mRNA and do not undergo displacement alongside the ribosome during translocation. As a result, the ribosome undergoes a shift in which the transfer RNA (tRNA) that was previously located in the P site is now positioned in the E site. Subsequently, this tRNA translocates into the cytoplasm, where it has the potential to be replenished with another amino acid (Kristensen et al., 2005). Translocation is a process that results in the repositioning of the transfer

RNA (tRNA) molecule, previously occupying the A site and bound to the elongating polypeptide chain, to the P site, thereby creating an available A site. Therefore, the sequential movement of each transfer RNA (tRNA) molecule through the ribosome during the elongation process can be succinctly described as follows: The sequence of events in protein synthesis involves the movement of the ribosome through different sites, namely the A site, P site, and E site, before returning to the cytoplasm. As previously mentioned, the initiator tRNA exhibits a unique characteristic whereby it directly binds to the P site without ever occupying the A site.

Following the process of translocation, it can be observed that the A site of the ribosome becomes unoccupied, thereby creating a state of readiness to accommodate the transfer RNA (tRNA) molecule that corresponds to the subsequent codon. The elongation cycle, as depicted in Figure 5.21, is a repetitive process. It involves the occupation of the A site by a charged tRNA along with its corresponding amino acid (Iadevaia et al., 2008).

5.5.4 Termination

The process of protein synthesis concludes with the translocation of the ribosome to a termination codon. The absence of transfer RNAs possessing anticodons that are complementary to the termination codons results in the exclusion of tRNA molecules from entering the A site of the ribosome upon encountering a termination codon. In contrast, proteins known as release factors exhibit binding affinity towards the ribosome. *Escherichia coli* possesses three distinct release factors, namely RF1, RF2, and RF3. Release factor 1 is responsible for the recognition of termination codons UAA and UAG, while release factor 2 specifically recognizes UGA and UAA. The formation of a complex between release factor 3 and GTP, followed by its binding to the ribosome, occurs. The release factors subsequently facilitate the hydrolysis of the transfer RNA in the P site from the polypeptide chain. This hydrolysis process leads to the conversion of guanosine triphosphate, which is bound to RF3, into guanosine diphosphate. Several supplementary factors contribute to the initiation of the tRNA's release from the P site, the subsequent release of the mRNA from the ribosome, and the subsequent dissociation of the ribosome. In eukaryotic cells, the process of translation termination is comparable, albeit with the involvement of two release factors: eRF1, responsible for recognizing all three termination codons, and eRF2, which interacts with GTP and facilitates the release of the polypeptide from the ribosome.

Recent studies have revealed that the termination of translation is facilitated by the release factors, which effectively conclude the process of protein synthesis by executing a final elongation cycle. In the present model, RF1 and RF2 exhibit comparable dimensions and morphology to transfer RNAs and are situated within the A site of the ribosome, akin to the amino acid–tRNA–EF–Tu–GTP complex's positioning during an elongation cycle. Release factor 3 exhibits structural similarities to EF-G and subsequently facilitates the translocation of RF1 and RF2 to the P site, along with the final tRNA to the E site. This process resembles the mechanism by which EF-G induces translocation. When the transfer RNA (tRNA) molecules are removed from both the A site and the P site of the ribosome, the ribosome has the ability to undergo dissociation. The research findings additionally demonstrate that certain sequences within the ribosomal RNA are involved in the recognition process of termination codons (Stansfield et al., 1995).

5.6 RNA-RNA INTERACTIONS IN TRANSLATION

The translation process exhibits a significant presence of RNA–RNA interactions. In bacterial translation, the binding of the ribosome to mRNA is ensured by the pairing of the Shine-Dalgarno consensus sequence located at the 5' end of the mRNA with the 3' end of the 16S rRNA. The translation is hindered by mutations that modify the Shine-Dalgarno sequence, thereby disrupting the complementary base pairing between mRNA and rRNA. Mutations that occur in the rRNA and subsequently restore complementarity are able to facilitate the continuation of translation (Xue, 2022). RNA-RNA interactions occur between transfer RNAs (tRNAs) located in the A and P sites and ribosomal RNAs present in both the large and small subunits of the ribosome. In addition, the process of ribosome assembly involving the combination of the larger and smaller subunits may necessitate interactions between the 16S rRNA and the 23S rRNA. However, the involvement of ribosomal proteins in this process remains uncertain. The interaction between transfer RNAs (tRNAs) and messenger RNAs (mRNAs) occurs through the pairing of their respective codons and anticodons (Fabian & White, 2006).

5.6.1 Polyribosomes

Simultaneous translation of mRNA molecules by multiple ribosomes occurs in both prokaryotic and eukaryotic cells. The resultant configuration, characterized by multiple ribosomes attached to a messenger RNA molecule, is commonly referred to as a polyribosome. The ribosome sequentially binds to the ribosome-binding site located at the 5' terminus of the mRNA and proceeds in the direction of the 3' terminus. Consequently, the polypeptide linked to each ribosome gradually increases in length as the ribosome traverses the mRNA. In prokaryotic cells, the processes of transcription and translation occur concurrently, allowing for the attachment of multiple ribosomes to the

5' end of the mRNA molecule. In contrast, transcription is ongoing at the 3' end. In the past, it was commonly believed that transcription and translation did not occur simultaneously in eukaryotes due to the spatial separation of these processes. Transcription was understood to occur exclusively within the nucleus, while the translation was assumed to transpire solely in the cytoplasm. Nevertheless, recent research has provided evidence indicating that certain translation processes occur within the eukaryotic nucleus. Moreover, it has been suggested that transcription and translation might occur concurrently in cases where translation occurs within the nucleus, resembling the simultaneous nature observed in prokaryotes (Tajima et al., 2011).

5.6.2 The Post-translational Modifications of Proteins

Following the translation process, proteins in both prokaryotic and eukaryotic cells can potentially undergo posttranslational modifications. There exist various potential modifications. As previously stated, the formyl group or the complete methionine residue can be eliminated from the amino terminus of a protein (Walsh, 2010). Certain proteins undergo synthesis as larger precursor proteins, necessitating enzymatic cleavage and trimming processes to enable their functional maturation. In some cases, the activation process may necessitate the attachment of carbohydrates. Properly folding polypeptide chains is crucial for the optimal functioning of numerous proteins. While certain proteins can spontaneously fold into their correct shapes, others may necessitate the involvement of additional molecules known as molecular chaperones for initial correct folding. Post-translationally, it is common for the amino terminus of a protein to undergo acetylation in eukaryotic cells (Khoury et al., 2011).

5.6.3 Translation & Antibiotics

Antibiotics are pharmaceutical substances designed to eradicate microorganisms. In order to develop a potent antibiotic, it is crucial to ensure that the microorganism is effectively eradicated while minimizing any detrimental effects on the host organism. The selection of antibiotics necessitates careful consideration to ensure the targeted destruction of bacterial cells while minimizing harm to the eukaryotic cells of the host organism. Antibiotics often target the process of translation due to their vital role in all living organisms and the notable differences observed between bacterial and eukaryotic cells. An illustration of the dissimilarities between bacterial and eukaryotic ribosomes can be observed in their contrasting size and composition. Several antibiotics exhibit selective binding to bacterial ribosomes, thereby inhibiting different stages of translation (Lee et al., 2022). Tetracyclines are a group of antibiotics that selectively attach to the A site of bacterial ribosomes, thereby inhibiting the translocation of charged tRNAs. It is worth noting that these antibiotics do not exhibit any impact on eukaryotic ribosomes. Neomycin exhibits an affinity for the ribosome near the A site, thereby instigating translational inaccuracies, likely by disrupting the proper attachment of charged transfer RNAs to the A site. Chloramphenicol exhibits an affinity for the large subunit of the ribosome, thereby impeding the process of peptide-bond formation. Streptomycin exhibits an affinity for the small subunit of the ribosome, thereby impeding the initiation process. Conversely, erythromycin functions by obstructing translocation.

While chloramphenicol and streptomycin exhibit strong inhibitory effects on translation in bacteria, they do not demonstrate the same inhibitory effects on translation in archaebacteria. The structural conformation of puromycin resembles the 3' terminus of a charged transfer RNA (tRNA), thereby facilitating its efficient binding to the A site of a ribosome and subsequent inhibition of tRNA entry (Yoo et al., 2016). The formation of a peptide bond is possible between the puromycin molecule located in the A site and an amino acid present on the tRNA situated in the P site of the ribosome. However, it is important to note that puromycin cannot bind to the P site, thereby preventing translocation. Consequently, this obstruction hinders the continuation of protein elongation. Due to the conserved structure of transfer RNA (tRNA) across various organisms, the translation process is impeded by puromycin.

Consequently, the cytotoxic effects of puromycin extend to eukaryotic cells, including bacteria, making it a potential agent in cancer therapy to eradicate tumor cells. Numerous antibiotics exert their mechanism of action by impeding distinct stages in the translation process, whereby diverse antibiotics exhibit varying impacts on various steps involved in protein synthesis. Due to their distinctive attributes, antibiotics are commonly employed in investigating protein synthesis mechanisms (Dahl & Rosenthal, 2008).

Remember

Through the examination of DNA, researchers are able to elucidate human migration patterns and reveal ancestral origins, thereby providing insights into our shared historical narrative.

SUMMARY

The chapter titled "The Genetic Code and Translation" provides an in-depth analysis of genetics, delving into the intricate molecular connections between the phenotype and genotype, the structural and functional aspects of proteins, and the complex mechanisms underlying the genetic code and translation processes.

The chapter primarily focuses on exploring the correlation between phenotype and genotype. The term "genotype" pertains to the genetic constitution of an organism, which is determined by the specific arrangement of genes within its DNA. The genetic code plays a significant role in determining the synthesis of proteins, which subsequently regulate a range of physical attributes and biological features collectively called the phenotype. This chapter explores the relationship between specific genetic variations and the resulting phenotypic differences, highlighting the significant impact that changes in the genetic code can have on an organism's traits and overall well-being.

The genotype-phenotype relationship is examined through the exploration of Lesch-Nyhan Syndrome. The etiology of this uncommon hereditary condition can be attributed to a mutation occurring in the HPRT1 gene, leading to an insufficiency of the hypoxanthine-guanine phosphoribosyltransferase (HPRT) enzyme. The deficiency of this particular enzyme gives rise to profound neurological and behavioral manifestations, including self-harming behaviors and impairments in motor function. The disorder above is a prime example of how a single genetic mutation can give rise to a significant phenotypic impact.

Proteins are crucial in cellular processes, serving as the primary agents responsible for executing various vital functions. The chapter explores the organization and operation of proteins, providing insight into how their three-dimensional structure governs their functions. Proteins consist of elongated polypeptide chains comprised of amino acids, and the specific arrangement of these amino acids is determined by the genetic code. Their shape and function are determined by the distinctive arrangement of amino acids, thereby demonstrating the direct molecular connection between genotype and phenotype.

The genetic code is a subject of great interest discussed in detail within the chapter. The genetic code refers to a predetermined set of regulations that dictate the process of converting nucleotide sequences found within DNA into distinct amino acids during the synthesis of proteins. The triplet code employs combinations of three nucleotides, codons, to encode individual amino acids. The genetic code is widely conserved across diverse living organisms and demonstrates a phenomenon known as degeneracy, whereby multiple codons can encode for a single amino acid. This characteristic allows for flexibility and redundancy during the process of protein synthesis.

The chapter provides a detailed description of the intricate process of protein synthesis, known as translation. The intricate molecular process occurs within ribosomes, where transfer RNA (tRNA) molecules interact with messenger RNA (mRNA) to facilitate the delivery of accurate amino acids to the developing protein sequence. The ribosome is responsible for interpreting the mRNA codons and facilitating the synthesis of peptide bonds, thereby assembling the amino acids in the appropriate order to produce a biologically active protein.

The chapter also examines RNA-RNA interactions during translation, emphasizing the significant contributions of different RNA molecules in the regulation and precise modulation of the translation process.

To summarize, the chapter titled "The Genetic Code and Translation" thoroughly examines the molecular connections between genotype and phenotype. Moreover, it provides valuable insights into particular genetic disorders, such as Lesch-Nyhan Syndrome, that serve as compelling illustrations of the significant impact of genotype on an organism's phenotype. By comprehending these essential biological mechanisms, scientists acquire invaluable insights into the intricacies of life and its molecular foundations.

REFERENCES

1. Koonin, E. V., & Novozhilov, A. S. (2009). Origin and evolution of the genetic code: the universal enigma. *IUBMB life*, *61*(2), 99-111.
2. Ambrogelly, A., Palioura, S., & Söll, D. (2007). Natural expansion of the genetic code. *Nature chemical biology*, *3*(1), 29-35.
3. Neumann, H. (2012). Rewiring translation–genetic code expansion and its applications. *FEBS letters*, *586*(15), 2057-2064.
4. Bossi, L., & Roth, J. R. (1980). The influence of codon context on genetic code translation. *nature*, *286*(5769), 123-127.
5. Rodin, A. S., Szathmáry, E., & Rodin, S. N. (2011). On origin of genetic code and tRNA before translation. *Biology Direct*, *6*, 1-24.
6. Yarian, C., Townsend, H., Czestkowski, W., Sochacka, E., Malkiewicz, A. J., Guenther, R., ... & Agris, P. F. (2002). Accurate translation of the genetic code depends on tRNA modified nucleosides. *Journal of Biological Chemistry*, *277*(19), 16391-16395.
7. Novozhilov, A. S., Wolf, Y. I., & Koonin, E. V. (2007). Evolution of the genetic code: partial optimization of a random code for robustness to translation error in a rugged fitness landscape. *Biology Direct*, *2*(1), 1-24.
8. Hoffmann, G. W. (1974). On the origin of the genetic code and the stability of the translation apparatus. *Journal of Molecular Biology*, *86*(2), 349-362.
9. Wong, J. T. F., Ng, S. K., Mat, W. K., Hu, T., & Xue, H. (2016). Coevolution theory of the genetic code at age forty: pathway to translation and synthetic life. *Life*, *6*(1), 12.
10. Vetsigian, K., Woese, C., & Goldenfeld, N. (2006). Collective evolution and the genetic code. *Proceedings of the National Academy of Sciences*, *103*(28), 10696-10701.
11. Shandell, M. A., Tan, Z., & Cornish, V. W. (2021). Genetic code expansion: a brief history and perspective. *Biochemistry*, *60*(46), 3455-3469.
12. Knight, R. D., Freeland, S. J., & Landweber, L. F. (1999). Selection, history and chemistry: the three faces of the genetic code. *Trends in biochemical sciences*, *24*(6), 241-247.
13. Fournier, G. P., Neumann, J. E., & Gogarten, J. P. (2010). Inferring the ancient history of the translation machinery and genetic code via recapitulation of ribosomal subunit assembly orders. *PLoS One*, *5*(3), e9437.
14. Wolf, Y. I., & Koonin, E. V. (2007). On the origin of the translation system and the genetic code in the RNA world by means of natural selection, exaptation, and subfunctionalization. *Biology Direct*, *2*, 1-25.

15. Kato, Y. (2019). Translational control using an expanded genetic code. *International Journal of Molecular Sciences*, *20*(4), 887.
16. Sun, F. J., & Caetano-Anollés, G. (2008). Evolutionary patterns in the sequence and structure of transfer RNA: a window into early translation and the genetic code. *PLoS One*, *3*(7), e2799.
17. Wiltschi, B., & Budisa, N. (2007). Natural history and experimental evolution of the genetic code. *Applied microbiology and biotechnology*, *74*, 739-753.
18. Di Giulio, M. (1989). Some aspects of the organization and evolution of the genetic code. *Journal of molecular evolution*, *29*, 191-201.
19. Freeland, S. J., & Hurst, L. D. (1998). Load minimization of the genetic code: history does not explain the pattern. *Proceedings of the Royal Society of London. Series B: Biological Sciences*, *265*(1410), 2111-2119.
20. Osawa, S., Jukes, T. H., Watanabe, K., & Muto, A. (1992). Recent evidence for evolution of the genetic code. *Microbiological reviews*, *56*(1), 229-264.
21. Fu, R., Ceballos-Picot, I., Torres, R. J., Larovere, L. E., Yamada, Y., Nguyen, K. V., ... & Jinnah, H. A. (2014). Genotype–phenotype correlations in neurogenetics: Lesch-Nyhan disease as a model disorder. *Brain*, *137*(5), 1282-1303.
22. Genao, M. T., & Torres, R. J. (2014). From genotype to phenotype: Clinical variability in Lesch-Nyhan disease. The role of epigenetics. *Revista Clínica Española (English Edition)*, *214*(8), 461-465.
23. Ceballos-Picot, I., Le Dantec, A., Brassier, A., Jaïs, J. P., Ledroit, M., Cahu, J., ... & Pinson, B. (2015). New biomarkers for early diagnosis of Lesch-Nyhan disease revealed by metabolic analysis on a large cohort of patients. *Orphanet journal of rare diseases*, *10*, 1-11.
24. Nyhan, W. L., Bakay, B., Connor, J. D., Marks, J. F., & Keele, D. K. (1970). Hemizygous expression of glucose-6-phosphate dehydrogenase in erythrocytes of heterozygotes for the Lesch-Nyhan syndrome. *Proceedings of the National Academy of Sciences*, *65*(1), 214-218.
25. Sampat, R., Fu, R., Larovere, L. E., Torres, R. J., Ceballos-Picot, I., Fischbach, M., ... & Jinnah, H. A. (2011). Mechanisms for phenotypic variation in Lesch–Nyhan disease and its variants. *Human Genetics*, *129*, 71-78.
26. Andresen, B. S., Bross, P., Udvari, S., Kirk, J., Gray, G., Kmoch, S., ... & Gregersen, N. (1997). The molecular basis of medium-chain acyl-CoA dehydrogenase (MCAD) deficiency in compound heterozygous patients: is there correlation between genotype and phenotype? *Human molecular genetics*, *6*(5), 695-707.

27. Horowitz, N. H., & Leupold, U. (1951, January). Some recent studies bearing on the one gene-one enzyme hypothesis. In *Cold Spring Harbor Symposia on Quantitative Biology* (Vol. 16, pp. 65-74). Cold Spring Harbor Laboratory Press.
28. Hickman, M., & Cairns, J. (2003). The centenary of the one-gene one-enzyme hypothesis. *Genetics*, *163*(3), 839-841.
29. Rashevsky, N. (1962). Abstract mathematical molecular biology: II some relational consequences of the "one gene-one enzyme" hypothesis. *The bulletin of mathematical biophysics*, *24*, 327-334.
30. Yanofsky, C. (2005). The Favorable Features of Tryptophan Synthase for Proving Beadle and Tatum's One Gene–One Enzyme Hypothesis. *Genetics*, *169*(2), 511-516.
31. Robinson, A. D. (1974). An evaluation of Garrod's contribution to the one gene-one enzyme hypothesis. *BioScience*, *24*(6), 357-358.
32. Smith, S. (1994). The animal fatty acid synthase: one gene, one polypeptide, seven enzymes. *The FASEB journal*, *8*(15), 1248-1259.
33. Stark, G. R. (1977). Multifunctional proteins: one gene—more than one enzyme. *Trends in Biochemical Sciences*, *2*(3), 64-66.
34. Wang, H., Caruso, L. V., Downie, A. B., & Perry, S. E. (2004). The embryo MADS domain protein AGAMOUS-Like 15 directly regulates expression of a gene encoding an enzyme involved in gibberellin metabolism. *The Plant Cell*, *16*(5), 1206-1219.
35. Lacks, S., & Hotchkiss, R. D. (1960). A study of the genetic material determining an enzyme activity in pneumococcus. *Biochimica et biophysica acta*, *39*(3), 508-518.
36. Burn, J. E., Hurley, U. A., Birch, R. J., Arioli, T., Cork, A., & Williamson, R. E. (2002). The cellulose-deficient Arabidopsis mutant rsw3 is defective in a gene encoding a putative glucosidase II, an enzyme processing N-glycans during ER quality control. *The Plant Journal*, *32*(6), 949-960.
37. Awai, K., Kakimoto, T., Awai, C., Kaneko, T., Nakamura, Y., Takamiya, K. I., ... & Ohta, H. (2006). Comparative genomic analysis revealed a gene for monoglucosyldiacylglycerol synthase, an enzyme for photosynthetic membrane lipid synthesis in cyanobacteria. *Plant Physiology*, *141*(3), 1120-1127.
38. Johnson, L. N., & Barford, D. (1993). The effects of phosphorylation on the structure and function of proteins. *Annual review of biophysics and biomolecular structure*, *22*(1), 199-232.
39. Ellis, J., & Jones, D. (1998). Structure and function of proteins controlling

strain-specific pathogen resistance in plants. *Current opinion in plant biology*, *1*(4), 288-293.

40. Helmreich, E. J., & Hofmann, K. P. (1996). Structure and function of proteins in G-protein-coupled signal transfer. *Biochimica et Biophysica Acta (BBA)-Reviews on Biomembranes*, *1286*(3), 285-322.
41. Schumann, R. R., Leong, S. R., Flaggs, G. W., Gray, P. W., Wright, S. D., Mathison, J. C., ... & Ulevitch, R. J. (1990). Structure and function of lipopolysaccharide binding protein. *Science*, *249*(4975), 1429-1431.
42. Clary, D. O., & Wolstenholme, D. R. (1985). The mitochondrial DNA molecule of Drosophila yakuba: nucleotide sequence, gene organization, and genetic code. *Journal of Molecular Evolution*, *22*, 252-271.
43. Sánchez, R., Grau, R., & Morgado, E. (2006). A novel Lie algebra of the genetic code over the Galois field of four DNA bases. *Mathematical biosciences*, *202*(1), 156-174.
44. Kwon, I., Kirshenbaum, K., & Tirrell, D. A. (2003). Breaking the degeneracy of the genetic code. *Journal of the American Chemical Society*, *125*(25), 7512-7513.
45. Hornos, J. E. M., Hornos, Y. M., & Forger, M. (1999). Symmetry and symmetry breaking: an algebraic approach to the genetic code. *International Journal of Modern Physics B*, *13*(23), 2795-2885.
46. Nirenberg, M. (2004). Historical review: Deciphering the genetic code–a personal account. *Trends in biochemical sciences*, *29*(1), 46-54.
47. Lenstra, R. (2014). Evolution of the genetic code through progressive symmetry breaking. *Journal of theoretical biology*, *347*, 95-108.
48. Tejedor, J. R., & Valcárcel, J. (2010). Breaking the second genetic code. *Nature*, *465*(7294), 45-46.
49. Antoneli, F., & Forger, M. (2011). Symmetry breaking in the genetic code: Finite groups. *Mathematical and computer modelling*, *53*(7-8), 1469-1488.
50. Hornos, J. E. M., & Hornos, Y. M. (1993). Algebraic model for the evolution of the genetic code. *Physical Review Letters*, *71*(26), 4401.
51. Mat, W. K., Xue, H., & Wong, J. T. F. (2010). Genetic code mutations: the breaking of a three billion year invariance. *PLoS One*, *5*(8), e12206.
52. Martin, R. G. (1984). A revisionist view of the breaking of the genetic code. In *NIH: an Account of Research in its Laboratories and Clinics*. Academic Press. (Vol. 2, pp. 281-296).
53. Li, F., Zhang, H., Sun, Y., Pan, Y., Zhou, J., & Wang, J. (2013). Expanding the genetic code for photoclick chemistry in *E. coli*, mammalian cells, and *A. thaliana*. *Angewandte Chemie International Edition*, *52*(37), 9700-9704.

54. Boch, J., Scholze, H., Schornack, S., Landgraf, A., Hahn, S., Kay, S., ... & Bonas, U. (2009). Breaking the code of DNA binding specificity of TAL-type III effectors. *Science*, *326*(5959), 1509-1512.
55. Greenough, B. S. (2023). Breaking the genetic code. In *The Moral Authority of Government*. Routledge. (Vol. 2, pp. 267-269).
56. Subramaniam, A. R., Pan, T., & Cluzel, P. (2013). Environmental perturbations lift the degeneracy of the genetic code to regulate protein levels in bacteria. *Proceedings of the National Academy of Sciences*, *110*(6), 2419-2424.
57. Lehmann, J., & Libchaber, A. (2008). Degeneracy of the genetic code and stability of the base pair at the second position of the anticodon. *Rna*, *14*(7), 1264-1269.
58. Araya, A., Krauskopf, M., & Siddiqui, M. A. Q. (1975). Selective changes in methionine tRNA species during development of the posterior silkglands of *Bombyx mori*. *Biochemical and Biophysical Research Communications*, *67*(3), 924-934.
59. Alkatib, S., Scharff, L. B., Rogalski, M., Fleischmann, T. T., Matthes, A., Seeger, S., ... & Bock, R. (2012). The contributions of wobbling and superwobbling to the reading of the genetic code. *PLoS Genetics*, *8*(11), e1003076.
60. Rozov, A., Demeshkina, N., Khusainov, I., Westhof, E., Yusupov, M., & Yusupova, G. (2016). Novel base-pairing interactions at the tRNA wobble position crucial for accurate reading of the genetic code. *Nature Communications*, *7*(1), 10457.
61. Sieber, P., Platzer, M., & Schuster, S. (2018). The definition of open reading frame revisited. *Trends in Genetics*, *34*(3), 167-170.
62. Slightom, J. L., Durand-Tardif, M., Jouanin, L., & Tepfer, D. (1986). Nucleotide sequence analysis of TL-DNA of *Agrobacterium rhizogenes* agropine type plasmid. Identification of open reading frames. *Journal of Biological Chemistry*, *261*(1), 108-121.
63. Falkner, F. G., & Moss, B. E. R. N. A. R. D. (1988). *Escherichia coli* gpt gene provides dominant selection for vaccinia virus open reading frame expression vectors. *Journal of Virology*, *62*(6), 1849-1854.
64. Swart, E. C., Serra, V., Petroni, G., & Nowacki, M. (2016). Genetic codes with no dedicated stop codon: context-dependent translation termination. *Cell*, *166*(3), 691-702.
65. Effros, M., Visweswariah, K., Kulkarni, S. R., & Verdú, S. (2002). Universal lossless source coding with the Burrows Wheeler transform. *IEEE Transactions on Information Theory*, *48*(5), 1061-1081.

66. Imamoto, F. (1973). Diversity of regulation of genetic transcription: I. Effect of antibiotics which inhibit the process of translation on RNA metabolism in Escherichia coli. *Journal of Molecular Biology*, *74*(2), 113-136.
67. Kimball, S. R., & Jefferson, L. S. (2006). New functions for amino acids: effects on gene transcription and translation. *The American journal of clinical nutrition*, *83*(2), 500S-507S.
68. Jefferson, L. S., & Kimball, S. R. (2003). Amino acids as regulators of gene expression at the level of mRNA translation. *The Journal of Nutrition*, *133*(6), 2046S-2051S.
69. Mader, S., Lee, H., Pause, A., & Sonenberg, N. (1995). The translation initiation factor eIF-4E binds to a common motif shared by the translation factor eIF-4 gamma and the translational repressors 4E-binding proteins. *Molecular and cellular biology*.7(6), 540-545.
70. Proud, C. G. (2004). mTOR-mediated regulation of translation factors by amino acids. *Biochemical and Biophysical Research Communications*, *313*(2), 429-436.
71. Kozak, M. (2002). Pushing the limits of the scanning mechanism for initiation of translation. *Gene*, *299*(1-2), 1-34.
72. Sussman, J. K., Simons, E. L., & Simons, R. W. (1996). *Escherichia coli* translation initiation factor 3 discriminates the initiation codon in vivo. *Molecular microbiology*, *21*(2), 347-360.
73. Jackson, R. J., Hellen, C. U., & Pestova, T. V. (2010). The mechanism of eukaryotic translation initiation and principles of its regulation. *Nature reviews Molecular cell biology*, *11*(2), 113-127.
74. Touriol, C., Bornes, S., Bonnal, S., Audigier, S., Prats, H., Prats, A. C., & Vagner, S. (2003). Generation of protein isoform diversity by alternative initiation of translation at non-AUG codons. *Biology of the Cell*, *95*(3-4), 169-178.
75. Hyatt, D., Chen, G. L., LoCascio, P. F., Land, M. L., Larimer, F. W., & Hauser, L. J. (2010). Prodigal: prokaryotic gene recognition and translation initiation site identification. *BMC Bioinformatics*, *11*, 1-11.
76. Osterman, I. A., Evfratov, S. A., Sergiev, P. V., & Dontsova, O. A. (2013). Comparison of mRNA features affecting translation initiation and reinitiation. *Nucleic acids research*, *41*(1), 474-486.
77. Arai, H., Miyamoto, K. I., Taketani, Y., Yamamoto, H., Iemori, Y., Morita, K., ... & Takeda, E. (1997). A vitamin D receptor gene polymorphism in the translation initiation codon: effect on protein activity and relation to bone mineral density in Japanese women. *Journal of Bone and Mineral Research*, *12*(6), 915-921.

78. Macejak, D. G., & Sarnow, P. (1991). Internal initiation of translation mediated by the 5′ leader of a cellular mRNA. *Nature*, *353*(6339), 90-94.
79. Christian, B. E., & Spremulli, L. L. (2009). Evidence for an active role of IF3mt in the initiation of translation in mammalian mitochondria. *Biochemistry*, *48*(15), 3269-3278.
80. Pause, A., Methot, N., Svitkin, Y., Merrick, W. C., & Sonenberg, N. (1994). Dominant negative mutants of mammalian translation initiation factor eIF-4A define a critical role for eIF-4F in cap-dependent and cap-independent initiation of translation. *The EMBO journal*, *13*(5), 1205-1215.
81. Holcik, M., Sonenberg, N., & Korneluk, R. G. (2000). Internal ribosome initiation of translation and the control of cell death. *Trends in genetics*, *16*(10), 469-473.
82. Shalgi, R., Hurt, J. A., Krykbaeva, I., Taipale, M., Lindquist, S., & Burge, C. B. (2013). Widespread regulation of translation by elongation pausing in heat shock. *Molecular cell*, *49*(3), 439-452.
83. Rajkovic, A., & Ibba, M. (2017). Elongation factor P and the control of translation elongation. *Annual Review of Microbiology*, *71*, 117-131.
84. Iadevaia, V., Caldarola, S., Tino, E., Amaldi, F., & Loreni, F. (2008). All translation elongation factors and the e, f, and h subunits of translation initiation factor 3 are encoded by 5′-terminal oligopyrimidine (TOP) mRNAs. *Rna*, *14*(9), 1730-1736.
85. Spencer, P. S., Siller, E., Anderson, J. F., & Barral, J. M. (2012). Silent substitutions predictably alter translation elongation rates and protein folding efficiencies. *Journal of Molecular Biology*, *422*(3), 328-335.
86. Kristensen, R., Mona, T. O. R. P., Kosiak, B., & Holst-Jensen, A. (2005). Phylogeny and toxigenic potential is correlated in Fusarium species as revealed by partial translation elongation factor 1 alpha gene sequences. *Mycological Research*, *109*(2), 173-186.
87. Stansfield, I., Jones, K. M., Kushnirov, V. V., Dagkesamanskaya, A. R., Poznyakovski, A. I., Paushkin, S. V., ... & Tuite, M. F. (1995). The products of the SUP45 (eRF1) and SUP35 genes interact to mediate translation termination in *Saccharomyces cerevisiae*. *The EMBO journal*, *14*(17), 4365-4373.
88. Fabian, M. R., & White, K. A. (2006). Analysis of a 3′-translation enhancer in a tombusvirus: a dynamic model for RNA–RNA interactions of mRNA termini. *Rna*, *12*(7), 1304-1314.
89. Xue, Y. (2022). Architecture of RNA–RNA interactions. *Current Opinion in Genetics & Development*, *72*, 138-144.

90. Tajima, Y., Iwakawa, H. O., Kaido, M., Mise, K., & Okuno, T. (2011). A long-distance RNA–RNA interaction plays an important role in programmed− 1 ribosomal frameshifting in the translation of p88 replicase protein of Red clover necrotic mosaic virus. Virology, 417(1), 169-178.
91. Khoury, G. A., Baliban, R. C., & Floudas, C. A. (2011). Proteome-wide post-translational modification statistics: frequency analysis and curation of the swiss-prot database. *Scientific reports*, *1*(1), 90.
92. Walsh, G. (2010). Post-translational modifications of protein biopharmaceuticals. *Drug discovery today*, *15*(17-18), 773-780.
93. Dahl, E. L., & Rosenthal, P. J. (2008). Apicoplast translation, transcription and genome replication: targets for antimalarial antibiotics. *Trends in parasitology*, *24*(6), 279-284.
94. Lee, J. H., Lee, E. J., & Roe, J. H. (2022). uORF-mediated riboregulation controls transcription of whiB7/wblC antibiotic resistance gene. *Molecular Microbiology*, *117*(1), 179-192.
95. Yoo, J. S., Oh, G. S., Ryoo, S., & Roe, J. H. (2016). Induction of a stable sigma factor SigR by translation-inhibiting antibiotics confers resistance to antibiotics. *Scientific Reports*, *6*(1), 28628.

CHAPTER 6

GENES & CHROMOSOMES

Contents

6.1 INTRODUCTION

Genes and chromosomes stand as the weavers of a sophisticated genetic code that dictates the symphony of existence itself, and this code is contained inside the beautiful tapestry that is the diversity of life. The study of these genetic blueprints is at the heart of genetics. This study investigates the stability of chromosome complements, the dynamic processes of mitosis and meiosis, and the significant role that chromosomes play in transmitting hereditary traits. This chapter will take you on an exciting adventure into the world of genes and chromosomes, illuminating the secrets surrounding the organization, transmission, and inheritance of genetic information from one generation to another (Schmid et al., 2000).

A vital element of maintaining genetic order within living organisms is the consistency of chromosomal complements. As they multiply, organisms can keep their genetic makeup the same because of a process called mitosis, which is the process of cell division responsible for growth and repair. Mitosis guarantees that each new cell receives an identical set of chromosomes as the parent cell, which protects the genetic integrity of multicellular organisms and promotes the harmonious functioning of tissues and organs. Mitosis also ensures that each new cell receives identical chromosomes as the parent cell. The fact that mitosis ensures the continuation of the genetic code at every stage of development and regeneration demonstrates its crucial role in the orchestration of life (Lyons & Freeling, 2008).

Meiosis, on the other hand, is the stage upon which the intricate dance of genetic diversity is performed. Meiosis is a process that occurs only in reproductive cells and is necessary for the production of gametes, including sperm and eggs. In this intricate performance, homologous chromosomes hook up and engage in genetic recombination, which involves exchanging and rearranging genetic material. This process is called genetic recombination (Mizumoto et al., 2006). The result is a one-of-a-kind mix of genes contained within each gamete, contributing to an increase in genetic variety and heralding the complex transmission of characteristics from parents to children. The process of meiosis, which halves the number of chromosomes in each cell while at the same time guaranteeing that each new generation inherits a unique set of genetic characteristics, is essential to the continuation of life (Lichter et al., 1991).

Chromosomes, which serve as the building blocks of heredity, are critically important to the process by which genes are passed down from one generation to another. Genes are the functional units of heredity that hold the instructions for specific traits and features. They are stored within these chromosomal archives, the archives of chromosomes. Genes are passed down from generation to generation, gradually generating a mosaic of genetic inheritance with each new generation. Every person is unique because of the unique combination of genes they received from both of their parents. These genes allow each person to express the traits and qualities that were passed down to them from their ancestors (Berg, 1981).

This chapter dives into the fundamentals of heredity, namely Mendelian genetics, which provides a narrative that sheds light on the subject. It is possible to gain insight into the likelihood of acquiring particular characteristics based on the genetic makeup of one's parents by analyzing the inheritance patterns of features controlled by single genes that contain both dominant and recessive alleles. The study of genetic principles reveals the legacy that genes leave behind within families, the trends of inheritance that guide the incidence of genetic illnesses, and the complex tapestry of diversity that is woven within the genetic code (Caron et al., 2001). In addition to this, the chapter delves into the essential significance of chromosomes in the context of genetic illnesses. Genealogical abnormalities, which can affect both health and development, can be caused by minute changes in the structure or number of chromosomes. In order to diagnose genetic diseases and provide thorough genetic counseling to families, it is vital to have a solid understanding of the complexities involved with chromosomes (Vertès et al., 2005).

In conclusion, the chapter titled "Genes and Chromosomes" extends an open invitation to all of us to set sail on an adventure into the worlds of genetic codes, chromosomes, and the science of heredity. By investigating the dynamic processes of mitosis and meiosis, we can acquire insights into the balance that exists between genetic stability and diversity. The investigation of chromosomes as carriers of heredity reveals the one-of-a-kind symphony of genes, which orchestrates the complex mechanisms of transmission from generation to generation. The chapter enriches our admiration for the harmonious unity and diversity that genes and chromosomes provide to life on Earth, which contributes to our comprehension of the splendor of genetics and the majesty of nature's design (Williams-Blangero et al., 2002).

The human species possesses a complement of 23 pairs of chromosomes, resulting in a total of 46 chromosomes. In contrast, various other species exhibit dissimilar numbers of chromosomes.

6.2 THE STABILITY OF CHROMOSOMES COMPLEMENTS

The significance of the cell nucleus and its constituents was initially proposed in the 1840s when Carl Nägeli observed that the nucleus underwent division prior to other cellular components during cell division. This individual, identified as Nägeli, demonstrated a lack of comprehension of Mendel's scientific findings. Nägeli's initial discovery of nuclear division was accompanied by a failure to recognize its significance (Brinkley, 2001). He considered the cells in which he observed nuclear division to be abnormal. However, it was acknowledged during the 1870s that nuclear division is a characteristic that is universally present in cell division. The significance of the nucleus in the transmission of genetic traits was further emphasized by the revelation that the fusion of the nuclei of two gametes occurs during fertilization. A significant breakthrough occurred ten years later when chromosomes were discovered, becoming observable through light microscopy after being stained with basic dyes. Subsequently, it was discovered that chromosomes undergo a systematic segregation process during both cell division, leading to the formation of daughter cells and the division of reproductive cells, resulting in the formation of gametes. Ultimately, three significant patterns were discerned about the chromosome complement, which refers to the entirety of chromosomes in both plant and animal species (Sasai et al., 2004).

- The nucleus of every somatic cell consists of a predetermined and characteristic number of chromosomes specific to the given species.
- The nuclei of somatic cells typically contain chromosomes that are found in pairs. As an illustration, the chromosomal composition of human beings comprises 23 pairs, totaling 46 chromosomes (see Figure 6.1). In a similar vein, it can be observed that the genetic makeup of peas comprises 14 chromosomes, which are organized into 7 pairs. Cells that possess nuclei of this nature, which consist of two sets of chromosomes that are similar in composition, are referred to as diploid.
- The germ cells, also known as gametes, undergo fertilization to generate the diploid state of somatic cells (Sun et al., 2003).

Table 6.1. Somatic chromosome numbers of some organisms (Source: Francis, Creative Commons License)

Organism	Chromosome number	Organism	Chromosome number
Field horsetail	216	Yeast	32
Bracken Fern	116	Fruit fly	8
Giant sequoia	22	Nematode	11, 12
Macaroni wheat	28	House fly	12
Bread wheat	42	Scorpion	4

Fava bean	12	Geometrid moth	224
Garden pea	14	Common toad	22
Wall cress	10	Chicken	78
Corn	20	Mouse	40
Lily	12	Gibbon	44
Snapdragon	16	Human being	46

In the context of multicellular organisms originating from a single cell, the coexistence of the diploid chromosome count in somatic cells and the haploid chromosome count in germ cells signifies the occurrence of two distinct mechanisms of nuclear division. One of these processes, namely mitosis, is responsible for preserving the chromosome number, while the other process, known as meiosis, reduces the chromosome number by half. The subsequent sections will analyze these two processes (Narayan, 1991).

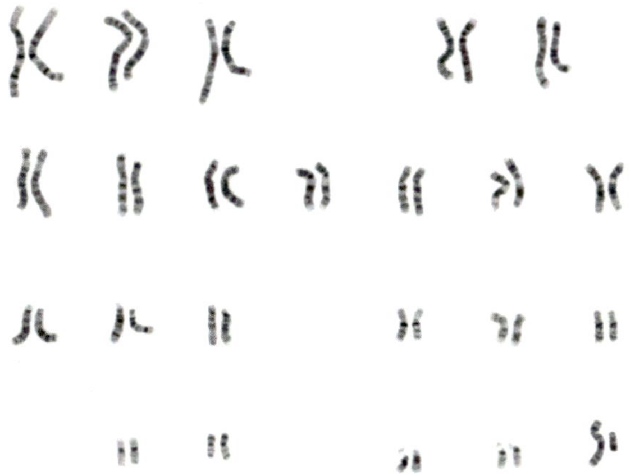

Figure 6.1. Complement of chromosomes of a male human. (Source: By Courtesy: National Human Genome Research Institute - Extracted image from http://www.genome.gov/glossary/resources/karyotype.pdf., Public Domain, https://commons.wikimedia.org/w/index.php?curid=583512, Creative Commons License)

Did You Know?

Under a microscope, chromosomes resemble an "X" shape, especially during cell division.

6.3 MITOSIS

Mitosis is a highly accurate mechanism of nuclear division that guarantees the distribution of an identical diploid set of chromosomes to each of the two resulting daughter cells, mirroring the diploid set found in the parent cell. Mitosis is typically accompanied by cytokinesis, a cellular process that results in the division of the cell itself, leading to the production of two daughter cells. The fundamental aspects of mitosis exhibit a consistent pattern across all organisms, thereby demonstrating a remarkable degree of uniformity in the underlying process (King et al., 1994).

- At the onset of nuclear division, each chromosome is duplicated.
- The process involves the longitudinal division of each chromosome into two identical halves, which subsequently become physically separated.
- During cellular division, each chromosome undergoes longitudinal division, forming two identical halves that subsequently separate from each other (McIntosh, 2016).

In a non-mitotic cell, the chromosomes are not discernible under the observation of a light microscope. The phase of the cell cycle, commonly referred to as interphase, is observed during this particular stage. During a specific phase of interphase known as S, the genetic material (DNA) contained within the chromosomes undergoes replication in anticipation of mitosis (as depicted in Figure 6.2). In this context, the letter "S" represents the process of DNA synthesis. The process of DNA replication is concomitant with the duplication of chromosomes. Preceding and succeeding the S phase exist distinct periods known as G1 and G2, during which DNA replication does not occur. The cell cycle, also known as the cellular life cycle, is typically delineated by three distinct interphase stages that are subsequently succeeded by mitosis, denoted as M. Consequently, the sequential arrangement of events follows the pattern of G1, S, G2, and M, as visually depicted in Figure 6.2. Cytokinesis, which involves partitioning the cytoplasm into two roughly equivalent portions containing a daughter nucleus, is encompassed within the M phase. The duration of a full cycle of life is variable across different cell types. In higher eukaryotes, a significant proportion of cells exhibit a temporal requirement ranging from 18 to 24 hours. The duration of the various periods in the cell cycle exhibits significant variation depending on the type of cell. Mitosis, which typically spans a duration of 30 minutes to 2 hours, is generally regarded as the briefest phase (Dunphy, 1994).

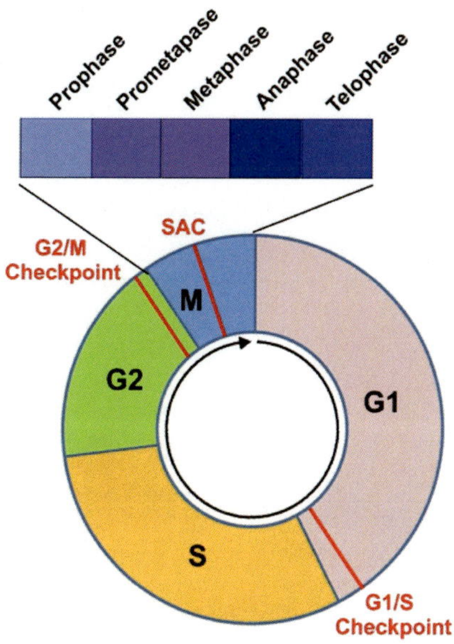

Figure 6.2. The 24-hour generation period of a normal mammalian cell's cell cycle as it grows in tissue culture. (Source: Wang Z. Regulation of Cell Cycle Progression by Growth Factor-Induced Cell Signaling. Cells. 2021; 10(12):3327. https://doi.org/10.3390/cells10123327, Creative Commons License)

The regulation of the cell cycle is primarily governed by genetic mechanisms. The control mechanisms seem to be fundamentally similar across all eukaryotic organisms. Two significant transitions exist in the cell cycle, namely the transition from the G1 phase to the S phase and the transition from the G2 phase to the M phase (as depicted in Figure 6.2). The G1/S and G2/M transitions are commonly referred to as "checkpoints" due to their temporally delayed characteristics until crucial processes have been successfully accomplished. At the G1/S checkpoint, DNA replication can only be initiated if either a certain amount of time has passed since the previous mitosis (in specific cell types) or if the cell has reached a certain size (in other cell types). The G2/M checkpoint necessitates the completion of DNA replication and the repair of any DNA damage before the initiation of the M phase (Mazia, 1961). Both major control points exhibit similar regulatory mechanisms and employ a specialized protein kinase, referred to as the p34 kinase subunit in Figure 6.2, which modulates the activity of target proteins through phosphorylation, involving the transfer of phosphate groups. The p34 kinase is classified as one of several protein kinases that play a crucial role in the regulation of various cellular processes. In order to undergo activation, the p34 polypeptide subunit necessitates association with multiple other polypeptide chains referred to as cyclins, owing to their periodic fluctuations in abundance that align with the progression of the cell cycle. At the G1/S transition, a specific group of cyclins binds with the p34 subunit, forming an active kinase. This active kinase is responsible for initiating DNA replication and other cellular processes that occur during the S phase. In a similar vein, during the G2/M checkpoint, a subsequent group of cyclins binds with the p34 subunit, resulting in the formation of the active kinase. This kinase then triggers the process of chromosome condensation, nuclear envelope disintegration, and cytoskeleton reorganization, all in anticipation of cytokinesis (Pickett-Heaps et al., 1982).

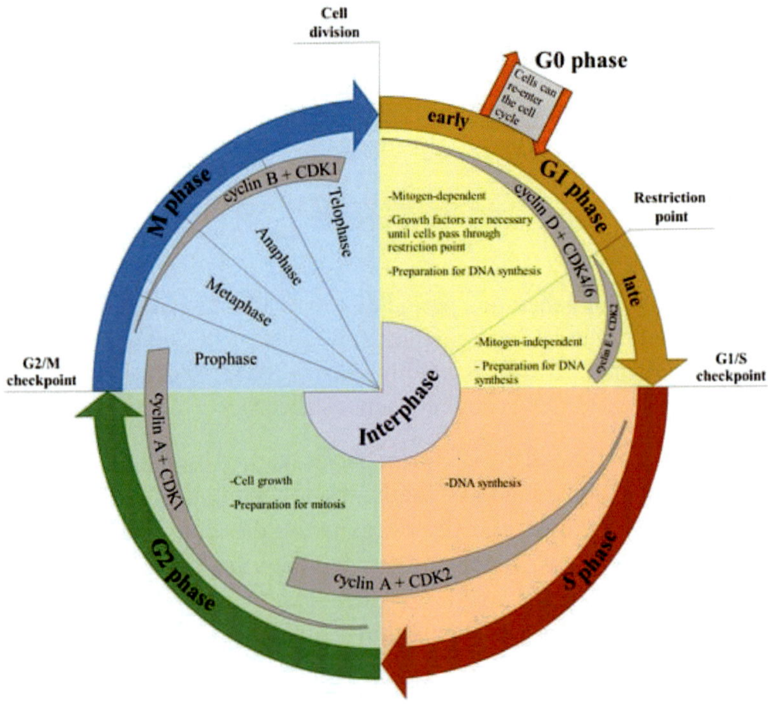

Figure 6.3. Mitosis in an organism (Source: Ligasová A, Frydrych I, Koberna K. Basic Methods of Cell Cycle Analysis. International Journal of Molecular Sciences. 2023; 24(4):3674. https://doi.org/10.3390/ijms24043674, Creative Commons License)

The essential characteristics of chromosome behavior during mitosis are depicted in Figure 6.3. The process of mitosis is traditionally categorized into four distinct phases: prophase, metaphase, anaphase, and telophase. If one encounters difficulty in recalling the sequence, it may be beneficial to employ the mnemonic device "peas make awful tarts" to jog one's memory. The stages exhibit the subsequent attributes (Gibcus et al., 2018):

- ***Prophase:*** During the interphase stage, the chromosomes exhibit an elongated filamentous structure and are not discernible as distinct entities under a light microscope. Except the existence of one or more prominent dark bodies known as nucleoli, the nucleus exhibits a diffuse and granular visual characteristic. The initiation of prophase is characterized by the compaction of chromosomes, resulting in the formation of discernible, slender filaments within the nucleus. Every chromosome is inherently composed of two longitudinally double subunits known as chromatids, which are closely associated with each other. The bipartite nature of each chromosome becomes apparent during the later stages of prophase. The formation of each pair of chromatids occurs as a result of the replication of a single chromosome during the S phase of interphase. The cohesion between chromatids within a chromosome pair is established at a distinct chromosomal site known as the centromere. During the prophase progression, the chromosomes undergo a transformation characterized by a reduction in length and an increase in thickness, which is attributed to a complex coiling process. During the final stage of prophase, the nucleoli undergo disappearance, and the nuclear envelope undergoes sudden disintegration (Tsuchiya & Lacefield, 2013).
- ***Metaphase:*** The mitotic spindle's formation occurs at the metaphase's onset. The spindle is a bipolar structure comprising bundles of microtubules that resemble fibers and span the cell from one spindle pole to the other. Multiple spindle fibers in the vicinity of the centromere region establish attachment with each individual chromosome. The anatomical feature that is functionally associated with the centromere and serves as the attachment site for spindle fibers is referred to as the kinetochore. Following the attachment of chromosomes to spindle fibers, they migrate towards the central region of the cell, ultimately aligning all kinetochores along an imaginary plane that is equidistant from the poles of the spindle. The hypothetical plane in question is commonly referred to as the metaphase plate. At the metaphase plate, the chromosomes become aligned and achieve their utmost level of contraction, facilitating counting and examining them for potential morphological variations. The accurate alignment of chromosomes is a critical regulatory checkpoint during the metaphase stage of both mitosis and meiosis (Wadsworth & Salmon, 1986). The delay in the completion of metaphase occurs when a chromosome is connected to a single pole of the spindle within a cell. By employing a micromanipulation needle

to seize a chromosome and exert a pulling force, it becomes possible to replicate the tension that the chromosome would encounter if it were connected on both ends. This mechanical tension facilitates the successful passage of the metaphase checkpoint, enabling the cell to progress into the subsequent division phase. The kinetochore is responsible for initiating the signal that facilitates chromosome alignment. The signal appears triggered by the dephosphorylation of specific proteins associated with the kinetochore. The function of the kinetochore is evidenced by the observation that the absence of a functional kinetochore resulting from targeted laser ablation does not cause a delay in the metaphase stage of cell division. The role of dephosphorylation is exemplified by employing an antibody that selectively interacts with certain kinetochore proteins exclusively in their phosphorylated state. The unattached kinetochores exhibit a robust affinity for the antibody, whereas the binding to the spindle attenuates the strength of the interaction. The reoccurrence of the antibody reaction with the kinetochore is observed in chromosomes that have undergone surgical detachment from the spindle. The metaphase checkpoint is successfully traversed, and cell division proceeds when the signaling mechanism ensures that all kinetochores are subjected to tension and properly lined up on the metaphase plate (Naruse, 2002).

- ***Anaphase:*** During the phase of anaphase, the centromeres undergo a longitudinal division, separating the two sister chromatids of each chromosome. These separated chromatids then proceed to migrate toward the opposite poles of the spindle structure. Upon the division of centromeres, it is acknowledged that each sister chromatid assumes the status of an independent chromosome. The movement of chromosomes is attributed, in part, to the gradual reduction in the length of the spindle fibers connected to the centromeres. This reduction exerts a force that pulls the chromosomes toward the poles in opposite directions. Following the conclusion of anaphase, the chromosomes are positioned in two distinct clusters located in close proximity to the opposite poles of the spindle. Each group possesses an equivalent number of chromosomes as initially present in the interphase nucleus (Kraft et al., 2003).
- ***Telophase:*** During the telophase stage of cell division, the chromosomes condense into compact groups and are enclosed by the formation of a nuclear envelope. Additionally, nucleoli, which are responsible for the production of ribosomes, are formed. Furthermore, the spindle apparatus, which aids in the separation of

chromosomes during earlier stages of cell division, disassembles and disappears. A gradual reversal characterizes the process of chromosome condensation until the chromosomes lose their distinct visibility. The two daughter nuclei gradually adopt a characteristic interphase morphology as the cell's cytoplasm undergoes division through the progressive formation of a deepening furrow along its periphery. During cellular division in plants, a novel cell wall is generated, which subsequently forms between the daughter cells, leading to their physical separation (Matsui et al., 2012).

Remember

Chromosome 1, the most extensive human chromosome, encompasses a staggering quantity of more than 200 million base pairs. Conversely, the Y chromosome, the smallest among the human chromosomes, comprises approximately 50 million base pairs.

6.4 MEIOSIS

Meiosis is a cellular process characterized by the creation of cells that possess a single copy of each chromosome pair found in the premeiotic cell. When a diploid cell, possessing two sets of chromosomes, undergoes the process of meiosis, it gives rise to four daughter cells. These daughter cells are genetically distinct from one another, and each contains a single set of chromosomes, known as haploid chromosomes. Meiosis is a cellular process that involves two consecutive nuclear divisions. Figure 6.4 provides a comprehensive overview of the fundamental aspects of the meiosis process (Bolcun-Filas & Handel, 2018).

This outline provides a comprehensive overview of the process of meiosis, along with an introductory explanation of its occurrence within a cellular context (Lee & Amon, 2003).

- Before the initial nuclear division, the chromosomes within each pair undergo a process of close association along their entire length (as depicted in Figure 6.4). The chromosomes that exhibit pairing are referred to as homologous chromosomes. Due to the prior replication of each member, a pair of homologous chromosomes is composed of two sister chromatids that are connected at the centromere. The association of the homologous chromosomes consequently results in the formation of a quadruple helical structure (Turner et al., 2006).
- During the initial nuclear division, the process of meiosis, the homologous chromosomes undergo segregation, with each member

of the pair moving towards opposite poles of the spindle, as depicted in Figure 6.4B. Two nuclei are generated, each consisting of a haploid complement of double-stranded chromosomes (as depicted in Figure 6.4C), with two chromatids present in each chromosome (Royo et al., 2010).

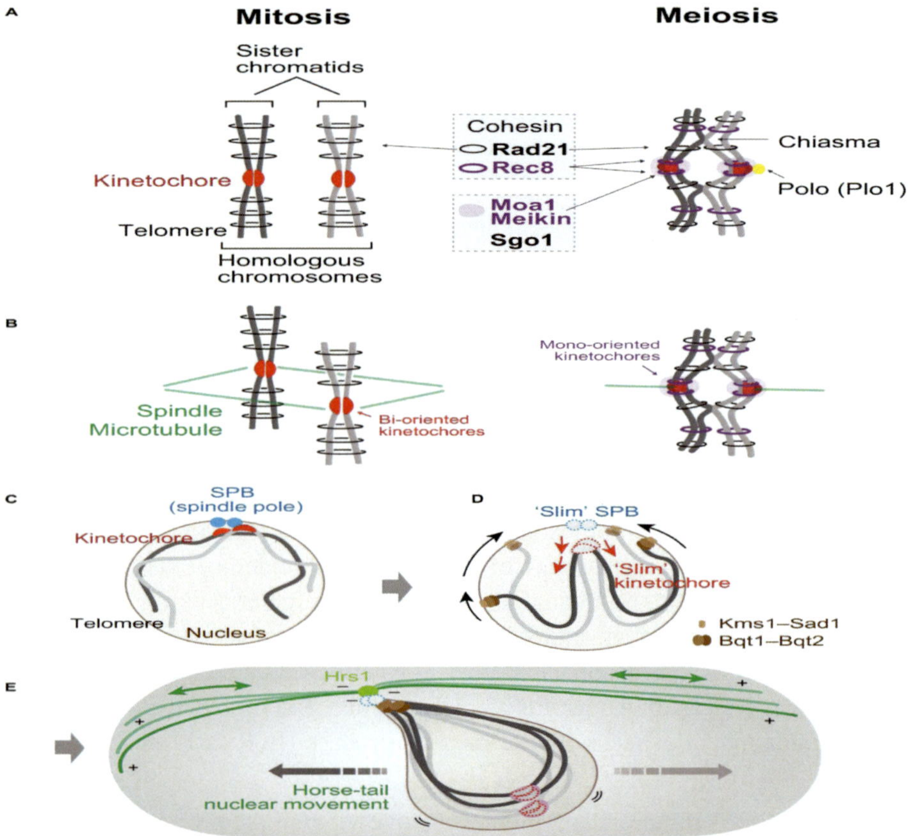

Figure 6.4. Overview of meiosis. (Source: Sato M, Kakui Y and Toya M (2021) Tell the Difference Between Mitosis and Meiosis: Interplay Between Chromosomes, Cytoskeleton, and Cell Cycle Regulation. Front. Cell Dev. Biol. 9:660322. doi: 10.3389/fcell.2021.660322, Creative Commons License)

- The subsequent nuclear division resembles a mitotic division without any chromosome replication. During cell division's metaphase stage, the chromosomes align along the metaphase plate. Subsequently, in the anaphase stage, the chromatids in each chromosome undergo separation, moving towards opposite daughter nuclei as depicted in Figure 6.4D. The overall outcome of the two stages of meiosis is the production of four haploid daughter nuclei. Each of these nuclei contains one sister chromatid from each pair of homologous chromosomes (Mahadevaiah et al., 2008).

The data presented in Figure 6.4 does not provide evidence for the occurrence of gene exchange between homologous chromosomes during the process of chromosome pairing. The process of genetic recombination during meiosis leads to the creation of chromosomes that contain segments derived from both homologous chromosomes, which are intermingled. The depiction of the exchanged chromosomes in Figure 6.4 involves the representation of segments with alternating colors. The process of exchange is a pivotal aspect of meiosis, and it will be analyzed in the subsequent section. Meiosis occurs in specialized cells known as meiocytes, encompassing primary oocytes and spermatocytes found in the gamete-producing tissues (Figure 6.5). Oocytes undergo maturation to develop into egg cells, while spermatocytes undergo a differentiation process to give rise to sperm cells. While the process of meiosis exhibits similarities across sexually reproducing organisms, it is noteworthy that in the female individuals of both animal and plant species, only one of the four resulting products undergoes development into a functional cell, with the remaining three undergoing disintegration. In animal biology, the outcome of meiosis gives rise to gametes, specifically sperm or eggs (Yi et al., 2013).

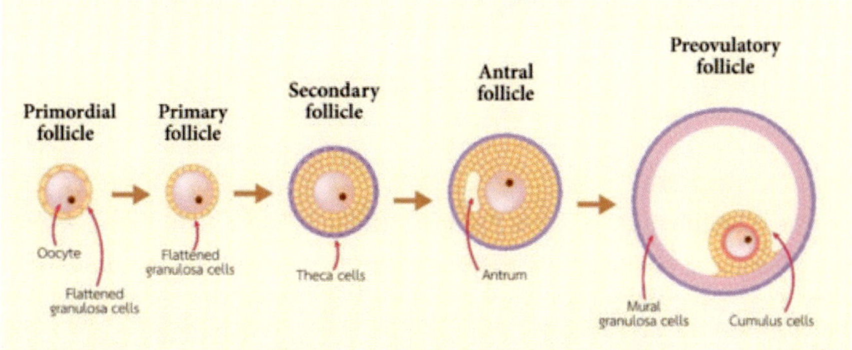

Figure 6.5. The life cycle of a typical oocyte (Source: Turathum B, Gao E-M, Chian R-C. The Function of Cumulus Cells in Oocyte Growth and Maturation and in Subsequent Ovulation and Fertilization. Cells. 2021; 10(9):2292. https://doi.org/10.3390/cells10092292, Creative Commons License)

In the context of plants, the situation exhibits a slightly greater complexity.

- The outcome of meiosis generally results in the production of spores, which subsequently undergo one or more rounds of mitotic divisions to generate a gametophyte organism with a haploid chromosome number. The production of gametes in the gametophyte occurs through the process of mitotic division of a haploid nucleus, as depicted in Figure 6.6.

- The fusion of haploid gametes results in the formation of a diploid zygote, which subsequently develops into the sporophyte plant (Khil et al., 2004).

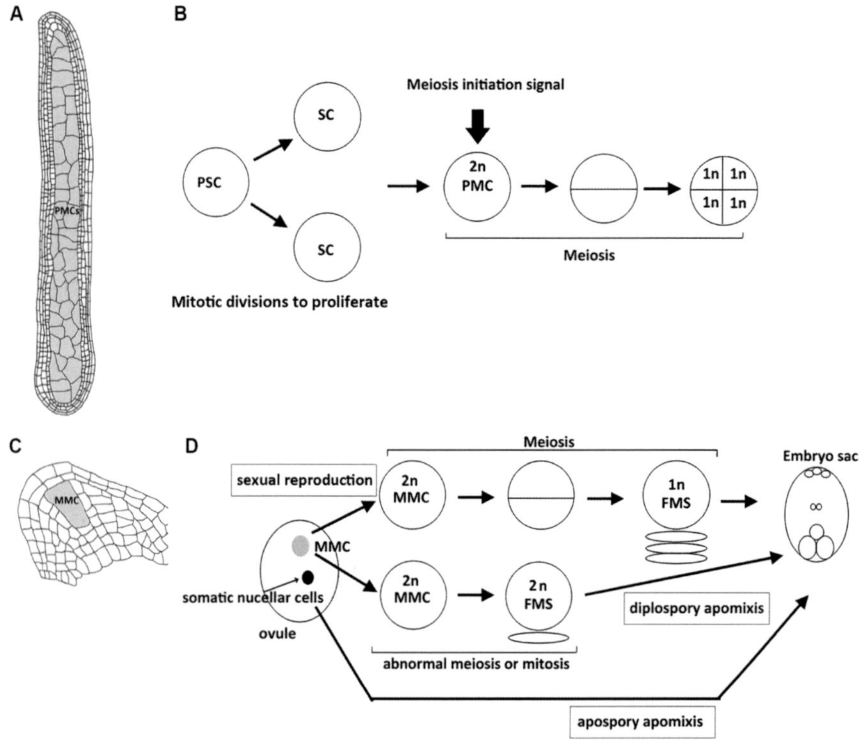

Figure 6.6. The life cycle of corn, Zea mays (Source: Wang C-JR and Tseng C-C (2014) Recent advances in understanding of meiosis initiation and the apomictic pathway in plants. Front. Plant Sci. 5:497. doi: 10.3389/fpls.2014.00497, Creative Commons License)

Meiosis, which can last days or weeks, is more complicated than mitosis. Figure 6.7 shows meiosis in a cell. Meiosis involves two nuclear divisions and one chromosome duplication. The first and second meiotic divisions, like mitosis, can be broken down into phases. This significant process occurs during the early division of the nucleus. The next section explains these events (Rasmussen & Holm, 1978).

6.4.1 The First Meiotic Division: Reduction

The initial stage of meiosis, known as meiosis I, is commonly referred to as the reductional division because it halves the chromosome count. The first meiotic division can be conceptually divided into four distinct stages, namely prophase

I, metaphase I, anaphase I, and telophase I, drawing an analogy with the process of mitosis. The stages of meiosis are typically more intricate in comparison to those of mitosis. The stages and substages can be observed and represented using Figures 6.7 and 6.8 (Vernet et al., 2011).

- **Prophase I:** Several important events occur during the first stage of meiosis, known as ***prophase I***. The prophase I stage, which spans multiple days in most complex organisms, is typically categorized into five distinct substages: leptotene, zygotene, pachytene, diplotene, and diakinesis. The terms above pertain to the visual characteristics exhibited by the chromosomes during each respective substage (Speed & Chandley, 1990).

During the ***leptotene*** stage, which originates etymologically from the term "thin thread," the chromosomes undergo their initial manifestation as elongated structures resembling threads. Electron microscopy enables the differentiation of sister chromatid pairs. During the initial phase of chromosome condensation, multiple dense granules manifest at irregular intervals along their length. The aforementioned localized contractions, chromomeres, exhibit distinct attributes in quantity, dimensions, and spatial arrangement within a specific chromosome (Börner et al., 2004).

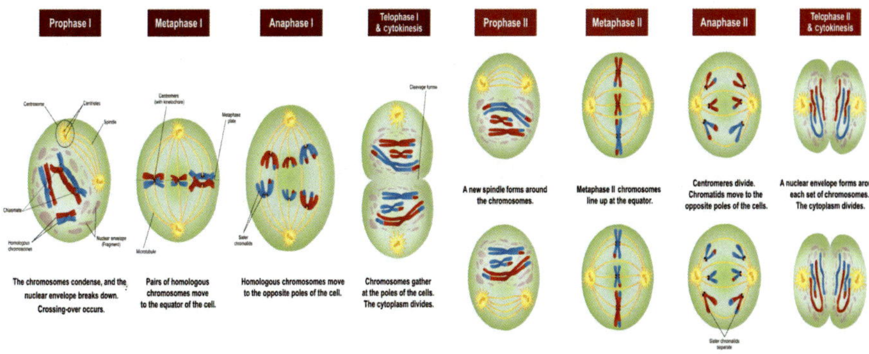

Figure 6.7. Stages of Meiosis (Source: By Ali Zifan - Own work using: information from Campbell Biology (10th Edition) by: Jane B. Reece & Steven A. Wasserman., CC BY-SA 4.0, https://commons.wikimedia.org/w/index.php?curid=49630204, Creative Commons License)

The *zygotene* phase is characterized by the process of lateral pairing, also known as synapsis, of homologous chromosomes, which initiates at the ends of the chromosomes. The term "zygotene" refers to the phenomenon of paired threads. During the progression of the pairing process, there is a specific and accurate association between chromomeres along the length of the chromosomes. The term "bivalent" describes a pair of synapsed homologous chromosomes (Murphy et al., 2014).

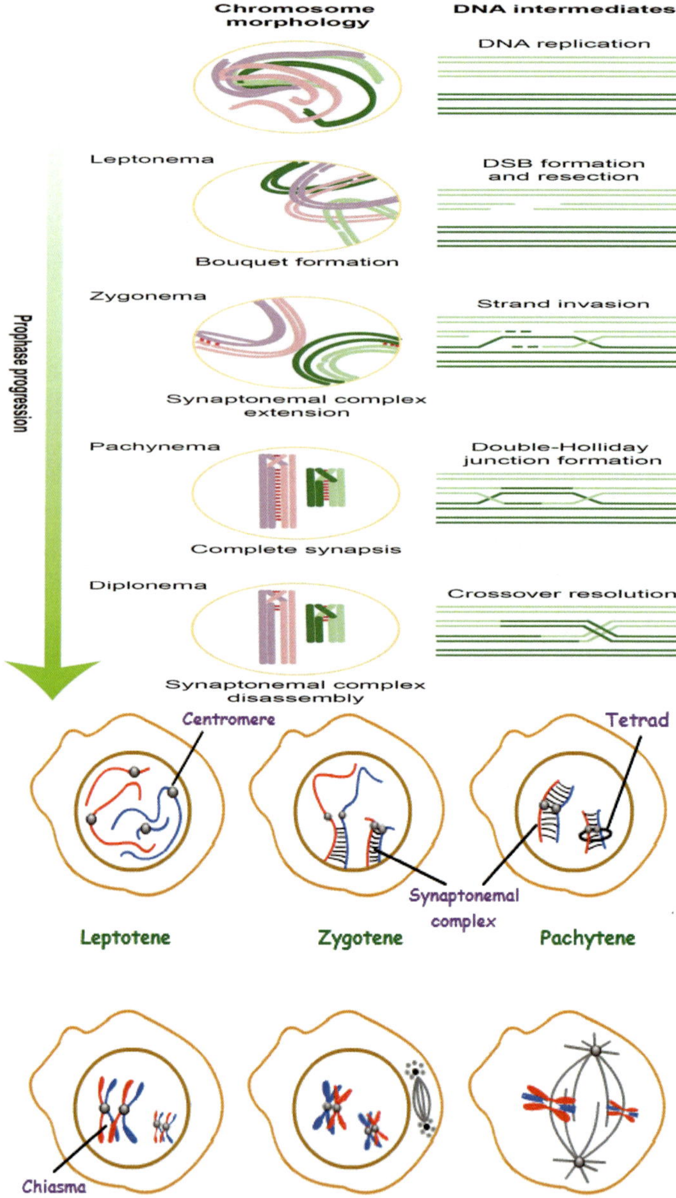

Figure 6.8. Stages of prophase 1 of Meiosis (Source: Kar FM and Hochwagen A (2021) Phospho-Regulation of Meiotic Prophase. Front. Cell Dev. Biol. 9:667073. doi: 10.3389/fcell.2021.667073, Creative Commons License)

During the ***pachytene*** stage, there is a continued process of chromosome condensation (Figure 6.8C).

The term "pachytene" is derived from the Greek words "pachys," meaning thick, and "tenos" meaning thread. The chromosomes undergo further condensation during this specific stage and become thicker, as depicted in Figure 6.7. During the late pachytene stage, it is occasionally observable that each bivalent, representing a pair of chromosomes, comprises a tetrad composed of four chromatids. However, the two sister chromatids of each chromosome typically exhibit close proximity. The occurrence of genetic exchange, known as crossing-over, is a significant event that transpires during the pachytene stage. However, the manifestation of crossing-over becomes evident only upon transitioning to the diplotene stage. The locations of exchange in Figure 6.7 are denoted by the intersections where chromatids of distinct colors intersect (Xu et al., 1995).

During the diplotene stage, the previously synapsed chromosomes commence the process of separation. The term "diplotene" refers to the state of chromosomes during meiosis, where they appear as double threads. This is evident in Figure 6.8, where the diplotene chromosomes are clearly observed to be double. Nevertheless, the homologous chromosomes persistently remain physically linked at specific intervals along their length due to the formation of cross-connections resulting from the process of crossing over. The formation of each cross connection, known as a chiasma (plural, chiasmata), occurs through the process of breakage and rejoining between chromatids that are not genetically identical. As depicted in Figure 6.9, the occurrence of a chiasma arises from the physical exchange that takes place between chromatids of homologous chromosomes (Carpenter, 1975).

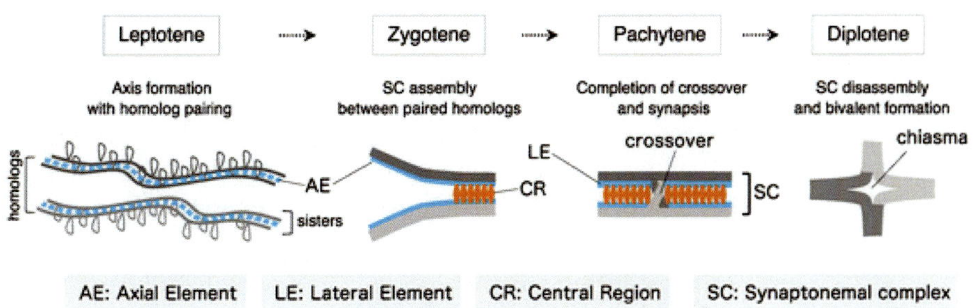

Figure 6.9. Chiasmata. (Source: Sakuno T, Hiraoka Y. Rec8 Cohesin: A Structural Platform for Shaping the Meiotic Chromosomes. Genes. 2022; 13(2):200. https://doi.org/10.3390/genes13020200, Creative Commons License)

The concluding stage of prophase I is known as *diakinesis*, during which the homologous chromosomes exhibit apparent repulsion towards each other, causing the segments that are not linked by chiasmata to separate. The term "diakinesis" refers to the process of moving apart. During this particular substage, the chromosomes undergo a process of maximum condensation, as depicted in Figure 6.8. The homologous chromosomes within a bivalent maintain their connection through the presence of at least one chiasma, which persists until the onset of the first meiotic anaphase. Towards the conclusion of the diakinesis phase, the commencement of spindle formation occurs, concomitant with the disintegration of the nuclear envelope (Clemons et al., 2013).

- ***Metaphase I*** is a stage in meiosis where homologous chromosomes align along the equatorial plane of the cell. The bivalents undergo alignment in which the centromeres of the two homologous chromosomes are situated on opposite sides of the metaphase plate (as depicted in Figure 6.10). As each pair of homologous chromosomes aligns at the metaphase plate, the orientation of their centromeres in relation to the spindle poles is randomized. As depicted in Figure 6.11, the bivalents resulting from nonhomologous pairs of chromosomes have the potential to align on the metaphase plate in two distinct orientations. The positioning of the centromeres plays a crucial role in determining the subsequent distribution of each member of the bivalent to the respective poles. Two alignments occur when each nonhomologous chromosome is heterozygous for two alleles. One alignment produces AB and ab gametes, whereas the other produces Ab and aB (Figure 6.11). Metaphase alignment is random. Therefore, the two types of alignment and the four gametes are equally likely. The alleles A, a, and B, b undergo independent assortment since the four gametes are distributed 1:1:1:1 (Solc et al., 2010).
- ***Anaphase I***: During this particular phase, homologous chromosomes, which consist of two chromatids connected by an undivided centromere, undergo separation and migrate towards opposite ends of the spindle structure (as depicted in Figure 6.10B). The cellular mechanism responsible for the segregation of alleles is the separation of chromosomes during anaphase (Salah & Nasmyth, 2000).
- ***Telophase I:*** Following the conclusion of anaphase I, a haploid complement of chromosomes, comprising one homolog from each bivalent, is positioned in close proximity to each pole of the spindle (as depicted in Figure 6.6). During telophase, the disassembly of the spindle occurs, and the subsequent events vary depending on the species. In some cases, a temporary nuclear envelope is formed

around each cluster of chromosomes, while in other cases, the chromosomes proceed to the second meiotic division with only a partial relaxation of their spiral structure (Kaczanowski et al., 1985).

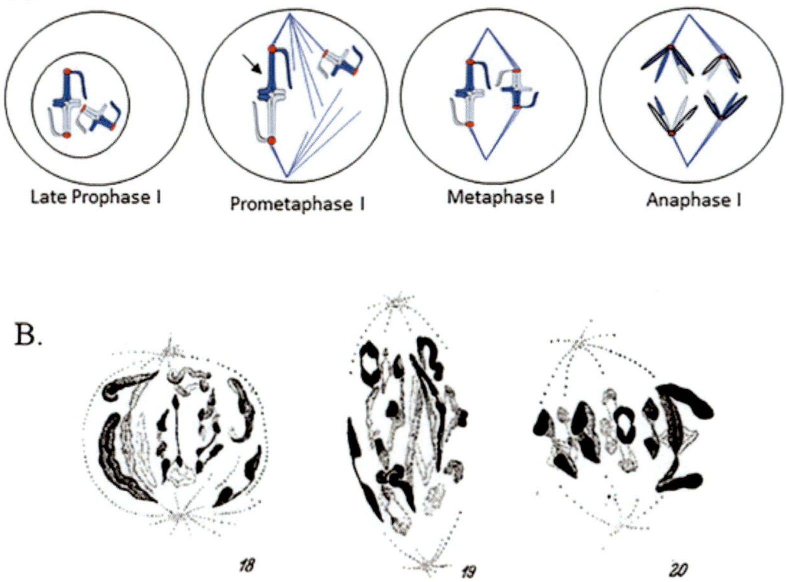

Figure 6.10. Later Meiotic stages (Source: Czekalski MA, Paliulis LV. The Pre-Metaphase Stretch: A Re-Examination. DNA. 2021; 1(1):3-12. https://doi.org/10.3390/dna1010002, Creative Commons License)

6.4.2 The Second Meiotic Division: Equation

The second stage of meiosis, known as meiosis II, is commonly referred to as the equational division due to the fact that the chromosome count remains constant in each cell both prior to and following this division. Certain species exhibit a direct transition of chromosomes from telophase I to prophase II, wherein condensation remains intact. Conversely, other species experience a short intermission between the two meiotic divisions, during which the chromosomes may undergo a partial "de-condensation" or uncoiling. Chromosome replication does not occur during the interphase between the two divisions. Thus, the chromosomes at the start of the second division are identical to those at the end of the first. After a brief prophase (prophase II) and the formation of second-division spindles, metaphase II aligns each nucleus' centromeres along the spindle's central plane. Anaphase II separates chromatids by longitudinally dividing centromeres. After anaphase II, each chromatid becomes a chromosome (Morrill et al., 1984).

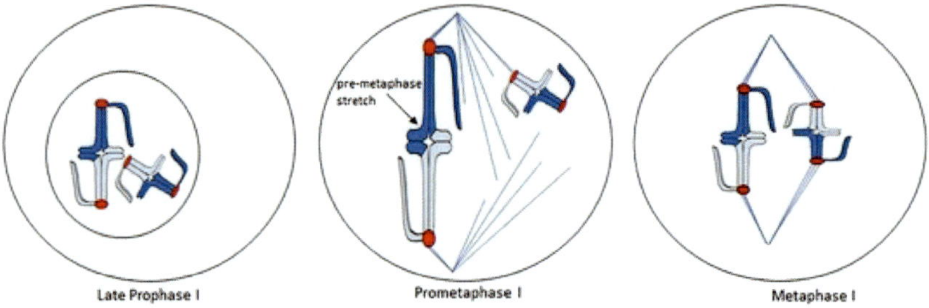

Figure 6.11. The occurrence of nonhomologous chromosomes randomly aligning at metaphase. (Source: Czekalski MA, Paliulis LV. The Pre-Metaphase Stretch: A Re-Examination. DNA. 2021; 1(1):3-12. https://doi.org/10.3390/dna1010002, Creative Commons License)

The occurrence of ***Telophase II*** is characterized by the chromosomes in the four haploid nuclei returning to the interphase state, along with the cytoplasmic division. Therefore, the second meiotic division bears a resemblance to a mitotic division on the surface (Takamiya et al., 1996).

Did You Know?

Telomeres, which are protective caps located at the ends of chromosomes, serve the crucial function of safeguarding genetic material from degradation during the process of replication.

6.5 CHROMOSOMES & HEREDITY

Following the reemergence of Mendel's seminal paper, a prevailing consensus swiftly emerged, positing that genes were indeed situated within the chromosomal structure. The most compelling evidence lies in the observation that Mendel's principles of segregation and independent assortment exhibit a striking resemblance to the behavior of chromosomes during the process of meiosis. However, the initial irrefutable evidence supporting the notion that genes are components of chromosomes was acquired through experiments focused on the transmission pattern of sex chromosomes. These chromosomes play a crucial role in determining the distinct sexes in certain plants and nearly all animals. In this section, an examination of these results will be conducted (Bridges, 1916).

6.5.1 Chromosomal Determination of Sex

The sex chromosomes deviate from the general principle that diploid organisms possess pairs of homologous chromosomes that are morphologically similar. Microscopic analysis conducted as early as 1891 revealed the absence of a homologous counterpart in one of the chromosomes found in males of certain insect species. The X chromosome, which was found in all somatic cells of males but only in half of their sperm cells, was referred to as an unpaired chromosome. The biological significance of these observations became evident when it was demonstrated that females of the same species possess a pair of X chromosomes. In species where females possess a pair of X chromosomes, males typically possess a single X chromosome in conjunction with a distinct morphological chromosome (Charlesworth, 2002).

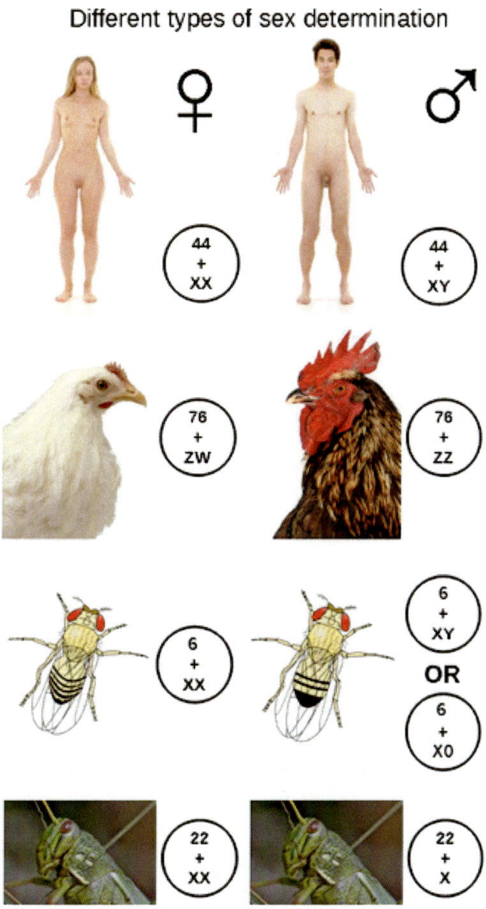

Figure 6.12. Chromosomal sex determination in mammals, insects, and other animals. (Source: By CFCF - Own work, CC BY-SA 3.0, https://commons.wikimedia.org/w/index.php?curid=29621037, Creative Commons License)

The Y chromosome, distinct from the X chromosome, undergoes pairing with the X chromosome during meiosis in males due to a shared region of homology between the X and Y chromosomes. The chromosomal constitution disparity between males and females serves as a chromosomal system for sex determination during the process of fertilization. While it is true that each egg cell in humans contains an X chromosome, it is important to note that only half of the sperm cells carry an X chromosome, while the remaining half carry a Y chromosome (Meisel, 2020).

The process of fertilization involves the fusion of an X-bearing egg with an X-bearing sperm, leading to the formation of a XX zygote, typically giving rise to a female individual. Conversely, when fertilization occurs between a Y-bearing sperm and an egg, an XY zygote is formed, which usually grows into a male individual (Figure 6.12). The outcome manifests as a distinctive interweaving arrangement of X chromosome inheritance, wherein males acquire their X chromosome exclusively from their mothers and subsequently pass it down solely to their female offspring (Pennell et al., 2018).

The designation of the homogametic sex is attributed to the female due to the production of a single type of gamete (specifically, the X-bearing gamete). Conversely, the heterogametic sex is assigned to the male due to the production of two distinct types of gametes (Marchal et al., 2003). In cases where the union of gametes during fertilization occurs randomly, it is anticipated that a sex ratio of 1:1 will be observed.

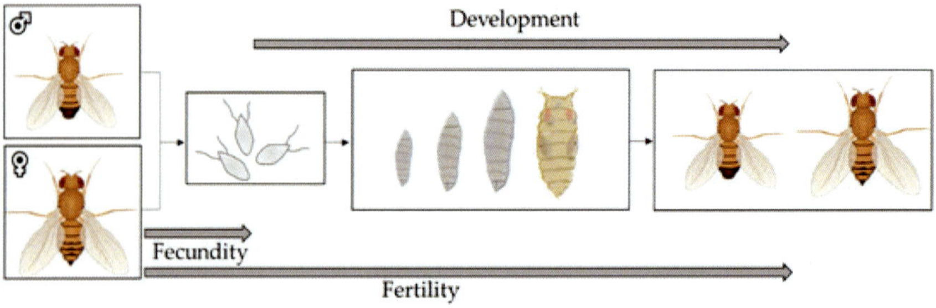

Figure 6.13. Drosophila melanogaster male and female (Source: Eickelberg V, Lüersen K, Staats S, Rimbach G. Phenotyping of Drosophila Melanogaster—A Nutritional Perspective. Biomolecules. 2022; 12(2):221. https://doi.org/10.3390/biom12020221, Creative Commons License)

This is because males produce an equal quantity of X-bearing and Y-bearing sperm. The X and Y chromosomes collectively comprise the sex chromosomes, a designation that sets them apart from autosomes, which refer to other pairs of chromosomes. The determination of early stages of female or male development is regulated by the sex chromosomes. However, the developmental process itself involves the contribution of numerous genes located throughout the entire set of chromosomes, including genes present in the autosomes. The X chromosome harbors numerous genes that possess functions that are not directly associated with sexual differentiation, as will be elucidated in the subsequent section. The Y chromosome in various organisms, including Homo sapiens, typically harbors a limited number of genes that are primarily associated with male determination (Kondo et al., 2009).

6.5.2 X-linked Inheritance

The study of a Drosophila gene for white eyes provided compelling evidence supporting the notion that genes are located within chromosomes, specifically the X chromosome. It is important to note that in Mendel's experimental crosses, the specific assignment of traits to the male and female parents did not have any bearing on the outcomes. The reciprocal crosses yielded identical outcomes. Thomas Hunt Morgan discovered one of the earliest deviations from this principle in 1910 during a preliminary investigation of a mutation observed in the fruit fly species *Drosophila melanogaster*, which resulted in the manifestation of white-colored eyes. The eye color of the wild type is characterized by a composite of red and brown pigments, resulting in a brick-red appearance (Dobyns et al., 2004).

White eyes can be attributed to specific combinations of autosomal genes that independently eliminate pigments. However, the white-eye mutation investigated by Morgan is characterized by a metabolic obstruction that simultaneously disrupts both pigments. The study conducted by Morgan commenced by observing the presence of a single male specimen exhibiting white eyes within a laboratory population of wild-type organisms that had been consistently bred for numerous generations. In a cross between a male exhibiting a particular trait and females with the wildtype phenotype, it was observed that all offspring in the first filial generation (F1) displayed the red eye phenotype, indicating the recessive nature of the allele responsible for white eyes. Morgan's observations of the F2 generation resulting from the crossbreeding of F1 males and females revealed the presence of 2459 red-eyed females, 1011 red-eyed males, and 782 white-eyed males. The white-eyed phenotype correlated with sex, as it was observed that all individuals displaying the white-eyed trait were of the male sex (Garn et al., 1965).

Conversely, the occurrence of white eyes was not limited solely to males. As an illustration, in the experiment, F1 females with red eyes resulting from the mating between wild type and white-eyed individuals were subjected to a backcross with their white-eyed fathers. The resulting offspring comprised both females with red eyes and white eyes, as well as males with red eyes and white eyes, in roughly equal proportions. A significant observation was made by mating female individuals with white eyes with males exhibiting the wild-type phenotype. All the female offspring exhibited the wild-type eye phenotype, while all the male offspring displayed the white-eye phenotype. The reciprocal crosses yielded divergent outcomes compared to the original cross between wildtype and white, wherein only wildtype females and males were observed (Meetei et al., 2004).

Morgan observed that conducting reciprocal crosses would produce disparate outcomes in the presence of the white eye allele on the X chromosome. The rationale behind this phenomenon lies in the distinct transmission patterns of the X chromosome by individuals of the male and female sexes. By Figure 6.12B, it is observed that males exclusively transmit their X chromosome solely to their female offspring, while females transmit one of their X chromosomes to the offspring of both genders. The normal chromosome complement of *Drosophila melanogaster* is depicted in Figure 6.14. Females possess a complement of XX chromosomes, while males possess XY chromosomes. It is important to note that the Y chromosome lacks a counterpart of the white gene. The designation of a gene on the X chromosome as X-linked is commonly employed (Depienne & Leguern, 2012).

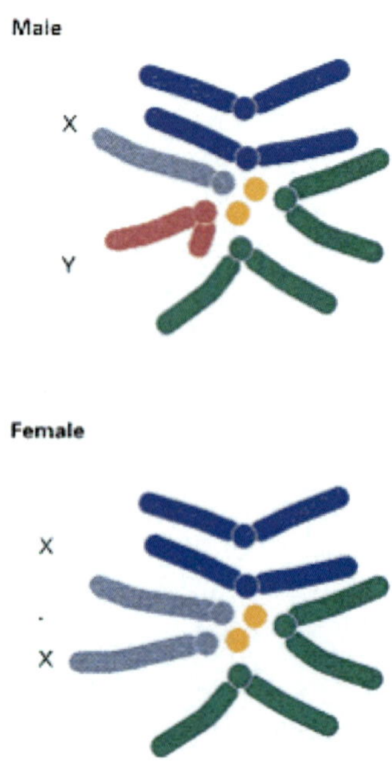

Figure 6.14. Drosophila melanogaster male and female diploid chromosomes. (Source: Mark, Creative Commons License)

The experimental design of a reciprocal crosses between wildtype Sbi: male and white individuals involved crossbreeding female Sbi (Cross A) mice with male Sbi (white) mice, followed by their subsequent mating with wild-type mice. The subject of interest in this study is the female individuals of the species Sbi (Cross B).

The symbols w, and w+ are used to represent the mutant and wildtype variants of the white gene found on the X chromosome. The genetic composition of a male with white eyes can be represented as wY, while that of a male with the typical eye color (wildtype) can be denoted as w+Y. Due to the recessive nature of the w allele, females with white eyes possess the genotype ww, while wild-type females can either be heterozygous w+w or homozygous w+w+.

The diagrams presented in Figure 6.15 explain the various phenotypic ratios that are observed in the offspring of the F1 and F2 generations resulting from the crosses. Subsequent discoveries in Drosophila have revealed the existence of numerous additional genes that adhere to the X-linked mode of inheritance. The salient features of X-linked inheritance can be succinctly summarized as follows (Androphy et al., 1985):

- Reciprocal crosses that lead to distinct phenotypic ratios between the sexes frequently suggest the presence of X linkage. For instance, in Drosophila, the reciprocal cross involving a red-eyed female and white-eyed male results in offspring that are exclusively red-eyed (as depicted in Figure 6.15). Conversely, when a white-eyed female is crossed with a red-eyed male, the progeny consists of red-eyed females and white-eyed males (Cross B, as illustrated in Figure 6.15).
- The F2 generation demonstrates that heterozygous females transmit each X-linked allele to roughly an equal proportion of their daughters and sons.
- Males who inherit an X-linked recessive allele express the recessive trait due to the absence of a wild-type counterpart of the gene on the Y chromosome. Males who are affected by the condition transmit the recessive allele exclusively to their female offspring, while their male offspring do not inherit the allele. In the context of genetic inheritance, it can be observed that males who do not exhibit any phenotypic effects possess the wild-type allele on their X chromosome (Gambineri et al., 2003).

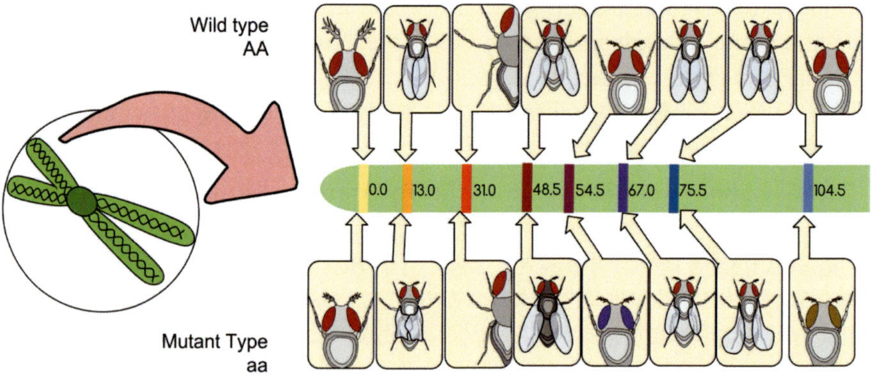

Figure 6.15. Drosophila F1 and F2 progenies' chromosomal interpretation (Source: By Twaanders17 - Own work, CC BY-SA 4.0, https://commons.wikimedia.org/w/index.php?curid=40694655, Creative Commons License)

Hemophilia A, a severe blood clotting disorder, is an illustrative example of a human trait that follows an X-linked inheritance pattern. A recessive allele determines this particular condition. Individuals affected by this condition exhibit a deficiency in a blood-clotting protein known as factor VIII, which is essential for normal clotting. Consequently, these individuals experience excessive and frequently life-threatening bleeding following an injury (Eksioğlu et al., 1996). Queen Victoria of England is renowned for being the progenitor of a well-known lineage affected by hemophilia. One of her offspring, namely Leopold, exhibited the medical condition of hemophilia, while two of her female offspring possessed the genetic trait of being heterozygous carriers of the gene above. Two of Queen Victoria's granddaughters were carriers of the gene, and through their marriages, they transmitted the gene to the royal families of Russia and Spain. The individual next in line to inherit the Russian monarchy of the Romanoffs, known as Tsarevich Alexis, was affected by the condition above. The individual in question acquired the gene in question through inheritance from his maternal ancestor, Tsarina Alexandra, who happened to be one of the granddaughters of Queen Victoria. In the course of the 1918 Russian Revolution, the Bolsheviks executed the Tsar, the Tsarina, Alexis, and their four sisters. Paradoxically, the contemporary lineage of the British royal family can be traced back to an ordinary offspring of Queen Victoria and is devoid of the ailment above (Gonçalves et al., 2014).

In certain organisms, there exists a reversal in the homogametic and heterogametic sexes, whereby the males possess a chromosomal composition of XX. In contrast, the females possess a chromosomal composition of XY. Avian species, certain reptilian and piscine organisms, as well as lepidopteran insects, exhibit this particular mode of sexual differentiation. The phenomenon of XX and XY reversal in the sexes leads to a contrasting pattern of nonreciprocal inheritance of X-linked genes. As an illustration, certain chicken breeds exhibit a phenotype known as "barred," wherein their feathers display alternating transverse bands of light and dark coloration. In the nonbarred phenotypes of other breeds, the feathers exhibit a uniform coloration. The genetics of true-breeding barred and nonbarred types are as follows (Hasstedt & Atkin, 1983):

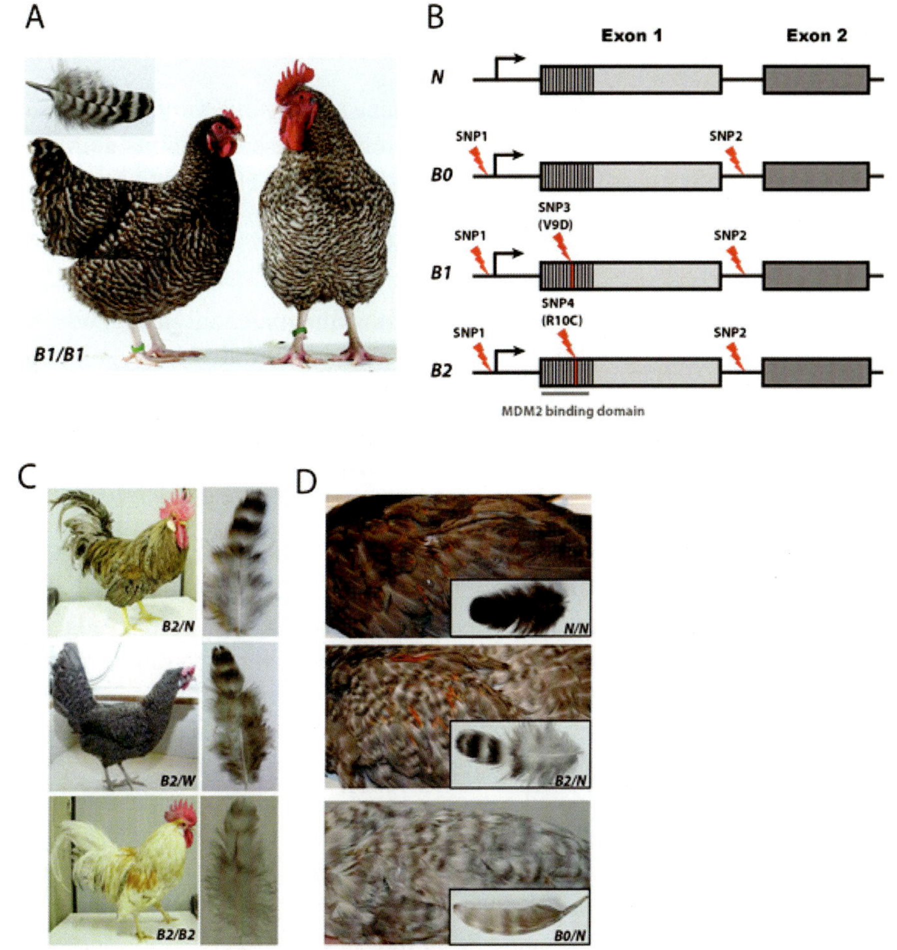

Figure 6.16. Alleles and phenotypes at the *Sex-linked barring* (*CDKN2A*) locus in chicken. (A) Female and male Coucou de Rennes chicken with separately depicted feather illustrating the iconic Sex-linked barring phenotype caused by the *B1* ***allele. (B)*** *Sex-linked barring* ***alleles and associated sequence variants. SNP1 and SNP2 are non-coding while SNP3 and SNP4 constitute non-synonymous changes in the region encoding the MDM2 binding domain. (C) Sex-linked dilution phenotype caused by the*** *B2* ***allele. Note how the homozygous male has an almost white***

appearance whereas the hemizygous female as well as the heterozygous male show a Sex-linked barring pattern. (D) Phenotype with individual feathers from *N /N*, *B2/N* ***and B0/N chicken. Photo credits: (A)*** Hervé Ronné, Ecomusée du pays de Rennes, (C) Susanne Kerje, (D) Dominic Wright and Doreen Schwochow-Thalmann. (Source: Schwochow Thalmann D, Ring H, Sundström E, Cao X, Larsson M, Kerje S, et al. (2017) The evolution of *Sex-linked barring* ***alleles in chickens involves both regulatory and coding changes in*** *CDKN2A*. PLoS Genet 13(4): e1006665. https://doi.org/10.1371/journal.pgen.1006665, Creative Commons License)

The findings of research suggest that the gene responsible for the expression of barring in chickens is located on the X chromosome and exhibits dominant inheritance. In order to differentiate the process of sex determination in birds, butterflies, and moths from the conventional XX-XY mechanism, these organisms occasionally exhibit a sex chromosome constitution denoted as WW in the homogametic sex and WZ in the heterogametic sex. Therefore, in avian species, as well as in butterflies and moths, the male individuals possess a chromosomal composition of WW, while the female individuals possess a chromosomal composition of WZ (Martin et al., 2021).

6.5.3 Non-disjunction As a Proof of the Chromosomes Theory of Heredity

The white mutation inheritance pattern and the X chromosome's genetic transmission supported the chromosome theory of heredity, which holds that genes are chromosomes. Drosophila experiments proved it. Calvin Bridges, a student of Morgan, made noteworthy observations regarding deviations from the anticipated inheritance patterns in crosses involving multiple X-linked genes. As an illustration, in the experimental setting where female Drosophila with white eyes was paired with male Drosophila with red eyes, the majority of the offspring exhibited the anticipated outcome of red-eyed females and white-eyed males. Nevertheless, approximately 0.05% of the F1 flies exhibited atypical characteristics, specifically either a female with white eyes or a male with red eyes. Bridges demonstrated that the occurrence of these uncommon exceptional offspring was attributed to the sporadic failure of the two X chromosomes in the maternal parent to segregate from one another during the process of meiosis, a biological event commonly referred to as nondisjunction. The outcome of X chromosome nondisjunction is the production of certain eggs containing two X chromosomes and others devoid of any X chromosome. The fertilization of these abnormal eggs is anticipated to yield four distinct classes of zygotes. Animals devoid of an X chromosome are undetectable due to the inviability of embryos lacking an X chromosome (Lamb et al., 1996).

Similarly, the majority of offspring possessing three X chromosomes experience early mortality during the developmental stage. The analysis of the chromosomes of the atypical offspring resulting from the white × wildtype cross revealed that the white-eyed females with exceptional traits possessed two X chromosomes in addition to a Y chromosome. Conversely, the red-eyed males with exceptional traits exhibited a single X chromosome but were devoid of a Y chromosome. The latter group, characterized by a sex-chromosome constitution designated as XO, exhibited sterility in males (Brush, 2002).

The previously mentioned experiments, along with other relevant studies, provided irrefutable evidence supporting the veracity of the chromosome theory of heredity (Safir, 1920).

Bridges provided evidence supporting the chromosome theory by demonstrating a direct correlation between extraordinary characteristics exhibited by chromosomes and the corresponding exceptional inheritance patterns of their genes. The experimental demonstration of the chromosome theory is widely regarded as one of the most significant and refined scientific investigations in the field of genetics (Sankaranarayanan, 1979).

6.5.4 Sex Determination in Drosophila

In the XX-XY mechanism of sex determination, it is widely recognized that the presence of the Y chromosome is correlated with the male phenotype. In certain organisms, such as *Homo sapiens*, this association arises due to the influence exerted by the Y chromosome on embryonic development, which manifests male sexual traits (Clough & Oliver, 2012). Drosophila exhibits a distinctive characteristic within organisms that possess an XX-XY system of sex determination, as the Y chromosome, despite its association with maleness, does not possess the ability to determine maleness. This phenomenon is exemplified by the discovery that in the species Drosophila, embryos with an XXY chromosomal configuration develop into females that are morphologically normal and capable of reproduction (Hediger et al., 2004). Meanwhile, embryos with an XO chromosomal configuration develop into males that are morphologically normal but unable to reproduce. The letter "O" is utilized in the formula XO to highlight the absence of a sex chromosome. The phenomenon of sterility observed in XO indicates that while the Y chromosome is not indispensable for male development, it plays a crucial role in male fertility. Specifically, the Drosophila Y chromosome encompasses six genes that are imperative for the proper formation of spermatozoa (Salz & Erickson, 2010).

The ratio between the number of X chromosomes and the number of sets of autosomes in an individual fly genetically influences the determination of sex in Drosophila. In the model organism Drosophila, a single set of autosomes in the haploid state comprises one copy of each of chromosomes 2, 3, and 4, which are classified as autosomes. Typically, diploid flies possess two sets of haploid autosomes, consisting of a homologous pair for each of the chromosomes 2, 3, and 4. There are two X chromosomes in females, while males have one X and one Y chromosome. The symbol A will be employed to denote a full set of autosomes in a haploid state (Baker & Ridge, 1980).

$$A = Chromosome2 + Chromosome3 + Chromosome4$$

In chromosomal composition, a typical male possesses the chromosomal complement XYAA. This results in an X/A ratio of 1 X: 2 A, or 1: 2, denoting the relative proportion of X chromosomes to sets of autosomes. Typical females possess a chromosomal composition of XXAA, resulting in an X/A ratio of 2 X : 2 A, which can also be expressed as 1 : 1. Male flies have X/A ratios smaller than 1:2, such as XAAA. On the other hand, female flies have X/A ratios greater than 1:1, like XXXAA. Intermediate X/A ratios, such as a ratio of 2:3, give rise to intersex individuals who exhibit certain characteristics of both sexes (Steinmann-Zwicky et al., 1990).

Sex-lethal (Sxl) gene regulates Drosophila sexual differentiation. Two protein isoforms of the Sxl gene vary depending on whether the messenger RNA contains a male-specific coding region. A feedback system controls the early embryo's Sxl protein level and gene expression. Low Sxl protein levels produce the male-specific protein, which suppresses gene expression (Du et al., 1998). The female isoform is produced at high Sxl protein concentrations, sustaining gene expression. The expression of specific genes is influenced by the X/A ratio, which in turn affects the production of certain gene products. These gene products play a crucial role in determining the quantity of Sxl protein available for the regulation of the Sxl gene. The genes responsible for perceiving the quantity of X chromosomes are referred to as numerator genes, as they ascertain the "numerator" component of the X/A ratio (Belote & Baker, 1982).

Conversely, the genes responsible for perceiving the number of autosomal sets are termed denominator genes, as they determine the "denominator" component of the X/A ratio. In the male population, characterized by a normal X/A ratio of 1:2, insufficient levels of Sxl protein lead to the suppression of the Sxl gene. Consequently, the absence of Sxl expression prompts the activation of the male pathway, which is considered the default route for sexual differentiation. In individuals of the female sex with a typical chromosomal composition (X/A = 1 : 1), the presence of an adequate amount of Sxl protein ensures the sustained expression of the Sxl gene. The sustained activation of the Sxl gene triggers a series of genetic processes, wherein each gene within the cascade regulates one or more downstream genes (Kan et al., 2017).

Consequently, this leads to the production of gene products specific to females and the suppression of gene products specific to males. In cases where the X/A ratio falls within the range of 1 : 2 to 1 : 1, certain genes that are unique to each sex become active, leading to the manifestation of an ambiguous sexual phenotype known as intersex. The Sxl protein is an RNA-binding protein that plays a crucial role in regulating the mRNA produced by certain sex-determining genes (Lence et al., 2016).

Remember

Certain organisms, such as bacteria, possess circular chromosomes as opposed to linear ones, a characteristic that is more prevalent among eukaryotes, which are organisms characterized by the presence of a nucleus within their cells.

SUMMARY

The chapter called "Genes and Chromosomes" delves into the fundamental principles of genetics, with a specific emphasis on the maintenance of chromosome complements, the mechanisms of mitosis and meiosis, and the pivotal significance of chromosomes in the transmission of hereditary traits. The chapter comprehensively explores the complexities inherent in these subjects, elucidating the organization, transmission, and intergenerational inheritance of genetic information.

Maintaining stable chromosome complements is a fundamental component of preserving genetic integrity in an organism. Cells employ the mechanisms of mitosis and meiosis to ensure the precise replication and equitable distribution of chromosomes during the process of cell division. Mitosis is the cellular process that facilitates cell proliferation and growth by producing genetically identical daughter cells that possess the same chromosome complement as the parent cell. The process above is of utmost importance in the growth, restoration, and sustenance of tissues in organisms composed of multiple cells. It plays a crucial role in safeguarding genetic material and upholding the constancy of chromosome counts within non-reproductive cells.

In contrast, meiosis is a distinct and specialized process of cellular division that takes place exclusively in reproductive cells, resulting in the production of gametes, namely sperm and eggs. During the process of meiosis, the homologous chromosomes align and engage in genetic recombination, resulting in the reorganization and exchange of genetic material. This process engenders genetic variability within the gametes, thereby enabling the transmission of a distinct amalgamation of genetic characteristics from each progenitor. The process of meiosis serves to achieve a halving of the chromosome number, thereby facilitating the preservation of chromosome complements throughout successive generations.

Chromosomes serve as the primary vehicles for transmitting genetic information, and their significance in the process of heredity is of utmost significance. The transmission of genetic material through gametes during sexual reproduction governs the inheritance of traits from parents to offspring. Genes, which are situated on chromosomes, function as the fundamental components of heredity, encompassing the directives for distinct traits and characteristics. The phenotypic variations observed in individuals can be attributed to the diverse assortment of genes inherited from both parents, reflecting the genetic contributions passed down from preceding generations.

The chapter provides an in-depth analysis of the genetic principles underlying heredity. It examines the patterns of inheritance elucidated by Mendelian genetics, which postulates that specific traits are controlled by individual genes possessing dominant and recessive alleles. These principles offer valuable insights into the probability of inheriting specific traits based on the genetic composition of parents and aid in predicting the occurrence of genetic disorders within familial contexts.

Moreover, the chapter underscores the importance of chromosomes in the inheritance of genetic disorders. Chromosomal aberrations, such as aneuploidy (the presence of an abnormal number of chromosomes) or chromosomal structural alterations, have the potential to give rise to genetic disorders and congenital anomalies. Comprehending these genetic aberrations is of utmost importance in the process of diagnosing and effectively managing these conditions, as well as in providing genetic counseling services to affected families.

In summary, the chapter on "Genes and Chromosomes" thoroughly examines the fundamental significance of chromosomes in the fields of genetics and heredity. This statement provides a comprehensive understanding of the mechanisms involved in mitosis and meiosis, which play crucial roles in maintaining the integrity of chromosome sets and promoting genetic variability. By comprehending the organization of chromosomes and the fundamental principles of heredity, researchers acquire invaluable knowledge regarding the intricacies of genetic inheritance. This knowledge contributes to a more profound comprehension of the diversity and interconnectedness of life on our planet. The acquisition of this knowledge is imperative for the progression of genetic research, medical diagnostics, and our understanding of the complex structure of living organisms.

REFERENCES

1. Androphy, E. J., Dvoretzky, I., & Lowy, D. R. (1985). X-linked inheritance of epidermodysplasia verruciformis: genetic and virologic studies of a kindred. *Archives of Dermatology*, *121*(7), 864-868.
2. Baker, B. S., & Ridge, K. A. (1980). Sex and the single cell. I. On the action of major loci affecting sex determination in *Drosophila melanogaster*. *Genetics*, *94*(2), 383-423.
3. Belote, J. M., & Baker, B. S. (1982). Sex determination in *Drosophila melanogaster*: analysis of transformer-2, a sex-transforming locus. *Proceedings of the National Academy of Sciences*, *79*(5), 1568-1572.
4. Berg, P. (1981). Dissections and reconstructions of genes and chromosomes. *Science*, *213*(4505), 296-303.
5. Bolcun-Filas, E., & Handel, M. A. (2018). Meiosis: the chromosomal foundation of reproduction. *Biology of Reproduction*, *99*(1), 112-126.
6. Börner, G. V., Kleckner, N., & Hunter, N. (2004). Crossover/noncrossover differentiation, synaptonemal complex formation, and regulatory surveillance at the leptotene/zygotene transition of meiosis. *Cell*, *117*(1), 29-45.
7. Bridges, C. B. (1916). Non-disjunction as proof of the chromosome theory of heredity (concluded). *Genetics*, *1*(2), 107.
8. Brinkley, B. R. (2001). Managing the centrosome numbers game: from chaos to stability in cancer cell division. *Trends in cell biology*, *11*(1), 18-21.
9. Brush, S. G. (2002). How theories became knowledge: Morgan's chromosome theory of heredity in America and Britain. *Journal of the History of Biology*, *35*(3), 471-535.
10. Caron, H., Schaik, B. V., Mee, M. V. D., Baas, F., Riggins, G., Sluis, P. V., ... & Versteeg, R. (2001). The human transcriptome map: clustering of highly expressed genes in chromosomal domains. *Science*, *291*(5507), 1289-1292.
11. Carpenter, A. T. (1975). Electron microscopy of meiosis in *Drosophila melanogaster* females: II: The recombination nodule—a recombination-associated structure at pachytene?. *Proceedings of the National Academy of Sciences*, *72*(8), 3186-3189.
12. Charlesworth, D. (2002). Plant sex determination and sex chromosomes. *Heredity*, *88*(2), 94-101.
13. Clemons, A. M., Brockway, H. M., Yin, Y., Kasinathan, B., Butterfield, Y.

S., Jones, S. J., ... & Smolikove, S. (2013). akirin is required for diakinesis bivalent structure and synaptonemal complex disassembly at meiotic prophase I. *Molecular biology of the cell*, *24*(7), 1053-1067.

14. Clough, E., & Oliver, B. (2012). Genomics of sex determination in Drosophila. *Briefings in functional genomics*, *11*(5), 387-394.
15. Depienne, C., & Leguern, E. (2012). PCDH19-related infantile epileptic encephalopathy: an unusual X-linked inheritance disorder. *Human mutation*, *33*(4), 627-634.
16. Dobyns, W. B., Filauro, A., Tomson, B. N., Chan, A. S., Ho, A. W., Ting, N. T., ... & Ober, C. (2004). Inheritance of most X-linked traits is not dominant or recessive, just X-linked. *American Journal of Medical Genetics Part A*, *129*(2), 136-143.
17. Du, C., McGuffin, M. E., Dauwalder, B., Rabinow, L., & Mattox, W. (1998). Protein phosphorylation plays an essential role in the regulation of alternative splicing and sex determination in Drosophila. *Molecular cell*, *2*(6), 741-750.
18. Dunphy, W. G. (1994). The decision to enter mitosis. *Trends in cell biology*, *4*(6), 202-207.
19. Eksioğlu, Y. Z., Scheffer, I. E., Cardenas, P., Knoll, J., DiMario, F., Ramsby, G., ... & Walsh, C. A. (1996). Periventricular heterotopia: an X-linked dominant epilepsy locus causing aberrant cerebral cortical development. *Neuron*, *16*(1), 77-87.
20. Gambineri, E., Torgerson, T. R., & Ochs, H. D. (2003). Immune dysregulation, polyendocrinopathy, enteropathy, and X-linked inheritance (IPEX), a syndrome of systemic autoimmunity caused by mutations of FOXP3, a critical regulator of T-cell homeostasis. *Current Opinion in Rheumatology*, *15*(4), 430-435.
21. Garn, S. M., Lewis, A. B., & Kerewsky, R. S. (1965). X-linked inheritance of tooth size. *Journal of Dental Research*, *44*(2), 439-441.
22. Gibcus, J. H., Samejima, K., Goloborodko, A., Samejima, I., Naumova, N., Nuebler, J., ... & Dekker, J. (2018). A pathway for mitotic chromosome formation. *Science*, *359*(6376), eaao6135.
23. Gonçalves, T. F., Gonçalves, A. P., Rodrigues, N. F., dos Santos, J. M., Pimentel, M. M. G., & Santos-Rebouças, C. B. (2014). KDM5C mutational screening among males with intellectual disability suggestive of X-Linked inheritance and review of the literature. *EuropeanJournal of Medical Genetics* , *57*(4), 138-144.
24. Hasstedt, S. J., & Atkin, C. L. (1983). X-linked inheritance of Alport syndrome: family P revisited. *American Journal of Human Genetics*, *35*(6), 1241.

25. Hediger, M., Burghardt, G., Siegenthaler, C., Buser, N., Hilfiker-Kleiner, D., Dübendorfer, A., & Bopp, D. (2004). Sex determination in *Drosophila melanogaster* and Musca domestica converges at the level of the terminal regulator doublesex. *Development genes and evolution*, *214*, 29-42.
26. Kaczanowski, A., Gaertig, J., & Kubiak, J. (1985). Effect of the antitubulin drug nocodazole on meiosis and post-meiotic development in *Tetrahymena thermophila*: Induction of achiasmatic meiosis. *Experimental cell research*, *158*(1), 244-256.
27. Kan, L., Grozhik, A. V., Vedanayagam, J., Patil, D. P., Pang, N., Lim, K. S., ... & Lai, E. C. (2017). The m6A pathway facilitates sex determination in Drosophila. *Nature Communications*, *8*(1), 15737.
28. Khil, P. P., Smirnova, N. A., Romanienko, P. J., & Camerini-Otero, R. D. (2004). The mouse X chromosome is enriched for sex-biased genes not subject to selection by meiotic sex chromosome inactivation. *Nature Genetics*, *36*(6), 642-646.
29. King, R. W., Jackson, P. K., & Kirschner, M. W. (1994). Mitosis in transition. *Cell*, *79*(4), 563-571.
30. Kondo, M., Nanda, I., Schmid, M., & Schartl, M. (2009). Sex determination and sex chromosome evolution: insights from medaka. *Sexual Development*, *3*(2-3), 88-98.
31. Kraft, C., Herzog, F., Gieffers, C., Mechtler, K., Hagting, A., Pines, J., & Peters, J. M. (2003). Mitotic regulation of the human anaphase-promoting complex by phosphorylation. *The EMBO Journal*, *22*(24), 6598-6609.
32. Lamb, N. E., Freeman, S. B., Savage-Austin, A., Pettay, D., Taft, L., Hersey, J., ... & Sherman, S. L. (1996). Susceptible chiasmate configurations of chromosome 21 predispose to non–disjunction in both maternal meiosis I and meiosis II. *Nature Genetics*, *14*(4), 400-405.
33. Lee, B. H., & Amon, A. (2003). Role of Polo-like kinase CDC5 in programming meiosis I chromosome segregation. *Science*, *300*(5618), 482-486.
34. Lence, T., Akhtar, J., Bayer, M., Schmid, K., Spindler, L., Ho, C. H., ... & Roignant, J. Y. (2016). m6A modulates neuronal functions and sex determination in Drosophila. *Nature*, *540*(7632), 242-247.
35. Lichter, P., Boyle, A. L., Cremer, T., & Ward, D. C. (1991). Analysis of genes and chromosomes by nonisotopic in situ hybridization. *Genetic Analysis: Biomolecular Engineering*, *8*(1), 24-35.
36. Lyons, E., & Freeling, M. (2008). How to usefully compare homologous plant genes and chromosomes as DNA sequences. *The Plant Journal*, *53*(4), 661-673.

37. Mahadevaiah, S. K., Bourc'his, D., de Rooij, D. G., Bestor, T. H., Turner, J. M., & Burgoyne, P. S. (2008). Extensive meiotic asynapsis in mice antagonises meiotic silencing of unsynapsed chromatin and consequently disrupts meiotic sex chromosome inactivation. *The Journal of Cell Biology*, *182*(2), 263-276.
38. Marchal, J. A., Acosta, M. J., Bullejos, M., Díaz De La Guardia, R., & Sánchez, A. (2003). Sex chromosomes, sex determination, and sex-linked sequences in Microtidae. *Cytogenetic and genome research*, *101*(3-4), 266-273.
39. Martin, H. C., Gardner, E. J., Samocha, K. E., Kaplanis, J., Akawi, N., Sifrim, A., ... & Hurles, M. E. (2021). The contribution of X-linked coding variation to severe developmental disorders. *Nature Communications*, *12*(1), 627.
40. Matsui, Y., Nakayama, Y., Okamoto, M., Fukumoto, Y., & Yamaguchi, N. (2012). Enrichment of cell populations in metaphase, anaphase, and telophase by synchronization using nocodazole and blebbistatin: a novel method suitable for examining dynamic changes in proteins during mitotic progression. *European Journal of Cell Biology*, *91*(5), 413-419.
41. Mazia, D. (1961). Mitosis and the physiology of cell division. In *The cell*. Academic Press. (Vol. 2, pp. 77-412).
42. McIntosh, J. R. (2016). Mitosis. *Cold Spring Harbor perspectives in biology*, *8*(9), a023218.
43. Meetei, A. R., Levitus, M., Xue, Y., Medhurst, A. L., Zwaan, M., Ling, C., ... & Joenje, H. (2004). X-linked inheritance of Fanconi anemia complementation group B. *Nature Genetics*, *36*(11), 1219-1224.
44. Meisel, R. P. (2020). Evolution of sex determination and sex chromosomes: a novel alternative paradigm. *BioEssays*, *42*(9), 1900212.
45. Mizumoto, Y., Kyo, S., Ohno, S., Hashimoto, M., Nakamura, M., Maida, Y., ... & Kiyono, T. (2006). Creation of tumorigenic human endometrial epithelial cells with intact chromosomes by introducing defined genetic elements. *Oncogene*, *25*(41), 5673-5682.
46. Morrill, G. A., Kostellow, A. B., Mahajan, S., & Gupta, R. K. (1984). Role of calcium in regulating intracellular pH following the stepwise release of the metabolic blocks at first-meiotic prophase and second-meiotic metaphase in amphibian oocytes. *Biochimica et Biophysica Acta (BBA)-Molecular Cell Research*, *804*(1), 107-117.
47. Murphy, S. P., Gumber, H. K., Mao, Y., & Bass, H. W. (2014). A dynamic meiotic SUN belt includes the zygotene-stage telomere bouquet and is disrupted in chromosome segregation mutants of maize (*Zea mays* L.). *Frontiers in Plant Science*, *5*, 314.

48. Narayan, R. K. J. (1991). Molecular organisation of the plant genome: Its relation to structure, recombination and evolution of chromosomes. *Journal of Genetics*, *70*, 43-61.
49. Naruse, Y. (2002). Mechanical vibration model for chromosomes in metaphase of mitosis and possible application to the interruption of cell division. *Biosystems*, *66*(1-2), 55-63.
50. Pennell, M. W., Mank, J. E., & Peichel, C. L. (2018). Transitions in sex determination and sex chromosomes across vertebrate species. *Molecular Ecology*, *27*(19), 3950-3963.
51. Pickett-Heaps, J. D., Tippit, D. H., & Porter, K. R. (1982). Rethinking mitosis. *Cell*, *29*(3), 729-744.
52. Rasmussen, S. W., & Holm, P. B. (1978). Human meiosis II. Chromosome pairing and recombination nodules in human spermatocytes. *Carlsberg Research Communications*, *43*, 275-327.
53. Royo, H., Polikiewicz, G., Mahadevaiah, S. K., Prosser, H., Mitchell, M., Bradley, A., ... & Turner, J. M. (2010). Evidence that meiotic sex chromosome inactivation is essential for male fertility. *Current Biology*, *20*(23), 2117-2123.
54. Safir, S. R. (1920). Genetic and cytological examination of the phenomena of primary non-disjunction in *Drosophila melanogaster*. *Genetics*, *5*(5), 459.
55. Salah, S. M., & Nasmyth, K. (2000). Destruction of the securin Pds1p occurs at the onset of anaphase during both meiotic divisions in yeast. *Chromosoma*, *109*(1-2), 27-34.
56. Salz, H., & Erickson, J. W. (2010). Sex determination in Drosophila: The view from the top. *Fly*, *4*(1), 60-70.
57. Sankaranarayanan, K. (1979). The role of non-disjunction in aneuploidy in man an overview. *Mutation Research/Fundamental and Molecular Mechanisms of Mutagenesis*, *61*(1), 1-28.
58. Sasai, K., Katayama, H., Stenoien, D. L., Fujii, S., Honda, R., Kimura, M., ... & Sen, S. (2004). Aurora-C kinase is a novel chromosomal passenger protein that can complement Aurora-B kinase function in mitotic cells. *Cell motility and the cytoskeleton*, *59*(4), 249-263.
59. Schmid, M., Nanda, I., Guttenbach, M., Steinlein, C., Hoehn, M., Schartl, M., ... & Mizuno, S. (2000). First report on chicken genes and chromosomes 2000. *Cytogenetic and Genome Research*, *90*(3-4), 169-218.
60. Solc, P., Schultz, R. M., & Motlik, J. (2010). Prophase I arrest and progression to metaphase I in mouse oocytes: comparison of resumption of meiosis and recovery from G2-arrest in somatic cells. *Molecular human reproduction*, *16*(9), 654-664.

61. Speed, R. M., & Chandley, A. C. (1990). Prophase of meiosis in human spermatocytes analysed by EM microspreading in infertile men and their controls and comparisons with human oocytes. *Human genetics*, *84*, 547-554.
62. Steinmann-Zwicky, M., Amrein, H., & Nöthiger, R. (1990). Genetic control of sex determination in Drosophila. *Advances in genetics*, *27*, 189-237.
63. Sun, X., Thrower, D., Qiu, J., Wu, P., Zheng, L., Zhou, M., ... & Shen, B. (2003). Complementary functions of the Saccharomyces cerevisiae Rad2 family nucleases in Okazaki fragment maturation, mutation avoidance, and chromosome stability. *DNA repair*, *2*(8), 925-940.
64. Takamiya, M., Watanabe, M., & Ono, K. (1996). Biosystematic studies on the genus Isoetes (Isoetaceae) in Japan. II. Meiotic behavior and reproductive mode of each cytotype. *American Journal of Botany*, *83*(10), 1309-1322.
65. Tsuchiya, D., & Lacefield, S. (2013). Cdk1 modulation ensures the coordination of cell-cycle events during the switch from meiotic prophase to mitosis. *Current Biology*, *23*(16), 1505-1513.
66. Turner, J. M., Mahadevaiah, S. K., Ellis, P. J., Mitchell, M. J., & Burgoyne, P. S. (2006). Pachytene asynapsis drives meiotic sex chromosome inactivation and leads to substantial postmeiotic repression in spermatids. *Developmental Cell*, *10*(4), 521-529.
67. Vernet, N., Mahadevaiah, S. K., Ojarikre, O. A., Longepied, G., Prosser, H. M., Bradley, A., ... & Burgoyne, P. S. (2011). The Y-encoded gene zfy2 acts to remove cells with unpaired chromosomes at the first meiotic metaphase in male mice. *Current Biology*, *21*(9), 787-793.
68. Vertès, A. A., Inui, M., & Yukawa, H. (2005). Manipulating corynebacteria, from individual genes to chromosomes. *Applied and Environmental Microbiology*, *71*(12), 7633-7642.
69. Wadsworth, P., & Salmon, E. D. (1986). Analysis of the treadmilling model during metaphase of mitosis using fluorescence redistribution after photobleaching. *The Journal of Cell Biology*, *102*(3), 1032-1038.
70. Williams-Blangero, S., VandeBerg, J. L., Subedi, J., Aivaliotis, M. J., Rai, D. R., Upadhayay, R. P., ... & Blangero, J. (2002). Genes on chromosomes 1 and 13 have significant effects on Ascaris infection. *Proceedings of the National Academy of Sciences*, *99*(8), 5533-5538.
71. Xu, L., Ajimura, M., Padmore, R., Klein, C., & Kleckner, N. (1995). NDT80, a meiosis-specific gene required for exit from pachytene in Saccharomyces cerevisiae. *Molecular and cellular biology*, *15*(12), 6572-6581.
72. Yi, K., Rubinstein, B., Unruh, J. R., Guo, F., Slaughter, B. D., & Li, R. (2013). Sequential actin-based pushing forces drive meiosis I chromosome migration and symmetry breaking in oocytes. *Journal of Cell Biology*, *200*(5), 567-576.

CHAPTER 7

GENE MUTATIONS & DNA REPAIR

Contents

7.1 INTRODUCTION

Within the expansive chronicle of life's genetic blueprint, the occurrence of gene mutations and subsequent DNA repair mechanisms constitutes a captivating segment, wherein narratives of genetic inheritances, peculiarities of nature, and the intricate interplay between constancy and diversity are unveiled. This chapter explores the complexities of gene mutations and the intricate processes involved in DNA repair, providing insight into their significant implications for various aspects of our world. These implications range from the consequences of the Chernobyl incident to the fundamental principles underlying genetic diversity and genetic illnesses (Pritchard et al., 2016).

The Genetic Legacy of Cherrnobyl serves as a poignant reminder of the profound impact of unleashed mutations. The nuclear disaster of 1986 resulted in the release of significant amounts of ionizing radiation into the surrounding environment, thereby causing the penetration of DNA molecules within living organisms (Antonarakis et al., 2018). The consequences of this incident had a significant impact beyond its immediate aftermath, as the introduction of gene mutations due to radiation exposure has had ongoing effects on the ecosystems surrounding Chernobyl, influencing various aspects such as plants, animals, and human beings. The historical significance highlights genetic mutations' profound influence on living organisms, encompassing immediate and enduring consequences. This underscores the fundamental importance of DNA repair mechanisms in preserving the stability and fidelity of genetic material (Brooks et al., 2008).

The central theme of the narrative revolves around the nature of mutations. This phenomenon possesses positive and negative aspects, as it serves as a driving force for evolution while occasionally resulting in catastrophic outcomes. Mutations are non-directed modifications in the DNA sequence arising from endogenous and exogenous influences. Certain mutations augment the variety of life forms and facilitate their ability to adapt to dynamic surroundings. Conversely, mutations can give rise to genetic disorders and ailments. The comprehension of mutation's complexities yields significant knowledge regarding the fundamental mechanisms of life, thereby providing indications of our lineage and the evolutionary trajectory (Wei et al., 2019).

This chapter explores the various types of mutations in genes that occur within the genetic code. Genetic variations encompass many alterations, from point mutations involving modifying a single base pair to more extensive structural mutations like deletions and insertions. These diverse genetic variations collectively contribute to the intricate fabric of genetic diversity. Moreover, certain mutations exhibit a silent phenotype, lacking any discernible effect, whereas others may have harmful or even advantageous consequences, providing benefits that facilitate survival and adaptation (Witkin, 1969). The root cause of mutation can arise from both exogenous and endogenous factors. Environmental factors can induce mutations, including radiation, chemical contact, and ultraviolet light. Furthermore, genetic flux is influenced by cellular processes, including DNA replication errors and the mobility of transposable elements. The comprehension of the varied sources of mutations reveals the complex interplay between surroundings and the genetic code, which influences biological evolution (Yap et al., 2014).

Mutation rates determine the frequency at which genetic alterations take place. The variation rates can exhibit variability across species and even within distinct genomic regions. Examining mutation rates yields significant insights for disciplines including evolutionary biology, epidemiology, and the comprehension of genetic disease susceptibility (Hsiehchen et al., 2020).

However, amid potential disorder, the narrative of DNA repair arises as a protector of genetic equilibrium. Cells have complex repair systems that diligently and effectively restore DNA damage. From the mechanism of base excision repair to the intricate process of homologous recombination, these complex biological processes play a crucial role in maintaining the integrity of genetic information and safeguarding organisms against the harmful consequences of mutations. The chapter explores resilient repair systems and highlights their vital function in preserving genetic integrity (Leongamornlert et al., 2019).

Genetic Disorders and Defects The intricate interplay of DNA repair mechanisms reveals the profound consequences that arise from impaired repair processes. Several genetic conditions, such as xeroderma pigmentosum, arise due to aberrant DNA repair mechanisms, resulting in an augmented susceptibility to ultraviolet radiation and an elevated likelihood of developing skin cancer. Acquiring knowledge about these connections is crucial for comprehending the diagnosis, mitigation, and potential therapeutic approaches to hereditary diseases (Marshall et al., 2019).

The chapter "Gene Mutations and DNA Repair" delves into the captivating narrative surrounding genetic diversity and preservation, intricately intertwined by mutation mechanisms and DNA repair robustness. This chapter explores the enigmatic legacy of the Chernobyl incident and delves into the intricacies of genetic alterations. It aims to navigate the genetic landscape and shed light on mutations' profound influence on the diversity and complexity of life. Through investigating the complex mechanisms involved in DNA repair, valuable perspectives are obtained regarding the equilibrium necessary for the perpetuation of life. This process safeguards the invaluable genetic material, a common bond among all living entities (Schweizer et al., 2019).

For astronauts, DNA repair mechanisms are put to the test in space due to increased exposure to cosmic radiation. Understanding how space affects DNA can help protect future space travelers.

7.2 THE GENETIC LEGACY OF CHERNOBYL

On April 26, 1986, an explosion occurred at unit 4 of the Chernobyl nuclear power plant in northern Ukraine, resulting in the most severe nuclear catastrophe ever recorded. The detonation resulted in the detachment of a 2000-ton metallic panel that had been serving as a seal for the upper part of the reactor. Subsequently, several hundred tons of graphite were ignited, leading to uncontrolled combustion that persisted for ten days. The precise quantity of radiation discharged during the explosion and subsequent fire remains uncertain; however, a conservative approximation suggests a minimum of 100 million curies, comparable to the magnitude of a nuclear strike of moderate scale. A dispersion of radioactive particles emanated in a westerly and northerly direction from the impaired reactor, resulting in the deposition of hazardous radiation levels over an extensive area spanning thousands of square kilometers. Regions located as distantly as Germany and Norway experienced the impact, with even Japan and the United States observing quantifiable rises in radiation levels (Bojcun, 1991).

Following the incident, 31 individuals, primarily firefighters who bravely combatted the conflagration, succumbed to acute radiation sickness. Over 400,000 workers were subsequently engaged in interring radioactive and chemical debris resulting from the incident and enclosing the remnants of the inoperative reactor within a structure made of steel and concrete, commonly referred to as a sarcophagus. Many workers experience adverse health conditions, including immune suppression, heightened cancer incidence, and reproductive disorders. Radiation is widely recognized as a mutagenic agent that induces DNA damage. Over 13,000 children residing near Chernobyl were subjected to the radioactive isotope iodine-131, with a significant number experiencing exposures 400 times higher than the maximum annual radiation exposure limit for individuals employed in the nuclear sector. The incidence of thyroid cancer among the pediatric population in Ukraine has escalated to ten times the pre-Chernobyl era rates. Chromosomal aberrations have been observed in the cellular composition of numerous individuals close to the Chernobyl nuclear disaster site during the incident. Furthermore, there has been a notable rise in the incidence of congenital abnormalities within the affected population (Petryna, 2004).

To investigate germ-line mutations, specifically, those heritable and passed on to subsequent generations, geneticists conducted a study wherein they obtained blood samples from 79 families residing in districts heavily contaminated due to the Chernobyl accident. The families above consisted of individuals born in 1994 who had not been exposed to radiation yet potentially harbored genetic mutations inherited from their parents. The researchers analyzed DNA sequences obtained from both parents and their offspring, enabling them to detect potential germ-line mutations.

The observed germ-line mutation rate in these clans was twice as high compared to a control group of families in Britain. In addition, a positive correlation was observed between the mutation rate and the surface radiation level. Specifically, families where the parents had resided in districts with higher contamination levels, exhibited higher mutation rates than those from districts with lower contamination levels. This chapter focuses on the phenomenon of DNA infidelity, exploring the mechanisms underlying the occurrence of errors in genetic instructions and the subsequent repair processes that take place.

The Chernobyl disaster serves as a poignant example of the relationship between mutations, specifically those caused by radiation, and the adverse consequences that can arise from DNA damage. The initial focus of this study involves a concise analysis of various categories of mutations, encompassing their phenotypic consequences, potential mechanisms of suppression, and rates of occurrence. The subsequent section delves into the mechanisms by which mutations arise spontaneously during replication and subsequent processes and how chemicals and radiation can induce mutations (Marples, 1997).

In conclusion, an examination is conducted on the process of DNA repair and the various diseases that manifest as a result of impaired DNA repair mechanisms. It is important to remember throughout this chapter that mutations, as per their definition, are heritable alterations in the sequence of DNA, necessitating their transmission to subsequent generations. A mutation is a biological phenomenon that necessitates two conditions: firstly, a modification in the structure of a DNA molecule, and secondly, the replication of this altered genetic information (Becker, 2006).

7.3 THE NATURE OF MUTATION

DNA is an exceptionally stable molecule that undergoes replication with remarkable precision; however, alterations in DNA structure and replication errors do manifest. A mutation is characterized as a heritable alteration in genetic information, which can manifest in the form of cells generated through cellular division or distinct organisms generated through reproduction (Srinivas & Patnaik, 1994).

7.3.1 The Importance of Mutations

Mutations play a dual role in the perpetuation of life and the instigator of significant distress. A mutation is the fundamental origin of genetic diversity, acting as the primary catalyst for evolutionary processes. The presence of mutations and the subsequent variation they introduce are crucial for the adaptive capacity of organisms in response to dynamic environmental conditions, thereby mitigating the risk of extinction. Conversely, most mutations exhibit deleterious consequences, attributing mutation as the underlying cause of numerous afflictions and disorders in the human population (Dzikiewicz-Krawczyk, 2008).

A significant portion of genetic research studies the inheritance of variants generated through mutation. The conduction of genetic crosses becomes insignificant when all individuals possess identical homozygosity for the same alleles. Mutations play a significant role in genetic analysis, as they are valuable tools for studying genetics. A comprehensive collection of mutants is typically the starting point for addressing nearly any genetic inquiry. A significant portion of Gregor Mendel's achievements in elucidating the principles of inheritance can be attributed to his deliberate selection of specific variations of the garden pea. Likewise, Thomas Hunt Morgan and his students made notable discoveries regarding fundamental genetic principles by examining mutated fruit flies (Figure 7.1) (Gagnon & Delpire, 2013).

Mutations are additionally advantageous in the investigation of fundamental biological processes. Identifying and analyzing mutations impacting various constituents of a biological system frequently facilitates the comprehension of said system. The approach known as genetic dissection can be likened to understanding the functioning of an automobile by systematically dismantling various components and observing the effects. For instance, damaging the radiator leads to engine overheating, thereby indicating the role of the radiator in cooling the engine (Jia & Zhao, 2017). The perturbation of physiological function in individual organisms harboring specific genetic mutations can also serve as a valuable means to understand biological processes. Geneticists have initiated elucidating the intricate molecular mechanisms underlying development by examining mutations that impede different embryonic stages in Drosophila. Despite the seemingly simplistic nature of this approach in dissecting and discerning the functions of various components within a system, it possesses significant efficacy. It has been widely employed in biochemistry, developmental biology, and physiological and behavioral science. However, it is worth noting that this method is not advisable for comprehending automotive systems' mechanics (Pardiñas et al., 2018).

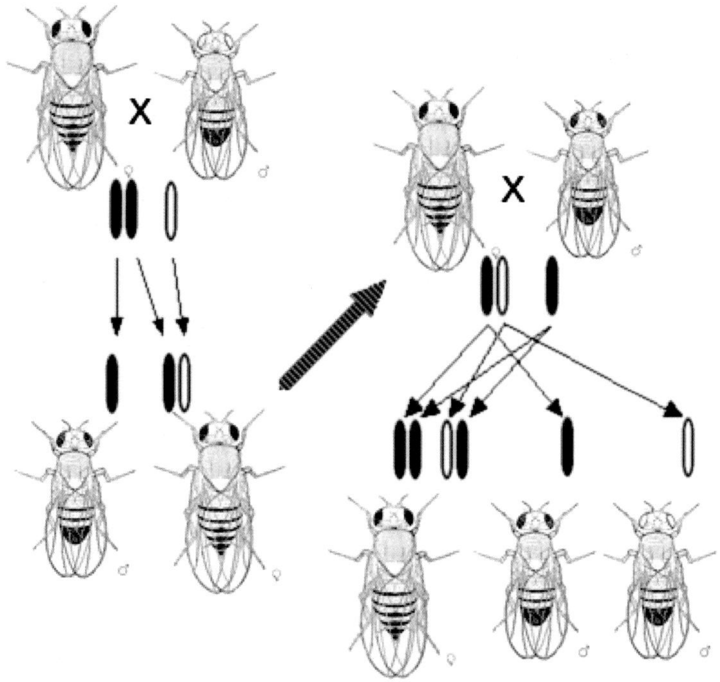

Figure 7.1. Morgan and his students discovered many principles of heredity by studying mutation in Drosophila melanogaster. Shown here are several common mutations (Source: By Electric goat - Own work, Public Domain, https://commons.wikimedia.org/w/index.php?curid=86924, Creative Commons License)

7.3.2 Categories of Mutations

Within multicellular organisms, a clear distinction can be made between two overarching classifications of mutations: somatic and germline. Somatic mutations occur in somatic tissues, which are non-reproductive and do not give rise to gametes (Figure 7.2). The transmission of these mutations to subsequent cells occurs via mitosis, resulting in the proliferation of a genetically homogeneous cell population, commonly referred to as a clone. The extent of a cellular clone harboring a somatic mutation within an individual organism is directly proportional to the timing of occurrence of the mutation during development. Due to the substantial number of cells found within a typical eukaryotic organism, somatic mutations are expected to be abundant. As an illustration, the human body is estimated to contain approximately 10^14 cells (Rockova et al., 2011). Suppose a mutation occurs at a frequency of one occurrence per million cell divisions, which is a commonly observed mutation rate. In that case, it can be inferred that thousands of millions of somatic mutations will arise in an individual. The impact of these mutations is contingent upon various factors, such as the specific cellular context in which they manifest and the particular stage of development at which they emerge. Numerous somatic mutations exhibit no discernible impact on the organism's phenotype, as the mutant cell's functionality, including the cell itself, is supplanted by unaffected cells. Nevertheless, the proliferation and dissemination of cells exhibiting a somatic mutation that triggers cellular division can occur. This particular mutation can confer a selective advantage to the cells, thereby serving as the fundamental mechanism underlying the development of all forms of cancer (Mrózek et al., 2007).

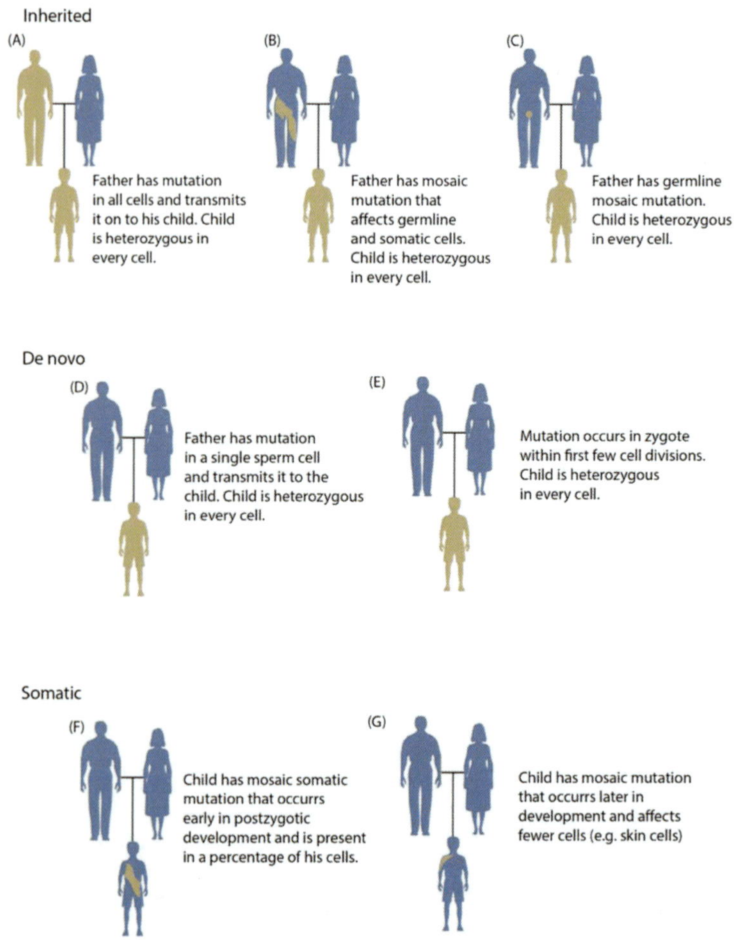

Figure 7.2. There are two basic classes of mutations: somatic mutations and germline mutations. (Source: By Donald Freed, Eric L. Stevens, and Jonathan Pevsner - Freed, Donald; Stevens, Eric; Pevsner, Jonathan (2014-12-11). "Somatic Mosaicism in the Human Genome". Genes. 5 (4): 1064–1094. doi:10.3390/genes5041064. ISSN 2073-4425. PMC 4276927. PMID 25513881., CC BY-SA 4.0, https://commons.wikimedia.org/w/index.php?curid=89635021, Creative Commons License)

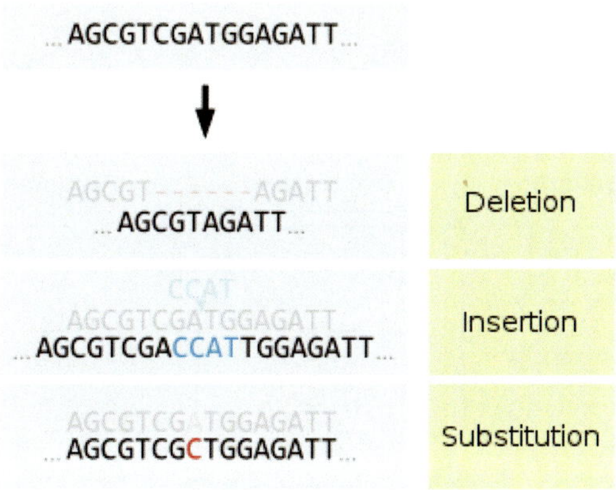

Figure 7.3. Three basic types of gene mutations are base substitutions, insertions, and deletions. (Source: https://commons.wikimedia.org/wiki/File:Deletion_Insertion_Substitution-en.svg, Creative Commons License)

> **Remember**
>
> In some organisms with short lifespans and high mutation rates, such as bacteria, evolution can occur rapidly. This phenomenon has significant implications for antibiotic resistance.

7.4 TYPES OF GENE MUTATIONS

There exist several methodologies for categorizing gene mutations. Certain classification schemes are derived from the characteristics of the phenotypic impact, specifically about whether the mutation modifies the order of amino acids of the protein and, if so, how it does so. Various schemes are predicated upon the causative agent responsible for the mutation, while alternative approaches center on the molecular characteristics of the aberration. The selection of an optimal scheme is contingent upon the underlying purpose of investigating the mutation (Cooper et al., 2011).

7.4.1 Base Substitutions

The most basic type of genetic mutation is a base substitution, where a single nucleotide in the DNA molecule is modified. The complementary nature of DNA strands means that any change in one nucleotide's base will cause a corresponding change in the base of its complementary nucleotide on the opposite strand during replication. As a result, a base substitution generally leads to the replacement of base pairs. The presence of two distinct types of base substitutions can be observed (Maki, 2002).

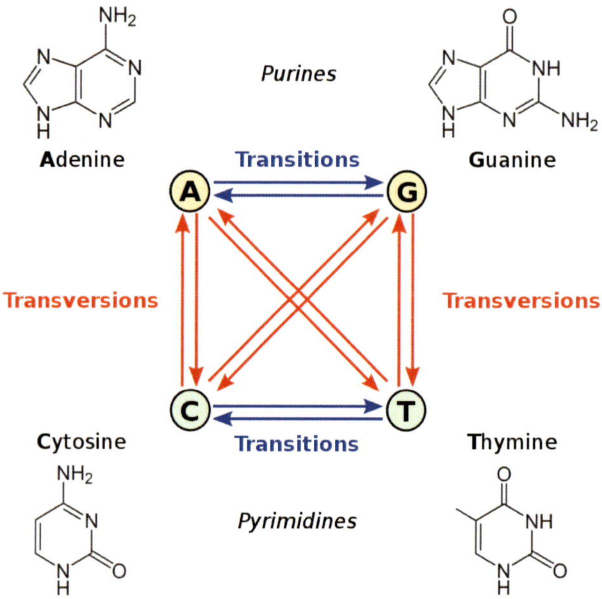

Figure 7.4. A transition is the substitution of a purine for a purine or a pyrimidine for a pyrimidine; a transversion is the substitution of a pyrimidine for a purine or a purine for a pyrimidine (Source: By Manudouz.Note: this work is an adapted version of Transitions-transversions-v4.svg created by Petulda.It has been transformed to show all possible transitions and transversions.It is distributed under the same license (Creative Commons Attribution-Share Alike 4.0 International license). - Own work, CC BY-SA 4.0, https://commons.wikimedia.org/w/index.php?curid=72412632, Creative Commons License)

7.4.2 Insertions & Deletions

The second important group of gene mutations consists of insertions and deletions. These mutations involve adding or removing one or more pairs of nucleotides. According to recent molecular analysis, it has been found that deletions and insertions occur more frequently than previously believed, challenging the notion that base substitutions are the most common form of mutation. Frameshift mutations are a type of genetic mutation that can happen when insertions or deletions in protein-coding sequences exist. These mutations cause alterations to the reading framework of the gene. The introduction or removal of a nucleotide usually causes a shift in the reading frame, resulting in changes to the sequence of amino acids encoded by codons following the mutation. Frameshift mutations have a significant effect on the phenotype because they have the potential to alter multiple amino acids. It is important to note that not all insertions and deletions lead to frameshift mutations. The reason for this is that codons, consisting of three nucleotides, can accommodate insertions and deletions that involve a multiple of three nucleotides without causing any disruption to the reading frame. Although the reading frame remains unchanged, adding or deleting one or more amino acids can still affect the phenotype. The genetic alterations mentioned earlier consist of in-frame deletions and insertions, respectively (Kim et al., 2013).

7.4.3 Expanding Trinucleotide Repeats

In the year 1991, a previously unidentified form of mutation was uncovered. The mutation above manifests within the FMR-1 gene, resulting in the development of fragile-X syndrome, widely recognized as the predominant hereditary factor contributing to cognitive impairment. The disorder is referred to as such due to the observation that in individuals with the condition, the extremity of the X chromosome is connected to the rest of the chromosome through a thin filament, as depicted in Figure 7.5. The FMR-1 gene harbors a series of contiguous repetitions of the CGG trinucleotide (Dowsing et al., 1999).

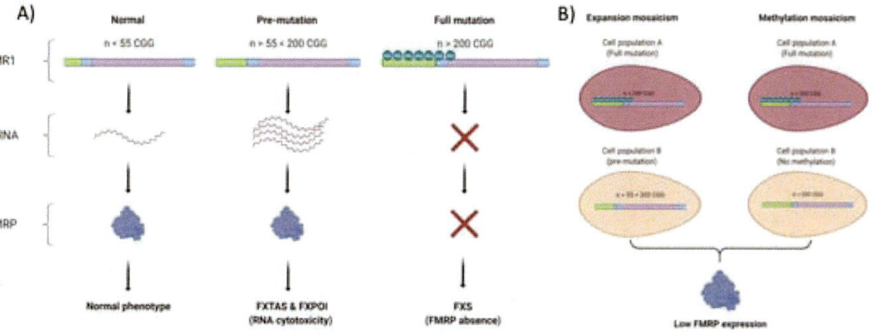

Figure 7.5. The fragile-X chromosome is associated with a characteristic constriction (fragile site) on the long arm (Source: Dionne O, Corbin F. An "Omic" Overview of Fragile X Syndrome. *Biology*. 2021; 10(5):433. https://doi.org/10.3390/biology10050433, Creative Commons License)

Trinucleotide repeat expansions have been identified in other pathological conditions affecting the human population. There is often a correlation between the quantity of trinucleotide repeat copies and the severity or age at which the disease manifests. The quantity of duplications of the repetitive sequence is also associated with the instability of trinucleotide repeats. Specifically, a higher number of repeats enhances the likelihood of further expansion. The instability in the genetic material gives rise to a phenomenon commonly referred to as anticipation, wherein diseases resulting from trinucleotide-repeat expansions progressively worsen with each successive generation. In certain cases, it is observed that there is a reduction in the number of trinucleotide repeats within a familial context, albeit less frequently. The precise mechanisms by which an augmentation in the number of trinucleotides gives rise to disease symptoms remain to be fully elucidated. In certain pathological conditions, such as Huntington's disease, the trinucleotide CAG expands within the coding region of a specific gene. This expansion leads to the production of a toxic protein that contains an excessive number of glutamine residues, which are the amino acids encoded by the CAG sequence (Paulson & Fischbeck, 1996).

In the case of additional diseases such as fragile-X syndrome and myotonic dystrophy, the repetitive sequence is located beyond the gene's coding region, necessitating an alternative mechanism of action. A particular medical condition, specifically a form of epilepsy infrequently observed, has recently been linked to an elongated repetition of a 12-base pair sequence (Snell et al., 1993). The inclusion of this particular repeat as an expanding trinucleotide is attributed to its repeat length being a multiple of three, despite not being a trinucleotide itself. The precise mechanism underlying the expansion of trinucleotide repeats remains uncertain.

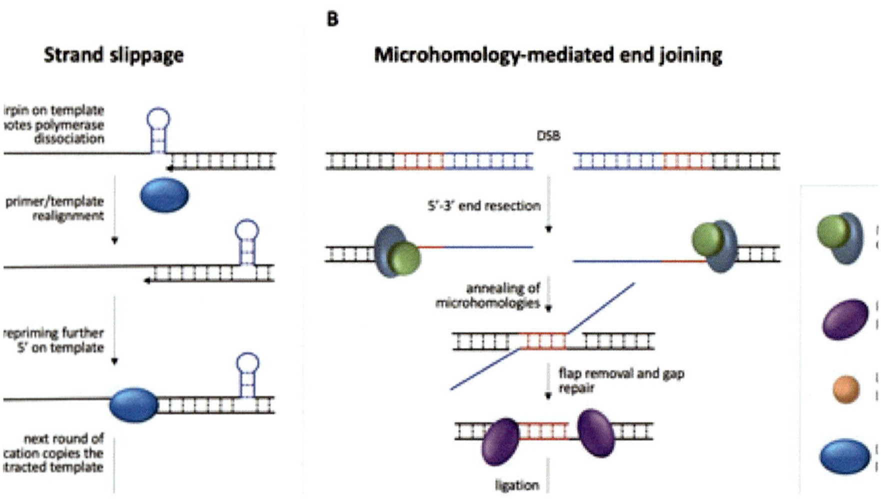

Figure 7.6. The number of copies of a trinucleotide may increase by strand slippage in replication. (Source: Hayward BE, Usdin K. Mechanisms of Genome Instability in the Fragile X-Related Disorders. *Genes*. 2021; 12(10):1633. https://doi.org/10.3390/genes12101633, Creative Commons License)

Two potential causes of expansion are strand slippage during DNA replication and crossing-over events between misaligned repeats. Certain trinucleotide repeats have been observed to possess single-stranded regions that exhibit a propensity to adopt hairpin conformations (as depicted in Figure 7.6) and other distinctive DNA structures. These particular structures have the potential to induce strand slippage during the process of replication, thereby impeding the identification and rectification of errors. Further elaboration on this topic can be found in the subsequent chapter, specifically in the section dedicated to mismatch repair (Bingham et al., 1995).

7.4.4 Phenotypic effects of mutations

Mutations exhibit a diverse range of phenotypic effects. The consideration of a mutation's effect necessitates the comparison of its phenotype to a reference phenotype, typically the wild-type phenotype. The wild-type phenotype represents the prevailing phenotype observed in the natural population of the organism. In nature, most *Drosophila melanogaster* exhibit red eye pigmentation, thus establishing red eyes as the wild-type eye color. Consequently, any alternative genetically selected eye color observed in fruit flies is classified as a mutant phenotype. A genetic alteration that modifies an organism's original observable characteristics is referred to as a forward mutation. In contrast, a reverse mutation, also known as a reversion, restores a mutant phenotype to the original wild-type state (Gillio et al., 1997).

Geneticists employ specialized terminology to delineate the phenotypic consequences of mutations. A missense mutation is a term used to describe a base substitution in the mRNA that leads to a codon modification, ultimately resulting in a different amino acid being incorporated into the protein. A non-sense mutation is characterized by altering a sense codon, responsible for specifying an amino acid, into a nonsense codon that leads to the termination of translation. In the event of an early occurrence of a nonsense mutation within the mRNA sequence, the resulting protein will probably undergo significant truncation, rendering it typically devoid of functionality. A silent mutation is a genetic alteration that affects a codon. A neutral mutation is a missense mutation that alters a protein's amino acid sequence while maintaining its functional integrity. Neutral mutations are observed when there is a substitution of one amino acid with another that possesses chemical similarities or when the specific amino acid in question has minimal impact on the overall functionality of the protein (Uchimura et al., 2015).

Loss-of-function mutations result in the full or partial lack of normal physiological or biochemical function. A mutation resulting in loss of function significantly modifies the protein's structure, impairing functionality (Steingrímsson et al., 1994). Alternatively, the mutation may manifest in regulatory regions, impacting the transcription, translation, or splicing processes associated with the protein. Loss-of-function mutations often exhibit a recessive pattern, requiring diploid individuals to possess two copies of the mutation (homozygous) to manifest the consequences of the functional protein's loss. On the contrary, a gain-of-function mutation results in the emergence of a completely novel characteristic or induces the manifestation of a trait in unsuitable tissues or during untimely stages of development. The expression of these mutations is commonly dominant. Another category of mutations is conditional mutations, which are only manifested under specific circumstances, alongside lethal mutations that result in untimely mortality (Dinges et al., 2001).

Nucleotide substitutions at RNA level

VariO:0308 **Missense variation** - Substitution in the mRNA coding region that leads to amino acid substitution in the coded protein.

VariO:0310 **Nonsense variation** - Substitution in the mRNA coding region that creates a terminator codon.

VariO:0318 **Synonymous (silent) variation** - Substitution in mRNA coding region that does not change the encoded amino acid.

VariO:0514 **Unsense variation** - Substitution in the mRNA coding region that affects gene expression and protein production without introducing a stop codon in the variation site.

VariO:0313 **Transition** - Change of a pyrimidine nucleotide, C or T (or U), into another pyrimidine nucleotide, or change of a purine nucleotide, A or G, into another purine nucleotide.

VariO:0315 **Purine transition** - A substitution of a purine, A or G, for another purine.

VariO:0314 **Pyrimidine transition** - A substitution of a pyrimidine, C or T (or U), for another pyrimidine.

VariO:0316 **Transversion** - Change of a pyrimidine nucleotide, C or T (or U), into a purine nucleotide, A or G, or vice versa.

Figure 7.7. Base substitutions can cause (a) missense, (b) nonsense, and (c) silent mutations. (Source: Vihinen M. When a Synonymous Variant Is Nonsynonymous. *Genes*. 2022; 13(8):1485. https://doi.org/10.3390/genes13081485, Creative Commons License)

7.4.5 Suppressor Mutations

A suppressor mutation refers to a specific type of genetic change that can mask or lessen the effects of another mutation. The abovementioned mutation can be differentiated from a reverse mutation, which occurs when the mutated site returns to the original wild-type sequence. This distinction is illustrated in Figure 7.8. A suppressor mutation occurs at a distinct site from the original mutation. As a result, an organism with a suppressor mutation is classified as a double mutant because it has both the primary and suppressor mutations. Nevertheless, even with these mutations, the organism exhibits the phenotype typical of an unmodified wild type (Panuzzo et al., 2020).

Suppressor mutations are classified by geneticists into two distinct categories: intragenic and intergenic. An intragenic suppressor refers to a specific type of genetic element found within the same gene where the mutation it suppresses is located. The suppressive effect of it is demonstrated through multiple mechanisms. The suppressor can alter a second nucleotide in the same codon originally impacted by the primary mutation. The modification leads to a codon that still codes for the same amino acid as the original codon, as shown in Figure 7.9. Intragenic suppressors can potentially reduce the impact of frameshift mutations by exerting suppressive effects. Suppose a one-base deletion occurs as the primary mutation. In that case, the initial reading frame can be restored by reintroducing a single base at a different location within the gene, as shown in Figure 7.9. There is a need to analyze the nucleotide sequence in DNA and the corresponding amino acids it encodes (Verlinsky et al., 2001):

DNA	AAA	TCA	CTT	GGC	GTA	CAA
Amino acids	Phe	Ser	Glu	Pro	His	Val

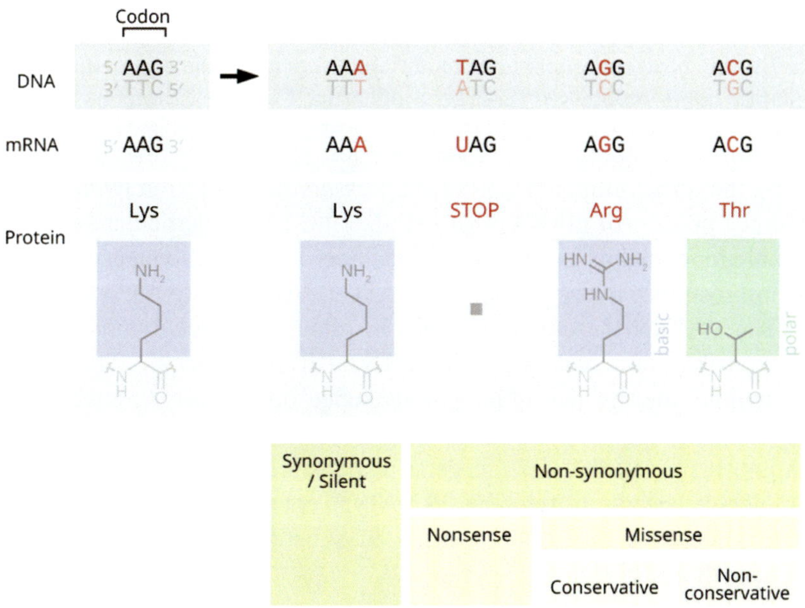

Figure 7.8. Point mutations (Source: By Jonsta247 - Own work, CC BY-SA 4.0, https://commons.wikimedia.org/w/index.php?curid=12481467, Creative Commons License)

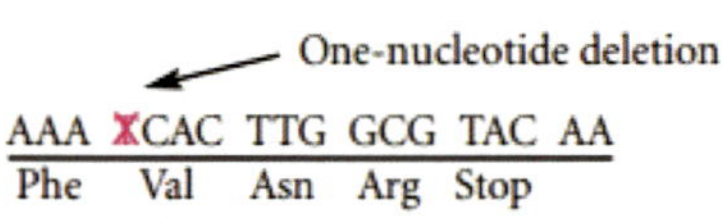

The restoration of the reading frame occurs when a suppressor mutation involving the addition of a single nucleotide to the third codon takes place.

However, it is important to note that this alteration leads to a discrepancy in two amino acids compared to those originally specified by the sequence (Ladeiro et al., 2008).

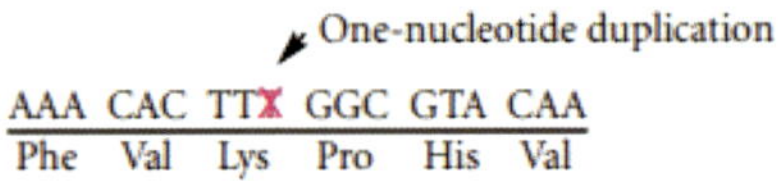

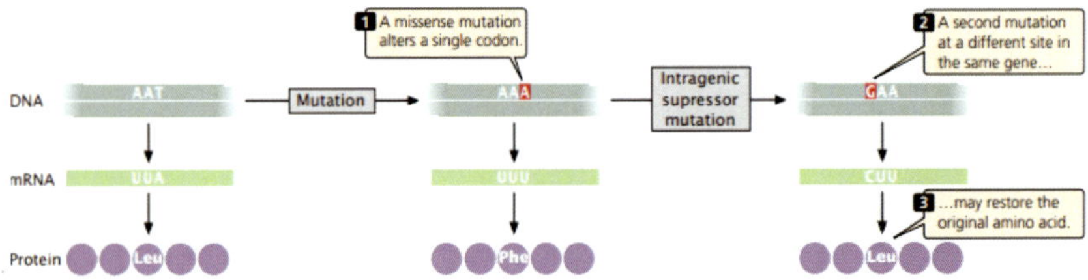

Figure 7.9. An intragenic suppressor mutation occurs in the same gene that contains the mutation being suppressed (Source: Siddhartha, Creative Commons License)

In contrast, intergenic suppressors manifest in a gene distinct from the one harboring the initial mutation. These suppressors may function by altering the translational process of mRNA. (Emr et al., 1981). This particular sequence is responsible for encoding the amino acid leucine. The given sequence undergoes a mutation resulting in the formation of ATC (UAG in mRNA), a termination codon. The ATC nonsense mutation may be alleviated through a mutation in a gene responsible for encoding a transfer RNA (tRNA) molecule. This mutation would involve altering the anticodon sequence on the tRNA, enabling it to form a complementary base pairing with the UAG termination codon. As an illustration, it is plausible for the gene responsible for encoding the transfer RNA molecule specific to tyrosine (tRNATyr), characterized by the anticodon AUA, to undergo a mutation resulting in substituting the anticodon AUC (Brachmann et al., 1998).

Consequently, this altered tRNA molecule could pair with the UAG stop codon. Instead of translation coming to a halt at the UAG codon, the amino acid tyrosine would be incorporated into the protein sequence, producing a complete protein. However, it should be noted that tyrosine would replace leucine in this context. The impact of this alteration would be contingent upon the amino acid's function within the protein's overall architecture. However, it is expected that the consequences would be less harmful than the premature termination of translation resulting from the nonsense mutation. Due to the presence of multiple copies of tRNA genes in the cells of numerous organisms, it is likely that unmutated copies of tRNATyr would persist, thereby enabling their continued recognition of tyrosine codons (Howell et al., 1991).

Intergenic suppressors may also exert their effects via interactions between genes. The interaction between polypeptide chains synthesized from two distinct genes can form a biologically active protein. A genetic mutation can potentially modify the polypeptide encoded by a specific gene, disrupting its interaction and subsequent loss of functionality, thereby preventing the production of a functional protein. A suppressor mutation occurring in the second gene has the potential to induce a compensatory alteration in its polypeptide, thereby reinstating the initial interaction (Huang et al., 1982).

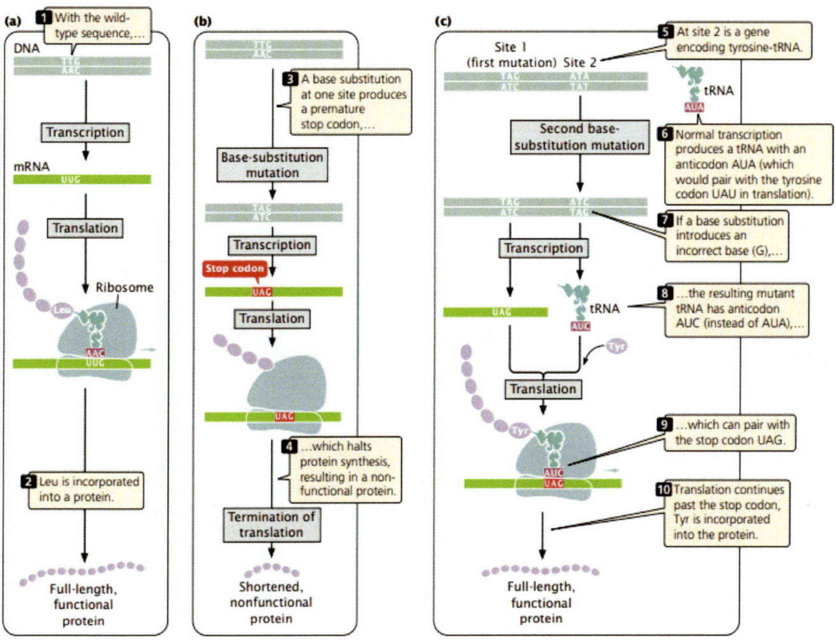

Figure 7.10. An intergenic suppressor mutation occurs in a different gene from the one bearing the original mutation (Source: Robert, Creative Commons License)

7.5 MUTATION RATES

The phenomenon in which a gene undergoes a transformation from its initial wild-type state to a mutant state is referred to as the mutation rate. The measurement of the rate is typically conducted by quantifying the frequency of mutations occurring within a defined biological entity, such as the number of mutations per cell gamete or per round of replication. The observed mutation rate for achondroplasia, a hereditary form of dwarfism, is approximately four mutations per 100,000 gametes, often denoted as 4 x 10 -5. Mutation frequency is a measure of the occurrence rate of a particular mutation type within a population of individual organisms. The available data indicate that achondroplasia exhibits a relatively low occurrence rate in the United States, estimated to be 2 x 10-4 in terms of prevalence. This data suggests that the prevalence of the genetic mutation linked to this condition is estimated to be around 1 in 20,000 individuals in the United States. There are three factors that exert an influence on mutation rates (Sniegowski et al., 2000).

The frequency of occurrences has an impact on the degree of reliance on primary alterations in DNA. The primary alterations in DNA can arise from either inherent molecular modifications or from exposure to environmental chemical or physical agents. The mutation rate may be subject to modulation based on the likelihood of repair mechanisms being activated in response to genetic modifications. A multitude of cellular mechanisms are possessed by the majority of cells, facilitating the repair of DNA that has undergone modifications. This mechanism ensures that the majority of alterations to the DNA sequence are rectified prior to the process of replication. In the event that the repair systems demonstrate efficacy, it is expected that the rates of mutation will be relatively low (Cutcutache et al., 2015).

Conversely, in the event of malfunctioning repair systems, it is anticipated that the rates of mutation will escalate. Certain mutations possess the capacity to enhance the mutation rate of additional genes. Mutations frequently arise within genes responsible for encoding the replication machinery or DNA repair enzymes, resulting in a range of observable outcomes. The calculation of mutation rates is influenced by the probability of detecting and documenting mutations. The utilization of DNA sequencing facilitates the identification and characterization of diverse forms of genetic alterations. The implementation of the sequencing process in real-world scenarios is frequently linked to substantial expenses. Consequently, an alternative methodology involves the identification of mutations through the observation of their phenotypic effects. Certain mutations may exhibit a seemingly elevated frequency due to their heightened detectability (Sniegowski et al., 1997).

The phenomenon of mutation rate variability can be observed in various organisms, as well as within organisms across different genes. Nevertheless, it is important to acknowledge that there are several overarching conclusions that can be inferred with respect to mutation rates. It is noteworthy to mention that the observed rates of spontaneous mutation in various studied organisms have been relatively low. The mutation rates observed in viral and bacterial genes generally fall within the range of 1 x 10^-8 to 1 x 10^-10 mutations per 10 billion cells. The mutation rates observed in the majority of eukaryotic genes typically exhibit a slightly elevated range, typically ranging from 1 to 10 mutations per million gametes (1 x 10-5 to 1 x 10-6). The elevated values observed in eukaryotes could potentially be attributed to the particular methodology employed for the computation of mutation rates. The rates specified are computed per gamete in eukaryotes, necessitating multiple cellular divisions for gamete production. In contrast to other organisms, prokaryotic cells and viruses exhibit mutation rates that are determined on a per-cell division basis (Nachman & Crowell, 2000).

There exists a notable disparity in mutation rates among various major taxonomic groups of organisms. The observed variations could potentially be accounted for by disparities in the capacity to rectify mutations, uneven levels of exposure to mutagens, or inherent discrepancies in the frequencies of spontaneously arising mutations. Variability in rates of variation is evident among various genes, even within a given species. The underlying factors contributing to the observed variability remain incompletely comprehended; however, it is established that specific DNA regions exhibit a higher propensity for undergoing mutations compared to others (Chen et al., 1998).

Did You Know?

In some cases, an individual's body can have different genetic mutations in various cells, resulting in genetic mosaicism. This phenomenon can lead to unique patterns in conditions like skin pigmentation.

7.6 CAUSES OF MUTATIONS

Mutations arise due to a combination of endogenous and exogenous influences. Spontaneous mutations refer to genetic alterations that arise from inherent changes in the structure of DNA. In contrast, induced mutations pertain to genetic changes caused by exposure to environmental chemicals or radiation (Zhu et al., 2013).

7.6.1 Spontaneous Replication Errors

DNA synthesis demonstrates remarkable precision, as errors during replication are remarkably low, with less than one billion inaccuracies. However, it is important to note that sporadic replication errors can sometimes occur. In the past, the prevailing belief was that tautomeric shifts, which involve the reorganization of protons within DNA bases, played a significant role in causing spontaneous replication errors. The base pairings proposed by Watson and Crick, which are adenine with thymine and cytosine with guanine, are formed between the most common forms of these bases. It is important to acknowledge that alternative base pairings can occur when the bases are in their less common tautomeric forms. According to Watson and Crick's proposal, mutations may occur due to tautomeric shifts. The proposition of explaining spontaneous replication errors as the prevailing framework was widely accepted for a significant amount of time. The available evidence does not support the idea that these rare tautomeric forms cause spontaneous mutations (Schaaper & Dunn, 1987).

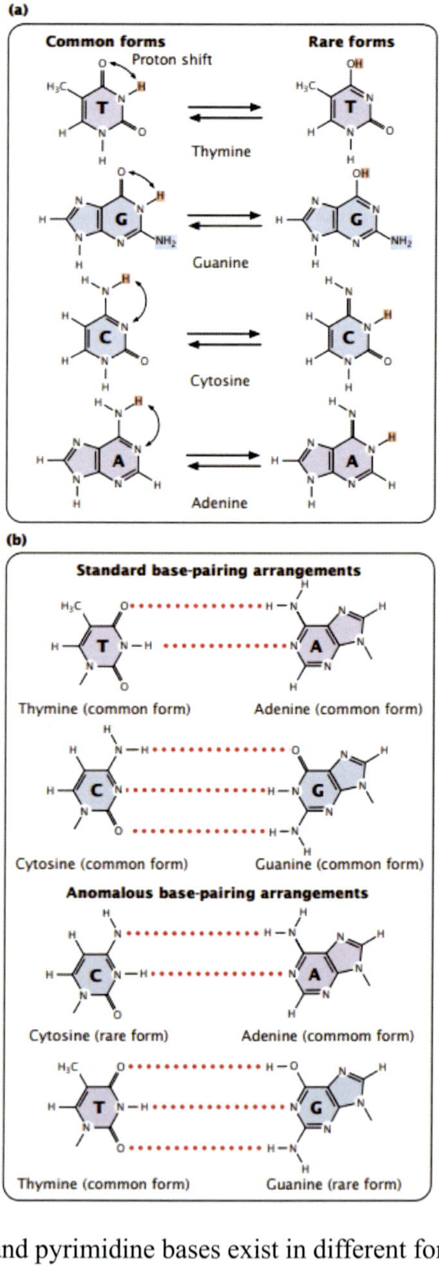

Figure 7.11. Purine and pyrimidine bases exist in different forms called tautomers. (Source: Carey, Creative Commons License)

Additionally, it is worth noting that recent research has revealed a lack of sufficient evidence to support these structures within DNA. Wobble interactions have the potential to facilitate mispairing due to the flexibility of the DNA helical structure. This flexibility allows for pairing between normal, protonated, and other base forms, as shown in Figure 7.12. The observation of these structures within DNA molecules has led to the current understanding that they have a significant impact on the occurrence of replication errors.

In the context of nucleotide chain synthesis, the occurrence of an error is attributed to the incorporation of a mismatched base. Assuming a scenario of replication, it is observed that thymine, which typically forms a base pair with adenine, exhibits a mispairing tendency with guanine due to the phenomenon known as wobble (as depicted in Figure 7.13). During the subsequent round of replication, the two bases that are not properly paired dissociate, and each acts as a template for synthesizing a novel strand of nucleotides (Lee et al., 2012).

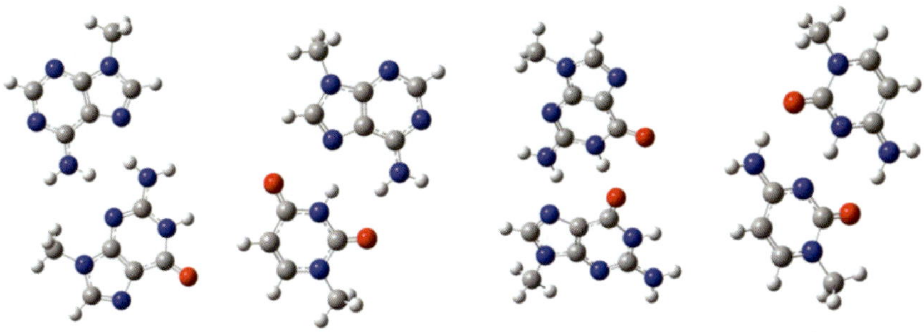

Adenine:Guanine *Trans* HS, found in many GNRA tetraloop

Adenine:Uracil *Trans* HW, found in many recurrent structural motifs

Guanine:Guanine *Cis* WH, found in G-Quadruplex of DNA or RNA and in many other motifs

Cytosine:Cytosine *Trans* WW, found in i-motif DNA and other forms. One of the Cyt base needs to be protonated to avoid electrostatic repulsion

Figure 7.12. Nonstandard base pairings can occur due to the flexibility in DNA structure. (Source: By Bhattasinp - Own work, CC BY-SA 4.0, https://commons.wikimedia.org/w/index.php?curid=85255263, Creative Commons License)

Spontaneous mutations resulting from small deletions and insertions can also arise during replication and traversing. Strand slippage can potentially manifest when a single nucleotide strand undergoes the formation of a miniature loop. If the nucleotides that are looped out occur on the strand that has been recently synthesized, it leads to the occurrence of an insertion. During the subsequent round of replication, the insertion will be integrated into both strands of the DNA molecule (Yoshiyama et al., 2001).

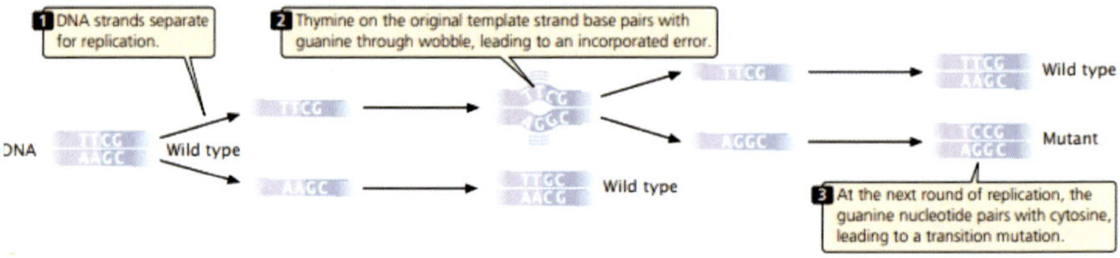

Figure 7.13. Wobble base pairing leads to a replicated error (Source: Nessa, Creative Commons License)

During normal crossing over, the homologous sequences of the two DNA molecules align with each other. This alignment does not cause any changes in the total number of nucleotides within either molecule. The occurrence of misalignment of DNA strands during pairing has the potential to cause unequal crossing over. This can result in the insertion of genetic material in one DNA molecule and the deletion of genetic material in the other, as depicted in Figure 7.15 (Prolla et al., 1998).

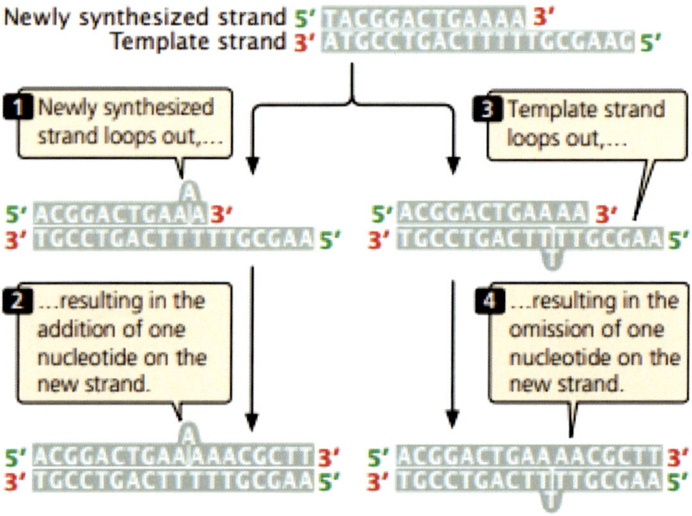

Figure 7.14. Insertions and deletions may result from strand slippage (Source: Nessa, Creative Commons License)

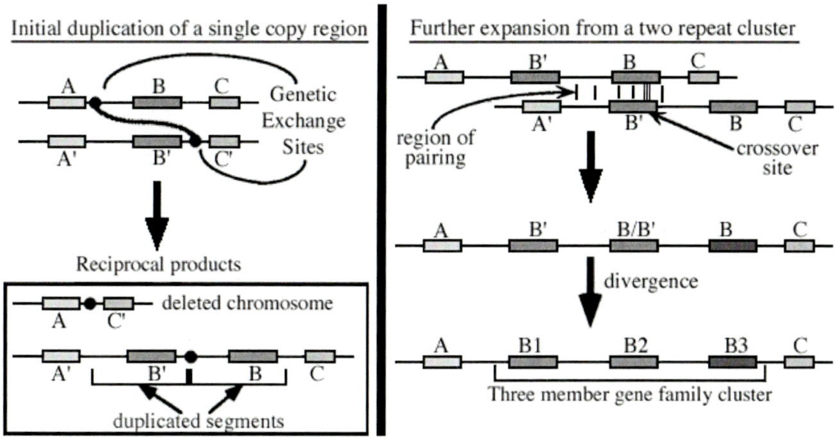

Figure 7.15. Unequal crossing over produces insertions and deletions. (Source: By Lee M. Silver - Own work, CC BY-SA 3.0, https://commons.wikimedia.org/w/index.php?curid=15030041, Creative Commons License)

Did You Know?

Dog breeds display a wide range of physical traits due to selective breeding and genetic mutations. Some breeds, like Dachshunds, have mutations associated with their unique body shapes.

7.6.2 Spontaneous Chemical Change

Mutations can occur not only due to spontaneous mutations during replication but also due to spontaneous chemical alterations in DNA. Depurination is a type of modification that involves the removal of a purine base from a nucleotide. Depurination is a process where the covalent bond between the purine and the 1'-carbon atom of the deoxyribose sugar is broken, forming an apurinic site. The website discusses a nucleotide that lacks its purine base. An apurinic site lacks the necessary information to act as a template for synthesizing a complementary base during replication. Without base-pairing constraints, the newly synthesized DNA strand may have an incorrect nucleotide, usually adenine, inserted across from the apurinic site (as depicted in Figure 7.16b). This can often lead to the incorporation of an error. The initial error is transformed into a replication error during the following replication process. Depurination is a common cause of spontaneous mutation in mammalian cells cultured in a laboratory. It results in the loss of approximately 10,000 purines per day. Deamination is a chemical process that naturally occurs within the DNA molecule. It involves the removal of an amino group (NH2) from a base. Deamination can occur spontaneously or as a consequence of exposure to mutagenic chemicals (Gates, 2009).

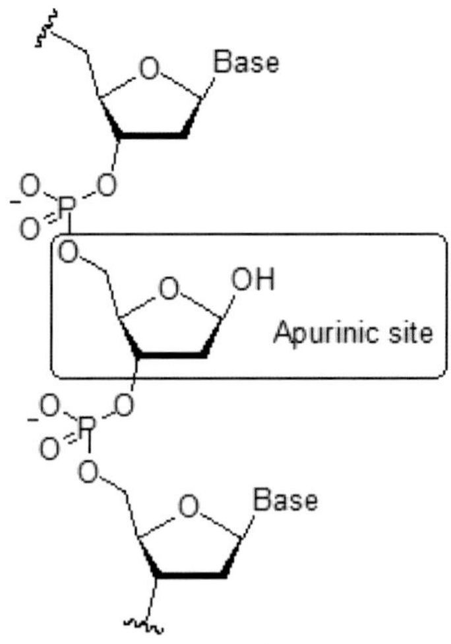

Figure 7.16. Depurination, the loss of a purine base from the nucleotide, produces an apurinic site (Source: By Chemist234 - Own work, CC BY-SA 3.0, https://commons.wikimedia.org/w/index.php?curid=23215145, Creative Commons License)

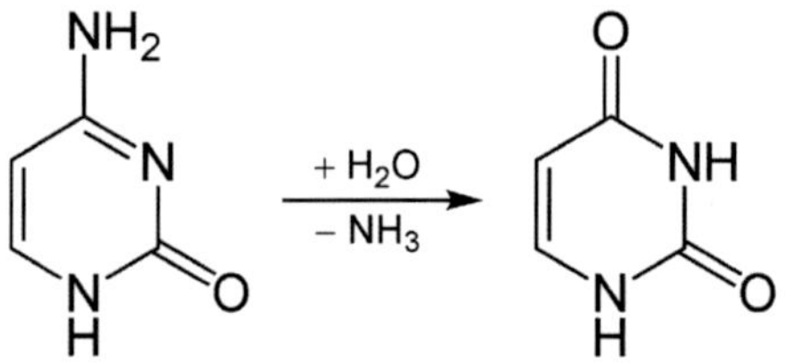

Figure 7.17. Deamination alters DNA bases. (Source: By Yikrazuul - Own work, Public Domain, https://commons.wikimedia.org/w/index.php?curid=3513581, Creative Commons License)

The process of deamination has the potential to modify the characteristics of base pairing. For instance, the deamination of cytosine results in the formation of uracil (as depicted in Figure 7.17), which then forms a complementary base pair with adenine during the replication process. Following a subsequent replication cycle, the nucleotide adenine will form a complementary base pair with thymine, replacing the original cytosine-guanine base pair (C.G) with a thymine-adenine base pair (T.A) (Seligmann, 2012). This alteration in the chemical composition represents a transition mutation. Enzymatic repair mechanisms are typically responsible for rectifying this mutation, wherein uracil is eliminated from DNA upon detection. The capacity to identify the outcome of cytosine deamination could potentially elucidate the reason for the presence of thymine, rather than uracil, in DNA. Certain cytosine bases in DNA occur naturally in a methylated state, specifically 5-methylcytosine. These methylated cytosine bases are converted upon deamination into thymine, as depicted in Figure 7.17. The pairing of thymine with adenine during replication is a well-established phenomenon. It has been observed that the deamination of 5-methylcytosine can modify the original C.G pair, resulting in a transformation to T.A (C. G→5mC.A→T.A) (Demerec, 1937). The inability of DNA repair systems to detect this alteration is attributed to its generation of a regular base. As a result, it is common for eukaryotic cells to experience C.G→T.Transitions (Novick & Szilard, 1950).

7.6.3 Chemically Induced Mutation

While it is true that numerous mutations occur spontaneously, several environmental agents can cause DNA damage, such as specific chemicals and radiation. A mutagen is any environmental factor that induces a notable increase in the mutation rate beyond the inherent or spontaneous rate. The initial identification of a chemical mutagen was accomplished by Charlotte Auerbach, an individual of German descent born into a Jewish family in 1899. After completing university studies in Berlin and engaging in research activities, she subsequently dedicated several years to instructing at diverse educational institutions in Berlin. In response to the escalating anti-Semitic climate in Nazi Germany, Auerbach decided to emigrate to Britain. Once there, she engaged in research about the study of mutant development in Drosophila. During this encounter, Auerbach had the opportunity to meet Herman Muller, a prominent researcher who had previously demonstrated the mutagenic effects of radiation. Muller proposed that Auerbach explore the possibility of generating mutants in Drosophila through chemical treatments. The initial endeavors she made yielded minimal outcomes. Several researchers conducted classified investigations about mustard gas, employed as a chemical weapon during World War I. These scientists observed that mustard gas elicited numerous physiological responses akin to radiation-induced ones (Gavazzi et al., 1987).

Auerbach was tasked with assessing the mutagenic properties of mustard gas. Auerbach conducted a research project in collaboration with pharmacologist J. M. Robson, focusing on investigating the impact of mustard gas on the species *Drosophila melanogaster*. The experimental conditions employed in this study were rudimentary. The liquid mustard gas was subjected to heat using a Bunsen burner situated on the rooftop of the pharmacology building. Subsequently, the flies were exposed to the gas within a spacious chamber. Following severe burns on her hands caused by exposure to gas, Auerbach delegated the task of conducting subsequent exposures to others while she focused on analyzing the flies. The research conducted by Auerbach and Robson provided evidence supporting the mutagenic properties of mustard gas. Their study demonstrated that exposure to mustard gas resulted in a decrease in the viability of gametes and an increase in the occurrence of mutations in the offspring of flies subjected to the gas. The delay in the publication of their findings until 1947 can be attributed to the fact that the research was conducted as a component of the covert war effort (Drinkwater & Klinedinst, 1986).

7.6.3.1 Base Analogs

A category of chemical mutagens comprises base analogs characterized by their structural resemblance to any of the four standard bases found in DNA. DNA polymerases cannot differentiate between these analogs and the standard bases. Consequently, if base analogs are present during the process of replication, they have the potential to be integrated into newly formed DNA molecules. An instance of a compound known as 5-bromouracil (5BU) can be observed as an analog of thymine. Its structure resembles thymine, except it possesses a bromine (Br) atom instead of a methyl group on the 5-carbon atom (Figure 7.18a). In the typical scenario, the nucleobase 5-bromouracil forms a complementary base pair with adenine, similar to the pairing between thymine and adenine. However, there are instances where 5-bromouracil erroneously pairs with guanine, resulting in a transition mutation. This mutation sequence, depicted in Figure 7.19, involves the progression from a thymine-adenine pair to a 5-bromouracil-adenine pair, followed by a 5-bromouracil-guanine pair, and ultimately leading to a cytosine-guanine pair (Freese, 1959).

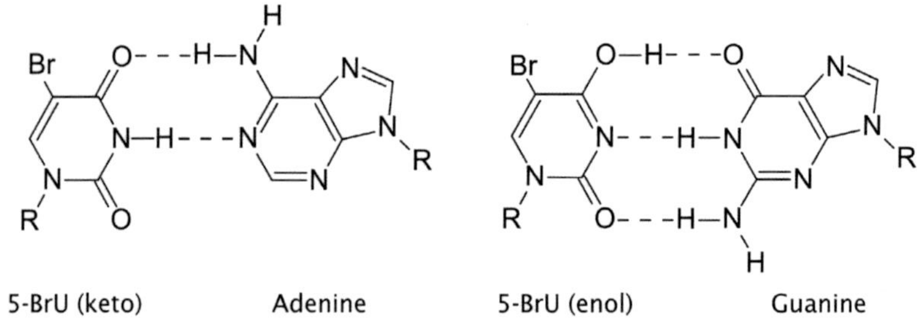

Figure 7.18. 5-Bromouracil (a base analog) resembles thymine, except it has a bromine atom in place of a methyl group on the 5-carbon atom. (Source: By Histidine - Own work, CC BY-SA 3.0, https://commons.wikimedia.org/w/index.php?curid=12872151, Creative Commons License)

Mispairing can result in incorporating 5-bromouracil into a recently synthesized DNA strand, specifically in a position opposite to guanine. During the subsequent round of replication, it is plausible that 5-bromouracil could form a base pair with adenine, thereby resulting in another transition (Kozmin et al., 1998).

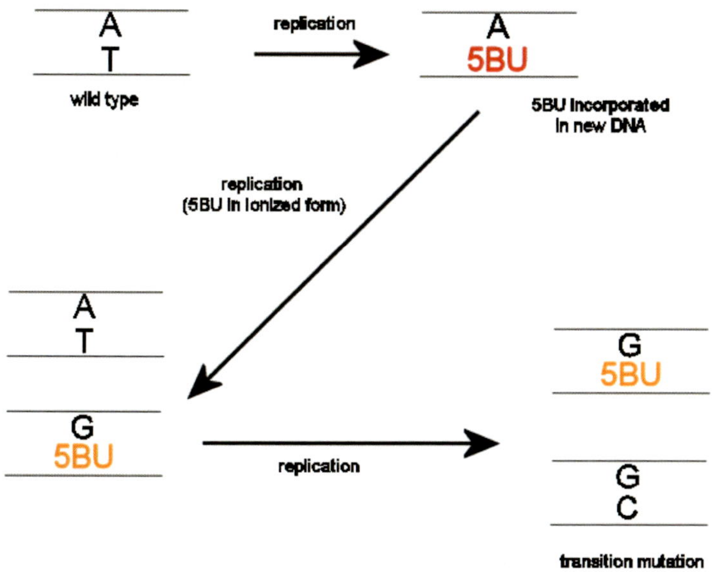

Figure 7.19. 5-Bromouracil can lead to a replicated error (Source: By Allen Gathman - Own work, CC BY-SA 2.5, https://commons.wikimedia.org/w/index.php?curid=34224265, Creative Commons License)

An additional chemical with mutagenic properties is 2-aminopurine (2AP), a base analog of adenine. Typically, the nucleobase 2-aminopurine forms complementary base pairs with thymine, but it can occasionally form incorrect base pairs with cytosine, resulting in a transition mutation (T.A→T.2AP→C.2AP→C.G). In an alternative scenario, the nucleotide 2-aminopurine (2AP) can be integrated into the nascent DNA strand by mispairing, specifically with cytosine. Subsequently, 2AP may form a base pair with thymine, resulting in a transition from cytosine-guanine (C-G) to cytosine-2AP (C-2AP), followed by a transition from thymine-2AP (T-2AP) to thymine-adenine (T-A). Both 5-bromouracil and 2-aminopurine can induce transition mutations. In the laboratory setting, it is possible to reverse mutations induced by base analogs through treatment with either the same analog or a different analog (Shelton et al., 2016).

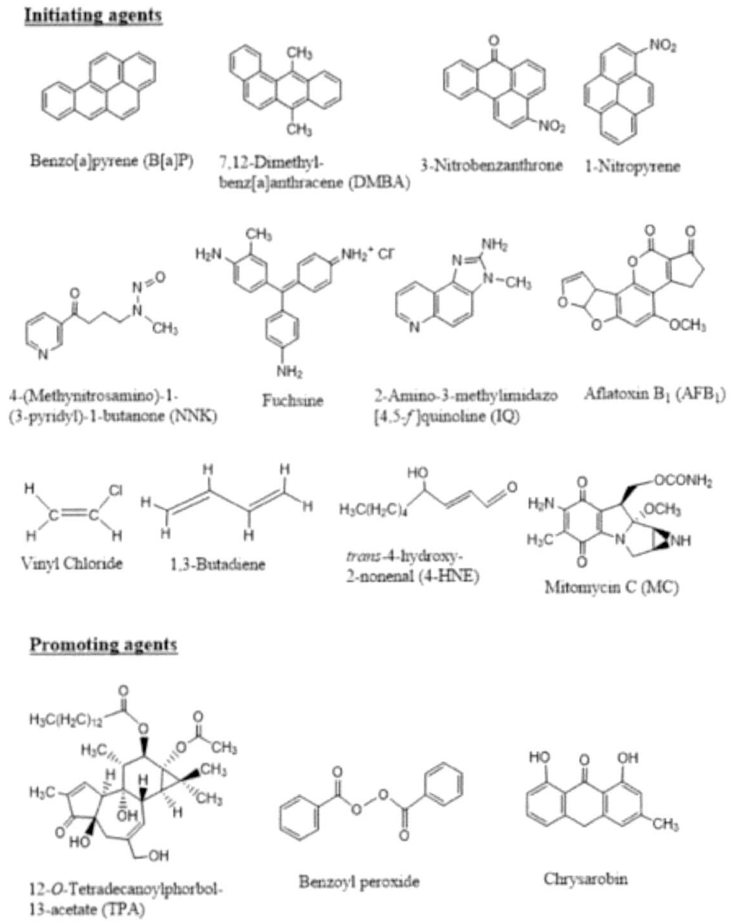

Figure 7.20. Chemicals may alter DNA bases. (Source: Basu AK. DNA Damage, Mutagenesis and Cancer. *International Journal of Molecular Sciences*. 2018; 19(4):970. https://doi.org/10.3390/ijms19040970, Creative Commons License)

Aside from its inherent spontaneity, as depicted in Figure 7.17, deamination can also be instigated by certain chemical agents. For example, nitrous acid deamination results in cytosine conversion to uracil. Subsequently, during the subsequent round of replication, uracil pairs with adenine, leading to a C.G→T.A transition mutation (refer to Figure 7.20). Nitrous acid induces the conversion of adenine to hypoxanthine, forming a base pair between hypoxanthine and cytosine. This base pairing alteration ultimately leads to a transition from thymine-adenine to cytosine-guanine. Nitrous acid is known to induce the deamination of guanine, resulting in the formation of xanthine. Xanthine can form base pairs with cytosine, similar to guanine (Lada et al., 2013). However, it is important to note that xanthine can also form base pairs with thymine, giving rise to a C.G→T.A transition. Nitrous acid is known to induce solely transition mutations, encompassing both C.G→T.A and T.A→C.G transitions. Consequently, applying nitrous acid enables the potential reversal of these mutations (Loakes et al., 2009).

Hydroxylamine: Hydroxylamine is a highly specialized mutagenic base modifier that introduces a hydroxyl group to cytosine, forming hydroxylaminocytosine (Figure 7.20). This transformation results in an elevated occurrence of an uncommon tautomer that forms a bond with adenine rather than guanine, resulting in transitions from C.G to T.A. Due to its specific interaction with cytosine, hydroxylamine cannot induce T.A. → C.G transitions. Consequently, this compound cannot reverse the mutations caused by hydroxylamine (Shcherbakova & Pavlov, 1996).

Oxidative reactions Reactive oxygen compounds, such as superoxide radicals, hydrogen peroxide, and hydroxyl radicals, are generated during aerobic metabolism and exposure to ozone, radiation, peroxides, and specific pharmaceutical substances (Ang et al., 2016).

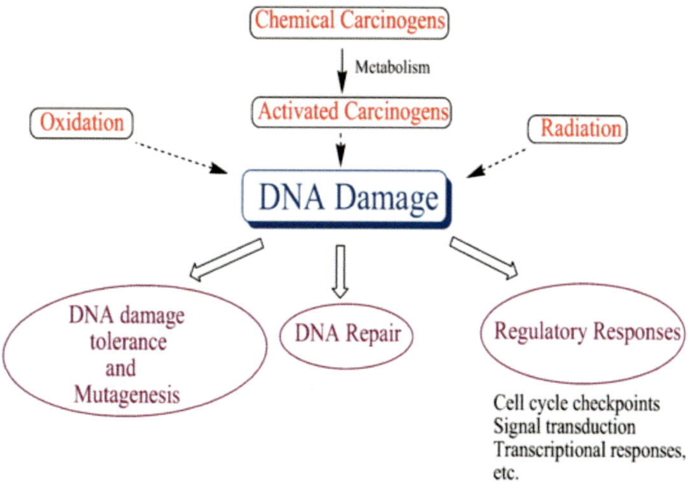

Figure 7.21. Oxidative radicals convert guanine into 8-oxy-7,8-dihydrodeoxyguanine, frequently mispairs adenine instead of cytosine, producing a C.G→T.A transversion (Source: Basu AK. DNA Damage, Mutagenesis and Cancer. *International Journal of Molecular Sciences*. 2018; 19(4):970. https://doi.org/10.3390/ijms19040970, Creative Commons License)

Intercalating agents, such as proflavin, acridine orange, and ethidium bromide, possess a similar size to that of a nucleotide. Mutations are generated through intercalation, wherein these entities position themselves between neighboring bases within the DNA molecule. This intercalation event leads to the distortion of the helical structure in three dimensions, resulting in single-nucleotide insertions and deletions during the replication process. Insertions and deletions often result in frameshift mutations, causing alterations to all amino acids located downstream of the mutation. Consequently, the mutagenic consequences of intercalating agents are typically significant. Intercalating agents can induce both insertions and deletions, thereby enabling them to counteract the consequences of their mutagenic alterations (Kozmin et al., 2000).

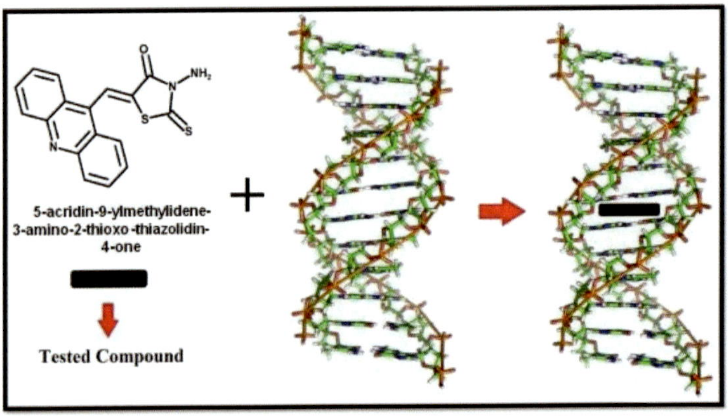

Figure 7.22. Intercalating agents such as proflavin and acridine orange insert themselves between adjacent DNA bases, distorting the helix's three-dimensional structure and causing single-nucleotide insertions and deletions in replication. (Source: Lafayette EA, Vitalino de Almeida SM, Da Rocha Pitta MG, Carneiro Beltrão EI, Gonçalves da Silva T, Olímpio de Moura R, Da Rocha Pitta I, De Carvalho LB Júnior, Do Carmo Alves de Lima M. Synthesis, DNA Binding and Topoisomerase I Inhibition Activity of Thiazacridine and Imidazacridine Derivatives. *Molecules*. 2013; 18(12):15035-15050. https://doi.org/10.3390/molecules181215035, Creative Commons License)

7.6.4 Radiation

In 1927, Herman Muller conducted an experiment that successfully demonstrated the ability to induce mutations in fruit flies through X-rays. The subsequent studies yielded findings demonstrating a significant increase in mutation rates across all organisms due to exposure to X-rays. Gamma, X-rays, and cosmic rays possess significant energy levels, enabling them to effectively traverse tissues and induce detrimental effects on DNA structures (Figure 7.23). Ionizing radiation dislodges electrons from atoms, transforming stable molecules into free radicals and reactive ions. These altered structures of bases and broken phosphodiester bonds in DNA are the subsequent effects of this process. Ionizing radiation often leads to breaks in double strands in DNA. Efforts to rectify these disruptions may result in chromosomal mutations (Sankaranarayanan, 1991).

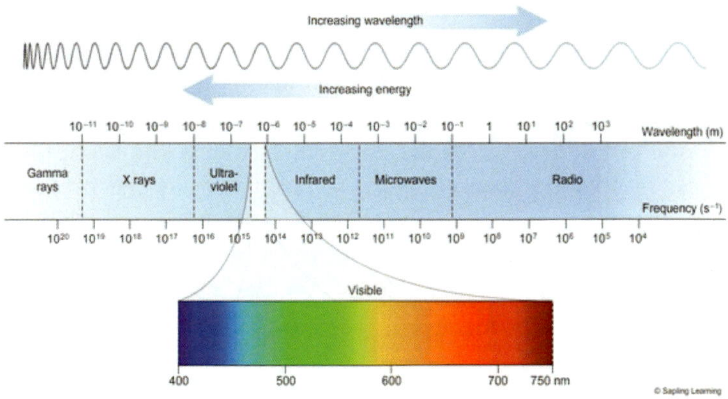

Figure 7.23. In the electromagnetic spectrum, as wavelength decreases, energy increases (Source: Gartshore A, Kidd M, Joshi LT. Applications of Microwave Energy in Medicine. *Biosensors*. 2021; 11(4):96. https://doi.org/10.3390/bios11040096, Creative Commons License)

Ultraviolet (UV) light possesses lower energy levels than ionizing radiation and does not induce the ejection of electrons or ionization. However, it still exhibits a significant mutagenic effect. The absorption of UV light by purine and pyrimidine bases leads to chemical bonds between neighboring pyrimidine molecules on the same DNA strand, generating pyrimidine dimers (Figure 7.24) (Wikonkal & Brash, 1999).

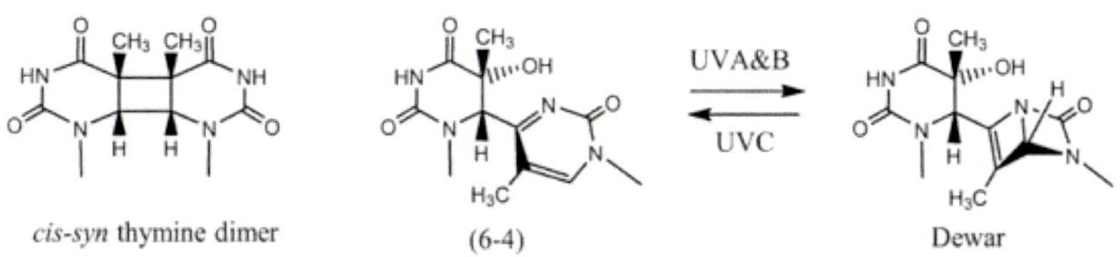

Figure 7.24. Pyrimidine dimers result from Ultraviolet light. (Source: Basu AK. DNA Damage, Mutagenesis and Cancer. *International Journal of Molecular Sciences*. 2018; 19(4):970. https://doi.org/10.3390/ijms19040970, Creative Commons License)

The presence of pyrimidine dimers hinders the replication process, inhibiting cell division and subsequent cell death. Consequently, UV light serves as a bactericidal agent and an efficient means of sterilization. For a mutation, a hereditary deviation in the genetic instructions, to transpire, it is imperative to surmount the replication block. How do bacteria and various other organisms undergo replication in the presence of thymine dimers (McGregor et al., 1999)?

Bacterial organisms can overcome obstacles in DNA replication caused by pyrimidine dimers and various forms of DNA damage by utilizing the SOS system. This system facilitates the resolution of replication blocks, albeit at the cost of introducing multiple errors and significantly augmenting the mutation rate. The ability of replication to occur in the presence of a block is attributed to the lack of strict adherence of the enzymes in the SOS system to the base-pairing rules. A potential consequence of replication is the preservation of cellular viability at the expense of compromising the fidelity of DNA synthesis. The SOS system exhibits a high degree of complexity, encompassing the products of a minimum of 25 genes. RecA, a protein, exhibits binding affinity towards the damaged DNA located at the obstructed replication fork, thereby undergoing activation.

The activation process facilitates the interaction between a protein known as LexA, which functions as a repressor of the SOS system. The RecA complex, when activated, triggers the process of self-cleavage in LexA, resulting in the elimination of its repressive activity. The process of inactivation facilitates the expression of additional SOS genes, which subsequently enable the progression of damaged DNA replication through the action of their respective products. The SOS system facilitates the incorporation of bases into a nascent DNA strand even without complementary bases on the template strand. However, these insertions give rise to many errors in the resulting base sequence (Efanov et al., 2018).

Eukaryotic cells possess a distinct DNA polymerase η (eta), specifically designed to circumvent the formation of pyrimidine dimers. Polymerase η prefers inserting adenine-adenine (AA) nucleotide pairs, specifically in the context of a pyrimidine dimer. This strategy appears rational as approximately two-thirds of pyrimidine dimers are specifically thymine dimers. Nevertheless, when an AA is inserted opposite a CT dimer, it leads to a transversion from C.G to A.T. Polymerase η is thus characterized as a polymerase with a propensity for errors (Perez et al., 2013).

Remember

Certain regions of the genome are more prone to mutations, known as mutational hotspots. These areas have higher mutation rates, and studying them can provide insights into the mutation process.

7.7 DNA REPAIR

The stability of DNA is continuously compromised by exposure to radiation, mutagenic chemicals, and spontaneous alterations. Despite the presence of these harmful agents, the frequency of genetic mutations remains notably low due to the proficient mechanisms of DNA repair. According to estimates, mutations are relatively rare, with less than one in a thousand DNA lesions undergoing this transformation, while most lesions are effectively repaired. There are many intricate pathways involved in the process of DNA repair; however, it is possible to make several overarching assertions regarding this fundamental biological mechanism. Most DNA repair mechanisms necessitate the presence of two nucleotide strands in DNA, as they typically involve the replacement of entire nucleotides. To accurately specify the base sequence, a template strand is required. DNA's inherent stability and replication efficiency are attributed to its complementary, double-stranded structure.

Additionally, this structure allows either strand to serve as a template for the necessary information to correct errors in the other strand. One additional characteristic of DNA repair is the presence of redundancy, which indicates that multiple repair pathways can address various types of DNA damage. The presence of redundancy in this context is evidence of the utmost significance of DNA repair for cellular survival, as it guarantees the rectification of nearly all errors. If an error eludes one repair mechanism, an alternative system will probably rectify it (Sancar et al., 2004).

7.7.1 Mismatch Repair

The replication process is highly precise, evidenced by only one error per billion nucleotides in each newly synthesized DNA copy. During the replication process, it is observed that there is an incorporation of mismatched bases into the newly synthesized DNA at a frequency ranging from approximately 10^-4 to 10^-5. Consequently, most errors during this stage are rectified and do not persist as permanent mutations. Certain corrections are implemented during the process of proofreading.

DNA polymerases can identify and rectify nucleotides that are not properly paired. Adding a nucleotide that does not match the template strand in a newly synthesized DNA strand leads to the stalling of the polymerase enzyme. Subsequently, the enzyme employs its 3'→5' exonuclease activity to retrogressively eliminate the erroneously incorporated nucleotide before resuming the polymerization process in the 5'→3' direction. Mismatch repair is a mechanism that corrects the presence of erroneously inserted nucleotides that manage to evade detection through proofreading (Ijsselsteijn et al., 2020). The distortion of the three-dimensional structure of DNA occurs due to incorrectly paired bases, which mismatch repair enzymes can detect. Furthermore, the mismatch-repair system identifies incorrectly paired bases and rectifies minor unpaired loops within the DNA. These loops may arise due to strand slippage during replication, as depicted in Figure 7.14. Certain trinucleotide repeats have the potential to generate secondary structures on the strand that is not paired with its complementary strand. This structural characteristic enables these repeats to evade detection by the mismatch-repair system. Upon identifying an incorporation error, the mismatch-repair enzymes excise the distorted segment from the recently synthesized strand. Subsequently, these enzymes replenish the resulting gap with fresh nucleotides, utilizing the original DNA strand as a template. For this strategy to be effective, the mismatch repair mechanism must be able to differentiate between the original and newly synthesized DNA strands. This distinction is crucial to accurately identify and remove the incorporation error while preserving the integrity of the original DNA strand (Kunz et al., 2009).

The identification of old and new strands in *E. coli* is accomplished by the proteins responsible for mismatch repair, which distinguish between them based on the occurrence of methyl groups on specific sequences of the old strand. Following the replication process, the adenine nucleotides present in the GATC sequence undergo methylation through the catalytic action of an enzyme known as Dam methylase. The methylation process is delayed, resulting in the immediate post-replication methylation of the pre-existing strand, while the newly synthesized strand remains unmethylated.

DNA polymerase and DNA ligase are enzymes that are responsible for the repair of DNA. Specifically, they play a crucial role in filling in gaps on the unmethylated strand by accurately pairing nucleotides. The repair of mismatches in eukaryotic cells resembles that observed in *Escherichia coli* (E. coli), albeit with the distinction that multiple proteins exhibit homology to MutS. In contrast, several others display homology to MutL. These proteins collaborate in various configurations to identify diverse integration errors, including mispaired bases and tiny unpaired loops. There is an absence of proteins associated with *E. coli* MutH in eukaryotic cells. The specific enzyme responsible for creating the nick in eukaryotic cells remains uncertain. The mechanisms underlying the identification of old and new DNA strands in eukaryotic cells remain unknown, as certain eukaryotes, including yeast and fruit flies, exhibit an absence of detectable DNA methylation (Seifert & Reichrath, 2006).

7.7.2 Direct Repair

Direct-repair processes do not substitute modified nucleotides but restore them to their initial (accurate) configurations. Photoreactivation of UV-induced pyrimidine dimers is one of the most extensively studied and well-defined direct repair mechanisms. The enzyme photolyase is present in both *E. coli* and certain eukaryotic cells. It utilizes energy obtained from light to disrupt the covalent bonds connecting the pyrimidines within a dimer. Direct repair is a mechanism that also effectively rectifies O6-methylguanine, an alkylation derivative of guanine. This particular form of guanine pairs with adenine, resulting in G.C→T.A transversions. The methyl group is eliminated from O6 -methylguanine by the enzyme O6 -methylguanine-DNA methyltransferase, restoring the base to guanine (see Figure 7.25) (Mishina et al., 2005).

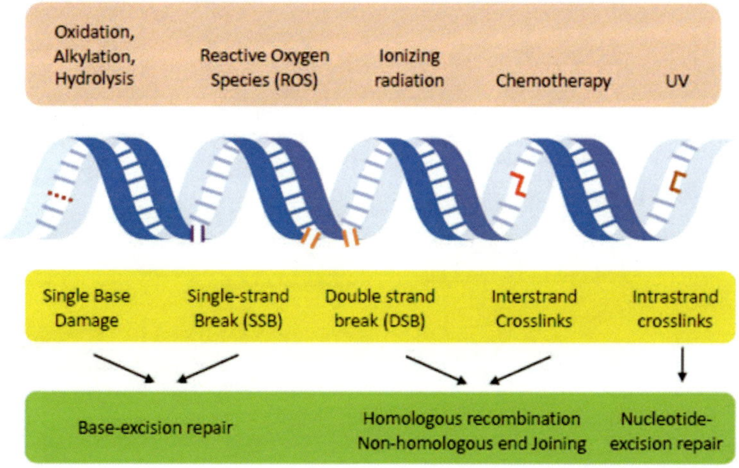

Figure 7.25. Direct repair changes nucleotides back into their original structures. (Source: Sohn J, Lee S-E, Shim E-Y. DNA Damage and Repair in Eye Diseases. *International Journal of Molecular Sciences*. 2023; 24(4):3916. https://doi.org/10.3390/ijms24043916, Creative Commons License)

7.7.3 Base-Excision Repair

The initial step of base-excision repair involves the excision of modified bases, followed by the subsequent replacement of the entire nucleotide. Removing modified bases from DNA is facilitated by a group of enzymes known as DNA glycosylases. These enzymes can identify and eliminate particular types of modified bases by breaking the bond that connects the base to the 1'-carbon atom of deoxyribose. One example of an enzyme known as uracil glycosylase can identify and eliminate uracil molecules arising from cytosine deamination. Various glycosylases can identify and interact with hypoxanthine, 7-methylguanine, 3-methyladenine, and other modified bases. Following the elimination of the base, an enzyme known as AP endonuclease initiates the cleavage of the phosphodiester bond, while additional enzymes facilitate the elimination of the deoxyribose sugar. Subsequently, the DNA polymerase incorporates fresh nucleotides into the unshielded 3'-OH moiety , substituting a segment of nucleotides on the impaired DNA strand. DNA ligase seals the nick in the phosphodiester backbone. This enzymatic action leads to restoring the original intact sequence, as shown in Figure 7.26 (Grin & Ishchenko, 2016).

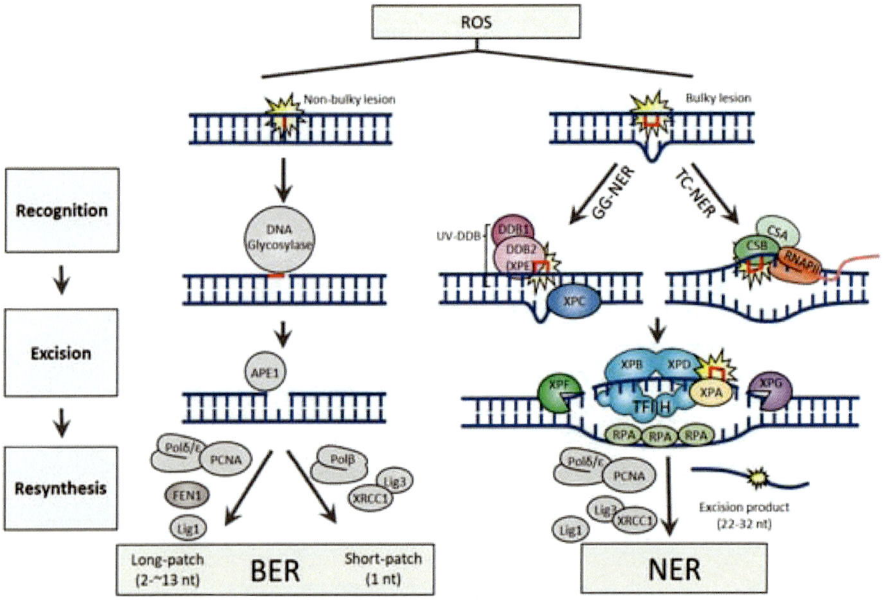

Figure 7.26. Base-excision repair excises modified bases and then replaces the entire nucleotide (Source: Lee T-H, Kang T-H. DNA Oxidation and Excision Repair Pathways. *International Journal of Molecular Sciences*. 2019; 20(23):6092. https://doi.org/10.3390/ijms20236092, Creative Commons License)

7.7.4 Nucleotide Excision Repair

The subsequent repair pathway under consideration is nucleotide excision repair, which is responsible for eliminating substantial DNA lesions that distort the double helix structure. These lesions include pyrimidine dimers or sizable hydrocarbons attached to the DNA. The process of nucleotide-excision repair exhibits a high degree of versatility, enabling it to rectify a wide range of DNA damage manifestations effectively. This repair mechanism in cells across all organisms, from bacteria to humans, holds significant importance. The nucleotide excision process is intricate, involving the participation of numerous genes in humans. Initially, a cluster of enzymes begins scrutinizing DNA to identify any alterations in its intricate three-dimensional structure. Upon detecting a distortion, supplementary enzymes facilitate the separation of the two nucleotide strands at the site of damage, while single-strand-binding proteins provide stability to the separated strands. Subsequently, the sugar-phosphate backbone of the impaired strand is enzymatically cleaved on both sides of the site of damage. A single incision is performed at a location that is 5 nucleotides preceding the site of damage, referred to as the 3' side. Another incision is made at a distance of either 8 nucleotides (in prokaryotes) or ranging from 21 to 23 nucleotides (in eukaryotes) following the damage, known as the 5' end. A portion of the compromised DNA strand is removed, resulting in the creation of a void. This gap is subsequently repaired through the enzymatic activity of DNA polymerase, which synthesizes new DNA to fill the gap. Finally, the repaired DNA is sealed by the action of DNA ligase (Rubbi & Milner, 2003).

7.7.5 Other Types Of DNA Repair

The previously mentioned DNA repair pathways are specifically tailored to address situations where damage occurs on a single strand of the DNA molecule. The preservation of the other strand is crucial as it serves as a reference during the DNA synthesis involved in the repair process. DNA damage can impact both strands of the DNA molecule, creating a more complex situation for the mechanisms involved in DNA repair. Ionizing radiation is a common cause of double-strand breaks in the DNA molecule. Homologous recombination is a commonly employed mechanism utilized to repair double-strand breaks. An interstrand cross-link is a form of damage that impacts both strands simultaneously. Duplex molecule linkage is a chemical phenomenon that occurs when the two strands of a molecule establish covalent bonds. Interstrand cross-links in cells are known to have a notable detrimental effect, as they impede the replication process. Chemotherapeutic agents such as mitomycin C, cisplatin, psoralen, and nitrogen mustard can cause interstrand cross-links. Nitrogen mustard, a chemical compound that bears structural similarities to mustard gas, was employed by Charlotte Auerbach to induce mutations in Drosophila. It holds the distinction of being the initial chemical agent utilized in chemotherapy treatment. The current understanding of the repair mechanisms for interstrand cross-links is limited and lacks comprehensiveness. According to the proposed model, it is suggested that double-strand breaks take place on both sides of the cross-link. These breaks are then repaired using the pathways responsible for repairing double-strand breaks (Featherstone & Jackson, 1999).

7.8 GENETIC DISEASES & FAULTY DNA REPAIR

Numerous human diseases are associated with abnormalities in DNA repair mechanisms. The diseases frequently correlate strongly with elevated occurrences of particular cancers due to impairments in DNA repair mechanisms, resulting in heightened mutation rates (Uziel et al., 2003).

One of the extensively researched human DNA repair illnesses is xeroderma pigmentosum (Figure 7.27), a rare autosomal recessive disorder characterized by atypical skin pigmentation and heightened susceptibility to sunlight. Individuals afflicted with this ailment exhibit a notable inclination towards developing skin cancer, with an occurrence rate ranging from one to two thousand times higher than unaffected individuals (McMurray, 2005). Sunlight comprises a substantial ultraviolet (UV) component, thereby leading to the generation of pyrimidine dimers within the DNA of cutaneous cells upon exposure. While it is true that human cells do not possess the enzyme photolyase, which is responsible for repairing pyrimidine dimers in bacteria, the majority of pyrimidine dimers in humans can still be effectively repaired through a process known as nucleotide excision repair. Nevertheless, individuals afflicted with xeroderma pigmentosum typically possess cells that exhibit deficiencies in nucleotide excision repair. Consequently, many pyrimidine dimers remain unrectified, increasing the risk of developing cancer (Barzilai et al., 2002).

Xeroderma pigmentosum can arise due to genetic abnormalities in multiple genes.

Cockayne syndrome and trichothiodystrophy, or brittle-hair syndrome, are two additional genetic disorders resulting from impairments in nucleotide excision repair. Individuals afflicted with either of these medical conditions do not demonstrate an elevated susceptibility to cancer, yet they manifest various developmental and neurological impairments. Both diseases are attributable to genetic mutations in the same gene set responsible for xeroderma pigmentosum. Several genes involved in nucleotide-excision repair produce proteins that contribute to replication and transcription initiation. It is plausible that these additional functions could potentially explain the developmental symptoms observed in individuals with Cockayne syndrome and trichothiodystrophy (Hanawalt & Spivak, 2008).

Hereditary non-polyposis colon cancer (HNPCC) is a genetic disorder resulting from impaired DNA repair mechanisms, leading to the development of colon cancer. This particular form of cancer is widely recognized as a prevalent hereditary cancer, constituting approximately 15% of all diagnosed cases of colon cancer. The available research suggests that Hereditary Nonpolyposis Colorectal Cancer (HNPCC) results from genetic mutations in the proteins responsible for executing the mismatch repair process (Jackson & Bartek, 2009).

The etiology of Li-Fraumeni syndrome can be attributed to genetic mutations occurring in the p53 gene, a pivotal regulator of the cell cycle. The protein synthesized by the p53 gene can impede cell division until the necessary repairs to damaged DNA have been completed. Additionally, it can directly initiate the process of DNA repair. The p53 gene product has the potential to induce apoptosis in cells that have incurred DNA damage, thereby inhibiting the transmission of their mutated genetic information. Individuals diagnosed with Li-Fraumeni syndrome manifest the presence of numerous distinct malignancies occurring in various tissues (Caldecott, 2008).

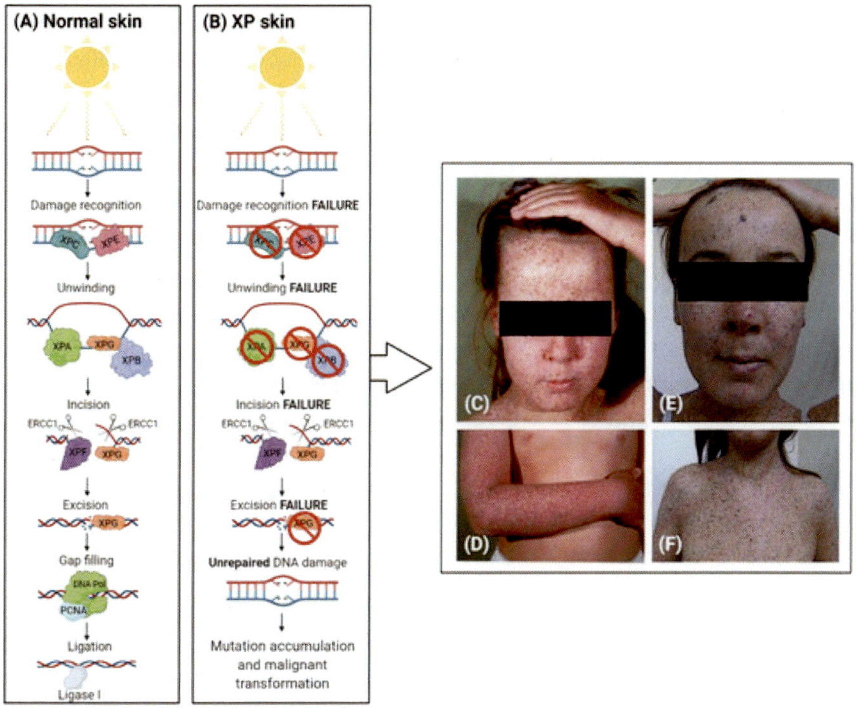

Figure 7.27. Xeroderma pigmentosum is a human disease that results from defects in DNA repair (Source: Piccione M, Belloni Fortina A, Ferri G, Andolina G, Beretta L, Cividini A, De Marni E, Caroppo F, Citernesi U, Di Liddo R. Xeroderma Pigmentosum: General Aspects and Management. *Journal of Personalized Medicine*. 2021; 11(11):1146. https://doi.org/10.3390/jpm11111146, Creative Commons License)

Remember

Repeating DNA sequences, called microsatellites, can undergo expansion or contraction due to mutations. These changes are linked to certain genetic disorders, like Huntington's disease.

SUMMARY

The chapter titled "Gene Mutations and DNA Repair" provides a comprehensive analysis of the genetic realm, delving into various aspects such as the aftermath of the Chernobyl nuclear catastrophe, the characteristics of mutations, different categories of gene mutations, factors contributing to mutations, rates of mutation occurrence, mechanisms involved in DNA repair, and the intricate relationship between defective DNA repair and genetic disorders.

The Genetic Legacy of Chernobyl serves as an important reminder of the enduring consequences of mutations. The nuclear incident of catastrophic proportions resulted in the emission of ionizing radiation, which subsequently caused genetic modifications in various forms of life. The enduring impact of this historical event persists in shaping the ecological system in the vicinity of Chernobyl, highlighting the significant role of mutations in influencing life forms and emphasizing the crucial role of DNA repair mechanisms in preserving genetic stability.

The fundamental essence of mutation lies at the core of the genetic discourse. Mutations are alterations in the DNA sequence that occur spontaneously, resulting from external and internal variables. While certain mutations contribute to the process of evolution and enhance genetic diversity, others have the potential to result in genetic disorders. The comprehension of mutation dynamics provides valuable insights into the evolutionary history of our species and the underlying mechanisms that contribute to the formation of genetic diversity.

The chapter explores various categories of gene mutations, encompassing point, and structural mutations. These genetic variations significantly contribute to the intricate and diverse fabric of genetic diversity. While certain mutations may not exhibit any discernible effects, others can have negative or positive consequences, playing pivotal roles in the adaptation process and ensuring the survival of organisms.

The causes of mutation can be attributed to various factors, including environmental exposures such as radiation and chemical compounds, as well as inner cellular processes, specifically errors occurring during DNA replication. The diverse origins of individuals serve to underscore the complex interplay between genetic factors and environmental influences, which collectively shape the course of an individual's life.

The rates of mutation exhibit variation across different species and genomic regions, thereby influencing the prevalence of genetic alterations. A comprehensive comprehension of mutation rates holds significant importance in evolutionary investigations, epidemiology, and the evaluation of genetic disease risks.

Despite the potential for genetic disorders, the chapter reveals the narrative of DNA Repair, an intricate and resilient collection of mechanisms that rectify DNA damage. From the mechanism of base excision repair to the intricate process of homologous recombination, these repair pathways play a crucial role in safeguarding the integrity of genetic information and protecting organisms from deleterious mutations.

This study investigates the relationship between faulty DNA repair processes and genetic diseases, elucidating how compromised repair mechanisms can contribute to the development of hereditary disorders. Conditions like xeroderma pigmentosum arise due to impaired DNA repair mechanisms, resulting in heightened susceptibility to UV rays and an escalated likelihood of developing skin cancer.

The chapter "Gene Mutations and DNA Repair" intricately integrates the complex aspects of genetic diversity and preservation. The enduring significance of the Chernobyl incident serves as a prime illustration of the consequences of mutations. At the same time, the robustness of DNA repair mechanisms acts as a protective shield for genetic material. Comprehending these mechanisms is imperative for the progression of medical investigation, the identification of genetic disorders, and the preservation of the genetic legacy that unifies all organisms.

REFERENCES

1. Ang, J., Song, L. Y., D'Souza, S., Hong, I. L., Luhar, R., Yung, M., & Miller, J. H. (2016). Mutagen synergy: hypermutability generated by specific pairs of base analogs. *Journal of Bacteriology*, *198*(20), 2776-2783.
2. Antonarakis, E. S., Lu, C., Luber, B., Liang, C., Wang, H., Chen, Y., ... & Luo, J. (2018). Germline DNA-repair gene mutations and outcomes in men with metastatic castration-resistant prostate cancer receiving first-line abiraterone and enzalutamide. *European Urology*, *74*(2), 218-225.
3. Barzilai, A., Rotman, G., & Shiloh, Y. (2002). ATM deficiency and oxidative stress: a new dimension of defective response to DNA damage. *DNA repair*, *1*(1), 3-25.
4. Becker, R. (2006). The legacy of Chernobyl: Palliative services making a difference in Belarus. *International Journal of Palliative Nursing*, *12*(7), 318-319.
5. Bingham, P. M., Scott, M. O., Wang, S., McPhaul, M. J., Wilson, E. M., Garbern, J. Y., ... & Fischbeck, K. H. (1995). Stability of an expanded trinucleotide repeat in the androgen receptor gene in transgenic mice. *Nature Genetics*, *9*(2), 191-196.
6. Bojcun, M. (1991). The legacy of Chernobyl. *New Scientist (London)*, *130*(1765), 30-35.
7. Brachmann, R. K., Yu, K., Eby, Y., Pavletich, N. P., & Boeke, J. D. (1998). Genetic selection of intragenic suppressor mutations that reverse the effect of common p53 cancer mutations. *The EMBO journal*, *17*(7), 1847-1859.
8. Brooks, P. J., Cheng, T. F., & Cooper, L. (2008). Do all of the neurologic diseases in patients with DNA repair gene mutations result from the accumulation of DNA damage?. *DNA repair*, *7*(6), 834-848.
9. Caldecott, K. W. (2008). Single-strand break repair and genetic disease. *Nature Reviews Genetics*, *9*(8), 619-631.
10. Chen, R. Z., Pettersson, U., Beard, C., Jackson-Grusby, L., & Jaenisch, R. (1998). DNA hypomethylation leads to elevated mutation rates. *Nature*, *395*(6697), 89-93.
11. Cooper, D. N., Bacolla, A., Férec, C., Vasquez, K. M., Kehrer Sawatzki, H., & Chen, J. M. (2011). On the sequence directed nature of human gene mutation: the role of genomic architecture and the local DNA sequence environment in mediating gene mutations underlying human inherited disease. *Human mutation*, *32*(10), 1075-1099.
12. Cutcutache, I., Suzuki, Y., Tan, I. B., Ramgopal, S., Zhang, S., Ramnarayanan, K., ... & Rozen, S. G. (2015). Exome-wide sequencing

shows low mutation rates and identifies novel mutated genes in seminomas. *European urology*, *68*(1), 77-83.

13. Demerec, M. (1937). Frequency of spontaneous mutations in certain stocks of *Drosophila melanogaster*. *Genetics*, *22*(5), 469.
14. Dinges, J. R., Colleoni, C., Myers, A. M., & James, M. G. (2001). Molecular structure of three mutations at the maize sugary1 locus and their allele-specific phenotypic effects. *Plant Physiology*, *125*(3), 1406-1418.
15. Dowsing, A. T., Yong, E. L., Clark, M., McLachlan, R. I., de Kretser, D. M., & Trounson, A. O. (1999). Linkage between male infertility and trinucleotide repeat expansion in the androgen-receptor gene. *The Lancet*, *354*(9179), 640-643.
16. Drinkwater, N. R., & Klinedinst, D. K. (1986). Chemically induced mutagenesis in a shuttle vector with a low-background mutant frequency. *Proceedings of the National Academy of Sciences*, *83*(10), 3402-3406.
17. Dzikiewicz-Krawczyk, A. (2008). The importance of making ends meet: mutations in genes and altered expression of proteins of the MRN complex and cancer. *Mutation Research/Reviews in Mutation Research*, *659*(3), 262-273.
18. Efanov, A. A., Brenner, A. V., Bogdanova, T. I., Kelly, L. M., Liu, P., Little, M. P., ... & Nikiforov, Y. E. (2018). Investigation of the relationship between radiation dose and gene mutations and fusions in post-Chernobyl thyroid cancer. *JNCI: Journal of the National Cancer Institute*, *110*(4), 371-378.
19. Emr, S. D., Hanley-Way, S., & Silhavy, T. J. (1981). Suppressor mutations that restore export of a protein with a defective signal sequence. *Cell*, *23*(1), 79-88.
20. Featherstone, C., & Jackson, S. P. (1999). Ku, a DNA repair protein with multiple cellular functions? *Mutation Research/DNA Repair*, *434*(1), 3-15.
21. Freese, E. (1959). The specific mutagenic effect of base analogues on phage T4. *Journal of Molecular Biology*, *1*(2), 87-105.
22. Gagnon, K. B., & Delpire, E. (2013). Physiology of SLC12 transporters: lessons from inherited human genetic mutations and genetically engineered mouse knockouts. *American Journal of Physiology-Cell Physiology*, *304*(8), C693-C714.
23. Gates, K. S. (2009). An overview of chemical processes that damage cellular DNA: spontaneous hydrolysis, alkylation, and reactions with radicals. *Chemical research in toxicology*, *22*(11), 1747-1760.

24. Gavazzi, G., Tonelli, C., Todesco, G., Arreghini, E., Raffaldi, F., Vecchio, F., ... & Sala, F. (1987). Somaclonal variation versus chemically induced mutagenesis in tomato (Lycopersicon esculentum L.). *Theoretical and Applied Genetics*, *74*, 733-738.
25. Gillio, A. P., Verlander, P. C., Batish, S. D., Giampietro, P. F., & Auerbach, A. D. (1997). Phenotypic consequences of mutations in the Fanconi anemia FAC gene: an International Fanconi Anemia Registry study. *Blood, The Journal of the American Society of Hematology*, *90*(1), 105-110.
26. Grin, I., & Ishchenko, A. A. (2016). An interplay of the base excision repair and mismatch repair pathways in active DNA demethylation. *Nucleic acids research*, *44*(8), 3713-3727.
27. Hanawalt, P. C., & Spivak, G. (2008). Transcription-coupled DNA repair: two decades of progress and surprises. *Nature reviews Molecular cell biology*, *9*(12), 958-970.
28. Howell, N., Kubacka, I., Xu, M., & McCullough, D. A. (1991). Leber hereditary optic neuropathy: involvement of the mitochondrial ND1 gene and evidence for an intragenic suppressor mutation. *American Journal of Human Genetics*, *48*(5), 935.
29. Hsiehchen, D., Hsieh, A., Samstein, R. M., Lu, T., Beg, M. S., Gerber, D. E., ... & Zhu, H. (2020). DNA repair gene mutations as predictors of immune checkpoint inhibitor response beyond tumor mutation burden. *Cell Reports Medicine*, *1*(3), 584-588.
30. Huang, B., Ramanis, Z., & Luck, D. J. (1982). Suppressor mutations in Chlamydomonas reveal a regulatory mechanism for flagellar function. *Cell*, *28*(1), 115-124.
31. Ijsselsteijn, R., Jansen, J. G., & de Wind, N. (2020). DNA mismatch repair-dependent DNA damage responses and cancer. *DNA repair*, *93*, 102923.
32. Jackson, S. P., & Bartek, J. (2009). The DNA-damage response in human biology and disease. *Nature*, *461*(7267), 1071-1078.
33. Jia, P., & Zhao, Z. (2017). Impacts of somatic mutations on gene expression: an association perspective. *Briefings in Bioinformatics*, *18*(3), 413-425.
34. Kim, D., Pertea, G., Trapnell, C., Pimentel, H., Kelley, R., & Salzberg, S. L. (2013). TopHat2: accurate alignment of transcriptomes in the presence of insertions, deletions and gene fusions. *Genome biology*, *14*(4), 1-13.
35. Kozmin, S. G., Pavlov, Y. I., Dunn, R. L., & Schaaper, R. M. (2000). Hypersensitivity of *Escherichia coli* Δ (uvrB-bio) mutants to 6-hydroxylaminopurine and other base analogs is due to a defect in molybdenum cofactor biosynthesis. *Journal of Bacteriology*, *182*(12), 3361-3367.

36. Kozmin, S. G., Schaaper, R. M., Shcherbakova, P. V., Kulikov, V. N., Noskov, V. N., Guetsova, M. L., ... & Pavlov, Y. I. (1998). Multiple antimutagenesis mechanisms affect mutagenic activity and specificity of the base analog 6-N-hydroxylaminopurine in bacteria and yeast. *Mutation Research/Fundamental and Molecular Mechanisms of Mutagenesis*, *402*(1-2), 41-50.
37. Kunz, C., Saito, Y., & Schär, P. (2009). DNA Repair in mammalian cells: Mismatched repair: variations on a theme. *Cellular and molecular life sciences*, *66*, 1021-1038.
38. Lada, A. G., Stepchenkova, E. I., Waisertreiger, I. S., Noskov, V. N., Dhar, A., Eudy, J. D., ... & Pavlov, Y. I. (2013). Genome-wide mutation avalanches induced in diploid yeast cells by a base analog or an APOBEC deaminase. *PLoS Genetics*, *9*(9), e1003736.
39. Ladeiro, Y., Couchy, G., Balabaud, C., Bioulac Sage, P., Pelletier, L., Rebouissou, S., & Zucman Rossi, J. (2008). MicroRNA profiling in hepatocellular tumors is associated with clinical features and oncogene/ tumor suppressor gene mutations. *Hepatology*, *47*(6), 1955-1963.
40. Lee, H., Popodi, E., Tang, H., & Foster, P. L. (2012). Rate and molecular spectrum of spontaneous mutations in the bacterium *Escherichia coli* as determined by whole-genome sequencing. *Proceedings of the National Academy of Sciences*, *109*(41), E2774-E2783.
41. Leongamornlert, D. A., Saunders, E. J., Wakerell, S., Whitmore, I., Dadaev, T., Cieza-Borrella, C., ... & Eeles, R. A. (2019). Germline DNA repair gene mutations in young-onset prostate cancer cases in the UK: evidence for a more extensive genetic panel. *European Urology*, *76*(3), 329-337.
42. Loakes, D., Gallego, J., Pinheiro, V. B., Kool, E. T., & Holliger, P. (2009). Evolving a polymerase for hydrophobic base analogues. *Journal of the American Chemical Society*, *131*(41), 14827-14837.
43. Maki, H. (2002). Origins of spontaneous mutations: specificity and directionality of base-substitution, frameshift, and sequence-substitution mutageneses. *Annual review of genetics*, *36*(1), 279-303.
44. Marples, D. R. (1997). The legacy of Chernobyl in 1997: impact on Ukraine and Belarus. *Post-Soviet geography and economics*, *38*(3), 163-170.
45. Marshall, C. H., Fu, W., Wang, H., Baras, A. S., Lotan, T. L., & Antonarakis, E. S. (2019). Prevalence of DNA repair gene mutations in localized prostate cancer according to clinical and pathologic features: association of Gleason score and tumor stage. *Prostate cancer and prostatic diseases*, *22*(1), 59-65.

46. McGregor, J. M., Crook, T., Fraser-Andrews, E. A., Rozycka, M., Crossland, S., Brooks, L., & Whittaker, S. J. (1999). Spectrum of p53 gene mutations suggests a possible role for ultraviolet radiation in the pathogenesis of advanced cutaneous lymphomas. *Journal of investigative dermatology*, *112*(3), 317-321.
47. McMurray, C. T. (2005). To die or not to die: DNA repair in neurons. *Mutation Research/Fundamental and Molecular Mechanisms of Mutagenesis*, *577*(1-2), 260-274.
48. Mishina, Y., Yang, C. G., & He, C. (2005). Direct repair of the exocyclic DNA adduct 1, N6-ethenoadenine by the DNA repair AlkB proteins. *Journal of the American Chemical Society*, *127*(42), 14594-14595.
49. Mrózek, K., Marcucci, G., Paschka, P., Whitman, S. P., & Bloomfield, C. D. (2007). Clinical relevance of mutations and gene-expression changes in adult acute myeloid leukemia with normal cytogenetics: are we ready for a prognostically prioritized molecular classification?. *Blood*, *109*(2), 431-448.
50. Nachman, M. W., & Crowell, S. L. (2000). Estimate of the mutation rate per nucleotide in humans. *Genetics*, *156*(1), 297-304.
51. Novick, A., & Szilard, L. (1950). Experiments with the chemostat on spontaneous mutations of bacteria. *Proceedings of the National Academy of Sciences*, *36*(12), 708-719.
52. Panuzzo, C., Signorino, E., Calabrese, C., Ali, M. S., Petiti, J., Bracco, E., & Cilloni, D. (2020). Landscape of tumor suppressor mutations in acute myeloid leukemia. *Journal of clinical medicine*, *9*(3), 802.
53. Pardiñas, A. F., Holmans, P., Pocklington, A. J., Escott-Price, V., Ripke, S., Carrera, N., ... & Walters, J. T. (2018). Common schizophrenia alleles are enriched in mutation-intolerant genes and in regions under strong background selection. *Nature Genetics*, *50*(3), 381-389.
54. Paulson, H. L., & Fischbeck, K. H. (1996). Trinucleotide repeats in neurogenetic disorders. *Annual review of neuroscience*, *19*(1), 79-107.
55. Perez, B. A., Ghafoori, A. P., Lee, C. L., Johnston, S. M., Li, Y. I., Moroshek, J. G., ... & Kirsch, D. G. (2013). Assessing the radiation response of lung cancer with different gene mutations using genetically engineered mice. *Frontiers in oncology*, *3*, 43811.
56. Petryna, A. (2004). Biological citizenship: The science and politics of Chernobyl-exposed populations. *Osiris*, *19*, 250-265.
57. Pritchard, C. C., Mateo, J., Walsh, M. F., De Sarkar, N., Abida, W., Beltran, H., ... & Nelson, P. S. (2016). Inherited DNA-repair gene mutations in men with metastatic prostate cancer. *New England Journal of Medicine*, *375*(5), 443-453.

58. Prolla, T. A., Baker, S. M., Harris, A. C., Tsao, J. L., Yao, X., Bronner, C. E., ... & Liskay, R. M. (1998). Tumour susceptibility and spontaneous mutation in mice deficient in Mlh1, Pms1 and Pms2 DMA mismatch repair. *Nature Genetics*, *18*(3), 276-279.
59. Rockova, V., Abbas, S., Wouters, B. J., Erpelinck, C. A., Beverloo, H. B., Delwel, R., ... & Valk, P. J. (2011). Risk stratification of intermediate-risk acute myeloid leukemia: integrative analysis of a multitude of gene mutation and gene expression markers. *Blood, The Journal of the American Society of Hematology*, *118*(4), 1069-1076.
60. Rubbi, C. P., & Milner, J. (2003). p53 is a chromatin accessibility factor for nucleotide excision repair of DNA damage. *The EMBO journal*, *22*(4), 975-986.
61. Sancar, A., Lindsey-Boltz, L. A., Ünsal-Kaçmaz, K., & Linn, S. (2004). Molecular mechanisms of mammalian DNA repair and the DNA damage checkpoints. *Annual review of biochemistry*, *73*(1), 39-85.
62. Sankaranarayanan, K. (1991). Ionizing radiation and genetic risks III. Nature of spontaneous and radiation-induced mutations in mammalian in vitro systems and mechanisms of induction of mutations by radiation. *Mutation Research/Reviews in Genetic Toxicology*, *258*(1), 75-97.
63. Schaaper, R. M., & Dunn, R. L. (1987). Spectra of spontaneous mutations in *Escherichia coli* strains defective in mismatch correction: the nature of in vivo DNA replication errors. *Proceedings of the National Academy of Sciences*, *84*(17), 6220-6224.
64. Schweizer, M. T., Antonarakis, E. S., Bismar, T. A., Guedes, L. B., Cheng, H. H., Tretiakova, M. S., ... & Pritchard, C. C. (2019). Genomic characterization of prostatic ductal adenocarcinoma identifies a high prevalence of DNA repair gene mutations. *JCO Precision Oncology*, *3*, 1-9.
65. Seifert, M., & Reichrath, J. (2006). The role of the human DNA mismatch repair gene hMSH2 in DNA repair, cell cycle control and apoptosis: implications for pathogenesis, progression and therapy of cancer. *Journal of molecular histology*, *37*, 301-307.
66. Seligmann, H. (2012). Coding constraints modulate chemically spontaneous mutational replication gradients in mitochondrial genomes. *Current Genomics*, *13*(1), 37-54.
67. Shcherbakova, P. V., & Pavlov, Y. I. (1996). $3' \rightarrow 5'$ exonucleases of DNA polymerases and δ correct base analog induced DNA replication errors on opposite DNA strands in *Saccharomyces cerevisiae*. *Genetics*, *142*(3), 717-726.

68. Shelton, J., Lu, X., Hollenbaugh, J. A., Cho, J. H., Amblard, F., & Schinazi, R. F. (2016). Metabolism, biochemical actions, and chemical synthesis of anticancer nucleosides, nucleotides, and base analogs. *Chemical Reviews*, *116*(23), 14379-14455.
69. Snell, R. G., MacMillan, J. C., Cheadle, J. P., Fenton, I., Lazarou, L. P., Davies, P., ... & Shaw, D. J. (1993). Relationship between trinucleotide repeat expansion and phenotypic variation in Huntington's disease. *Nature Genetics*, *4*(4), 393-397.
70. Sniegowski, P. D., Gerrish, P. J., & Lenski, R. E. (1997). Evolution of high mutation rates in experimental populations of E. coli. *Nature*, *387*(6634), 703-705.
71. Sniegowski, P. D., Gerrish, P. J., Johnson, T., & Shaver, A. (2000). The evolution of mutation rates: separating causes from consequences. *Bioessays*, *22*(12), 1057-1066.
72. Srinivas, M., & Patnaik, L. M. (1994). Adaptive probabilities of crossover and mutation in genetic algorithms. *IEEE Transactions on Systems, Man, and Cybernetics*, *24*(4), 656-667.
73. Steingrímsson, E., Moore, K. J., Lamoreux, M. L., Ferré-D'Amaré, A. R., Burley, S. K., Sanders Zimring, D. C., ... & Jenkins, N. A. (1994). Molecular basis of mouse microphthalmia (mi) mutations helps explain their developmental and phenotypic consequences. *Nature Genetics*, *8*(3), 256-263.
74. Uchimura, A., Higuchi, M., Minakuchi, Y., Ohno, M., Toyoda, A., Fujiyama, A., ... & Yagi, T. (2015). Germline mutation rates and the long-term phenotypic effects of mutation accumulation in wild-type laboratory mice and mutator mice. *Genome Research*, *25*(8), 1125-1134.
75. Uziel, T., Lerenthal, Y., Moyal, L., Andegeko, Y., Mittelman, L., & Shiloh, Y. (2003). Requirement of the MRN complex for ATM activation by DNA damage. *The EMBO journal*, *22*(20), 5612-5621.
76. Verlinsky, Y., Rechitsky, S., Verlinsky, O., Xu, K., Schattman, G., Masciangelo, C., ... & Kuliev, A. (2001). Preimplantation diagnosis for p53 tumour suppressor gene mutations. *Reproductive biomedicine online*, *2*(2), 102-105.
77. Wei, Y., Wu, J., Gu, W., Qin, X., Dai, B., Lin, G., ... & Ye, D. (2019). Germline DNA repair gene mutation landscape in Chinese prostate cancer patients. *European urology*, *76*(3), 280-283.
78. Wikonkal, N. M., & Brash, D. E. (1999, September). Ultraviolet radiation induced signature mutations in photocarcinogenesis. In *Journal of Investigative Dermatology Symposium Proceedings*. Elsevier. (Vol. 4, No. 1, pp. 6-10).

79. Witkin, E. M. (1969). Ultraviolet-induced mutation and DNA repair. *Annual Review of Genetics*, *3*(1), 525-552.

80. Yap, K. L., Kiyotani, K., Tamura, K., Antic, T., Jang, M., Montoya, M., ... & Nakamura, Y. (2014). Whole-exome sequencing of muscle-invasive bladder cancer identifies recurrent mutations of UNC5C and prognostic importance of DNA repair gene mutations on survival. *Clinical cancer research*, *20*(24), 6605-6617.

81. Yoshiyama, K., Higuchi, K., Matsumura, H., & Maki, H. (2001). Directionality of DNA replication fork movement strongly affects the generation of spontaneous mutations in Escherichia coli. *Journal of Molecular Biology*, *307*(5), 1195-1206.

82. Zhu, F., Gong, F., Lin, G., & Lu, G. (2013). DPY19L2 gene mutations are a major cause of globozoospermia: identification of three novel point mutations. *MHR: Basic science of reproductive medicine*, *19*(6), 395-404.

CHAPTER 8

THE CELL CYCLE

Contents

8.1 INTRODUCTION

The intricate interplay between genetic factors and the cell cycle serves as a cornerstone for comprehending the intricate mechanisms that govern the fundamental processes of life. Genetics is a fundamental factor in coordinating the complex progression of the cell cycle, starting from conception and continuing throughout the formation of a multicellular organism and beyond. This chapter aims to examine the intricate relationship between genetics and the cell cycle, focusing on elucidating the mechanisms that regulate cell division, maintain genetic stability, and the consequences of their disruption on human well-being and pathological conditions (Murray, 1995).

The cell cycle is a meticulously coordinated sequence of events that facilitates cellular growth, duplication of genetic material, and subsequent division into two progeny cells. This process holds significant importance in facilitating regular development, promoting tissue regeneration, and upholding the overall well-being of multicellular organisms. The cell cycle encompasses four primary phases, namely G1 (Gap 1), S (Synthesis), G2 (Gap 2), and M (Mitosis). The progression of each phase is rigorously regulated by a multitude of genetic elements, guaranteeing the seamless continuation of the cycle and rectifying any potential errors before the initiation of cell division (Hartwell, 1991).

Key regulatory molecules, such as cyclins and cyclin-dependent kinases (CDKs), play a central role in the genetic control of the cell cycle. These proteins function as molecular switches, facilitating the signaling of transitions between distinct cell cycle phases. Furthermore, the preservation of genome integrity is upheld by tumor suppressor genes such as p53, which can halt the progression of the cell cycle or initiate programmed cell death, known as apoptosis, upon identification of DNA damage. Conversely, oncogenes, such as Ras, have the potential to induce unregulated cellular proliferation and division in the event of their abnormal activation.

The relationship between genetics and the cell cycle is significant, and disruptions in this complex interaction can have serious ramifications. The occurrence of genetic mutations has the potential to impede the normal operation of genes responsible for regulating the cell cycle, resulting in unregulated cellular proliferation and the subsequent formation of cancer. In the present context, it is of utmost importance to comprehend the genetic underpinnings of the cell cycle to facilitate the advancement of precise therapeutic approaches and interventions aimed at addressing cancer and other disorders associated with the cell cycle (Baylin, 1997). In the forthcoming chapter, an in-depth examination will be conducted on the genetic foundations of the cell cycle, with a particular emphasis on elucidating the pivotal genes and proteins responsible for regulating its advancement. Additionally, an exploration will be conducted on the relationship between genetic mutations and the occurrence of cell cycle abnormalities, as well as the potential ramifications for human health. Furthermore, an examination will be conducted on the crucial significance of epigenetics in regulating the cell cycle, focusing on elucidating the mechanisms by which alterations in DNA methylation and histone modifications can impact the expression of genes throughout various phases of the cell cycle (Williams & Stoeber, 2012).

One of the crucial elements that will be analyzed is the function of cell cycle checkpoints. The control mentioned above points serve as crucial regulators that function as molecular gatekeepers, ensuring that the progression of the cell cycle occurs solely when specific prerequisites have been fulfilled. Checkpoints serve as surveillance mechanisms responsible for detecting and correcting any errors that may occur during DNA replication or chromosome segregation, ensuring that the cell does not progress to the subsequent phase until these errors have been resolved. Genomic instability, a characteristic feature of cancer cells, can arise due to dysfunction in these checkpoints (Menges et al., 2002).

In addition, we will investigate the intriguing phenomenon of mitosis, wherein a parental cell undergoes division to yield two daughter cells that possess identical genetic composition. This notable achievement entails the accurate allocation of genetic material, guaranteeing that every progeny cell obtains the appropriate assortment of chromosomes. Mitotic errors can result in chromosomal abnormalities, which have noteworthy implications for both genetic diversity and human health (Ruijtenberg & van den Heuvel, 2016).

Furthermore, we will explore the distinct mechanism of meiosis. This crucial process occurs during gametes (ova and spermatozoa) development and plays a vital role in sexual reproduction. The process of meiosis encompasses two successive rounds of cellular division, ultimately yielding four daughter cells that are genetically distinct from one another. The process mentioned above plays a pivotal role in generating genetic diversity within a given species and in transmitting genetic traits from parental individuals to their progeny (Lundberg & Weinberg, 1999).

In recent years, significant progress has been made in genetic testing, leading to a transformative impact on our comprehension of disorders associated with the cell cycle. Genetic testing enables the detection of distinct genetic mutations responsible for diverse irregularities in the cell cycle, thereby providing significant knowledge for diagnosing, predicting the course of the disease, and tailoring treatment approaches to individual patients. In this discourse, we shall deliberate upon the significance of genetic counseling and the ethical considerations encompassing genetic testing and its potential ramifications on individuals and their familial units (De Lichtenberg et al., 2005).

In conclusion, our focus will shift toward the prospective aspects of genetics and the cell cycle, delving into current research patterns and the potential implications of genetic breakthroughs in comprehending and managing diseases associated with the cell cycle.

The emergence of advanced technologies such as CRISPR-Cas9 gene editing has opened up limitless possibilities for precise therapeutic interventions (Nasmyth & Reed, 1980).

In summary, the complex interplay between genetic factors and the process of cell division represents a remarkable phenomenon that constitutes the fundamental essence of living organisms. Comprehending the genetic mechanisms that regulate the cell cycle is of utmost importance in elucidating the enigmas surrounding human development and well-being. Furthermore, it can potentially revolutionize the treatment of cell cycle-associated disorders, such as cancer. In this chapter, we aim to conduct a thorough investigation into the intriguing convergence of genetics and the cell cycle, thereby illuminating the intricacies of the fundamental processes underlying life (Papavasiliou & Schatz, 2000).

The cell cycle can be likened to a meticulously orchestrated performance, characterized by a sequence of meticulously timed processes that guarantee accurate cellular replication.

8.2 THE CELL CYCLE

The cell cycle regulates cell longevity. The process controls cell development, genetic replication, and cell division into two daughter cells. All living things depend on this process's complexity. Understanding the cell cycle has helped scientists understand cellular biology, genetics, and the causes of illnesses like cancer. This extensive study will examine the cell cycle's essential components and key events at each stage (Johnson & Walker, 1999).

8.2.1. Overview of the Cell Cycle

The cell cycle is a dynamic and uninterrupted process subdivided into discrete phases, each distinguished by specific events and activities. The phases are intricately linked and rigorously controlled, guaranteeing the appropriate synchronization and advancement of the cellular cycle. The cell cycle consists of distinct phases, which can be categorized as follows (Bodwell et al., 1998):

- **G1 Phase (Gap 1)**: After undergoing cell division, the cell proceeds to the G1 phase, which increases in size and synthesizes proteins essential for the subsequent phases. This stage is crucial for cellular growth and metabolic processes. Significantly, cells not currently undergoing division can withdraw from the cell cycle and transition into a non-dividing phase known as G0. In this state, cells can remain dormant until they receive external signals that prompt them to resume the cell cycle (Pramila et al., 2006).

- **S Phase (Synthesis)**: During this crucial stage, the DNA undergoes replication. The process of chromosome duplication within a cell leads to the formation of two sister chromatids that are identical and held together by a centromere. DNA replication is characterized by a high degree of coordination, as it incorporates various checkpoints to guarantee precise and error-free duplication of genetic material (Celis & Celis, 1985).
- **G2 Phase (Gap 2)**: Following the process of DNA replication, the cell proceeds to transition into the G2 phase. During this stage, the cellular entity synthesizes supplementary proteins and organelles in anticipation of the subsequent cellular division. Additionally, rigorous examinations are conducted to verify the integrity and accuracy of the replicated DNA, thereby ensuring its freedom from any potential errors or damage. If the DNA is healthy, the cell progresses to the subsequent phase (Natarajan et al., 1980).
- **M Phase (Mitosis):** The M phase represents the culmination of the cell cycle and is further subdivided into multiple sub-phases. Mitosis can be subdivided into four distinct phases: prophase, metaphase, anaphase, and telophase. Each phase is characterized by specific events contributing to the accurate segregation of genetic material. The process of mitosis guarantees the equitable distribution of a complete and identical set of chromosomes to each daughter cell.8.2.2. Cell Cycle Regulation.

The cell cycle is meticulously regulated to maintain genomic integrity and ensure cells divide appropriately. Key regulatory molecules known as cyclins and cyclin-dependent kinases (CDKs) control the progression from one phase to another. Cyclins are proteins that oscillate in concentration throughout the cell cycle, and CDKs are enzymes whose activity depends on the presence of specific cyclins. When a cyclin binds to its corresponding CDK, it activates the kinase, initiating the cellular events necessary to transition to the next phase (Grobelny et al., 2000).

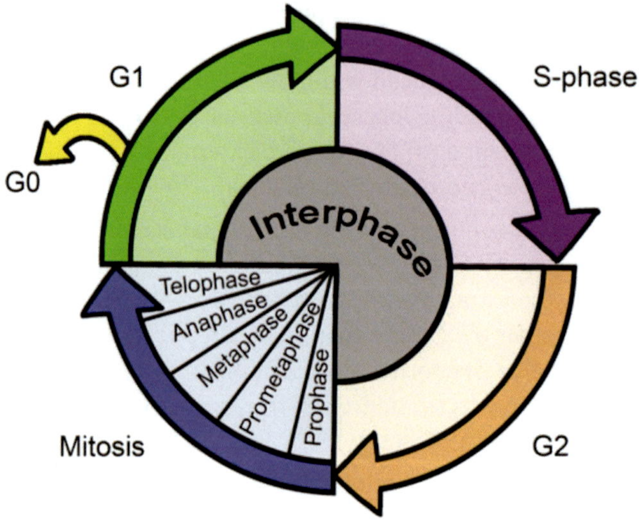

Figure 8.1. Cell cycle & its regulation (Source: Martí-Clúa J. Methods for Inferring Cell Cycle Parameters Using Thymidine Analogues. *Biology*. 2023; 12(6):885. https://doi.org/10.3390/biology12060885, Creative Commons License)

The G1 checkpoint, commonly known as the restriction point, is a significant milestone in the cell cycle. Currently, the cellular entity evaluates internal and external cues, thereby ascertaining the suitability of initiating DNA replication and cellular division. Cells that successfully traverse this particular checkpoint are obligated to proceed with the entirety of the cell cycle.

The G2-M checkpoint is a pivotal checkpoint that verifies the completion of DNA replication and the readiness of the cell to progress into mitosis. If errors or insufficient replication are identified, the cell cycle undergoes arrest to impede the development of genetically unstable progeny cells (Abrieu et al., 1998).

8.2.2. Cell Cycle Abnormalities and Genetic Mutations

The preservation of genetic material integrity within an organism relies on properly regulating the cell cycle. However, genetic mutations or the disturbance of genes regulating the cell cycle can lead to abnormal cell division, potentially contributing to the development of different diseases, such as cancer. For example, mutations in tumor suppressor genes, such as p53, can lead to the loss of cell cycle checkpoints, facilitating the growth of dysfunctional cells and the subsequent accumulation of additional genetic alterations. The progression of these events can culminate in the formation of malignant cells (Roncalli et al., 1998).

In contrast, oncogenes are genetic components that promote cellular proliferation and mitosis. The phenomenon of uncontrolled cellular proliferation, a recognized characteristic feature of cancer, can arise from mutations that activate oncogenes. For example, genetic modifications in the Ras gene can elicit continuous activation of the Ras protein, thus facilitating abnormal cellular proliferation.

In addition, dysfunctions in the DNA repair machinery can result in the accumulation of genetic mutations, thereby amplifying abnormalities in the cell cycle and contributing to the development of pathological disorders (MacPherson et al., 2003).

8.2.3. Genetic Regulation of Cell Cycle Checkpoints

Cell cycle checkpoints are crucial for genomic integrity. Cell cycle processes prevent cells with DNA damage or abnormal chromosomal counts.

G1 checkpoints depend on p53. The "guardian of the genome" p53 protein monitors DNA integrity and coordinates cellular responses to DNA damage. p53 protein induces cell cycle arrest to allow DNA repair after DNA damage. The p53 protein can cause apoptosis to kill damaged cells when repair is impossible. This prevents harmful mutations from spreading (Hu et al., 2001).

RB regulates cell cycle checkpoints. Retinoblastoma protein (RB) regulates the G1-S checkpoint, which prevents cells from initiating DNA synthesis (S phase) until they receive growth signals and are ready for replication.

The protein kinases Chk1 and Chk2 monitor DNA integrity and cell readiness for mitosis at the G2-M checkpoint. They ensure DNA replication and repair before mitosis (Nakanishi et al., 2006).

8.2.4. Epigenetics and the Cell Cycle

Epigenetics regulates the cell cycle with genetics. Chemical changes in DNA and histones can affect gene expression without changing the genetic coding. DNA methylation, histone acetylation, and histone methylation have been widely studied concerning cell cycle control.

DNA methylation attaches methyl groups to cytosine residues in the DNA molecule. This alteration often suppresses or inactivates genes. DNA methylation patterns affect cell cycle gene expression. DNA hypermethylation silences p16INK4A, allowing uncontrolled cell growth in many cancer cells.

Histone changes also regulate the cell cycle. Histone acetylation, which opens chromatin, makes gene regulatory areas more accessible and boosts gene expression. However, histone deacetylation can suppress genes and slow the cell cycle (Kamal et al., 2018).

8.2.5. Mitosis and Genetics

Mitosis is a cellular mechanism through which a parent cell partitions its replicated genetic material into two daughter cells with identical genetic content. The process of chromosome segregation is characterized by a high degree of coordination and precision.

In the prophase stage of cell division, the chromatin undergoes condensation, forming discernible chromosomes. Simultaneously, the nuclear envelope disassembles. The formation and organization of chromosomes is initiated by the mitotic spindle, comprised of microtubules (Weinert et al., 1994).

During the metaphase stage of cell division, the replicated chromosomes linearly arrange themselves along the equatorial plane of the cell, which is commonly known as the metaphase plate. This configuration guarantees that every daughter cell obtains an identical complement of chromosomes. During the anaphase phase, the sister chromatid separation process occurs, wherein the mitotic spindle facilitates the movement of these chromatids toward opposite poles of the cell. Ensuring the inheritance of a complete set of chromosomes is a crucial step (Sikorski et al., 1990).

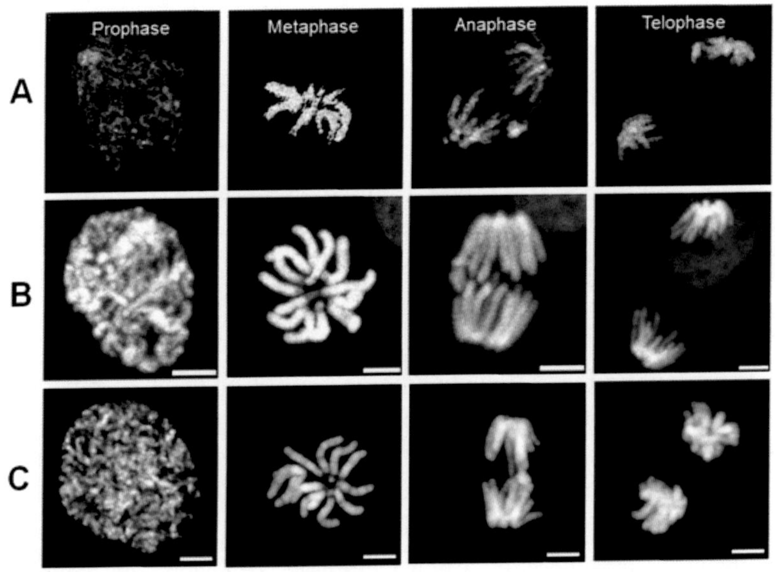

Figure 8.2. Mitosis phases (Source: Kim T-K, Lee B-W, Fujii F, Lee K-H, Lee S, Park Y, Kim JK, Lee S-W, Pack C-G. Mitotic Chromosomes in Live Cells Characterized Using High-Speed and Label-Free Optical Diffraction Tomography. *Cells*. 2019; 8(11):1368. https://doi.org/10.3390/cells8111368, Creative Commons License)

In the telophase stage, the two sets of chromosomes migrate towards opposite poles, and subsequently, a fresh nuclear envelope is established around each set. Subsequently, the cell proceeds with cytokinesis, a process in which the cytoplasm and organelles are partitioned between the resulting daughter cells. Mistakes occurring during mitosis can result in aneuploidy, a pathological state characterized by an atypical number of chromosomes within the progeny cells. Aneuploidy is a prevalent characteristic observed in cancer cells, which can play a role in the initiation and advancement of tumors (Russell & Nurse, 1987).

8.2.6. Meiosis and Genetic Variation

Mitosis is responsible for the accurate duplication and equitable allocation of genetic material to somatic cells. In contrast, meiosis is a specialized type of cellular division that generates haploid gametes (sperm and eggs) intended for sexual reproduction. Meiosis is a biphasic cellular process characterized by distinct stages of cell division (Biswas et al., 2021). During the first stage of meiosis, meiosis I, homologous chromosome pairing occurs, facilitating the exchange of genetic material through crossing over. The process of genetic recombination between homologous chromosomes contributes to the generation of genetic diversity, as the resulting combinations of alleles can give rise to unique traits in the progeny (Chowdhury et al., 2009). Meiosis II resembles mitosis as it encompasses the segregation of sister chromatids. In contrast to mitosis, the resultant progeny cells are haploid, possessing fewer chromosomes than somatic cells.

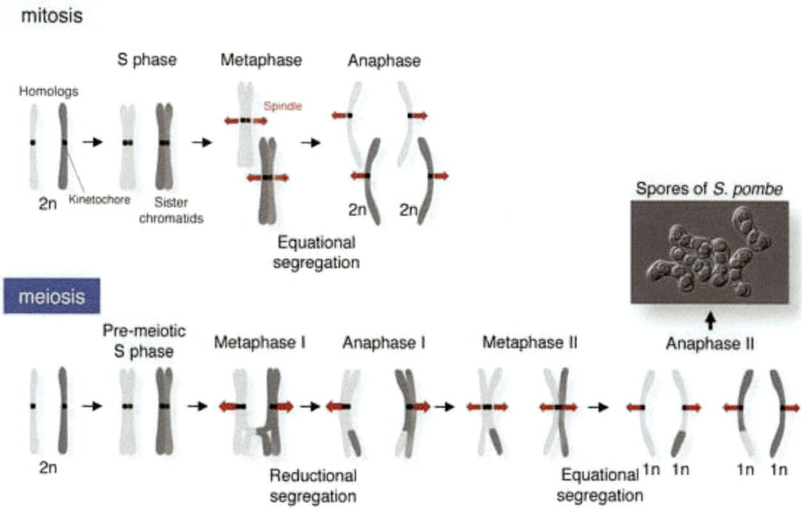

Figure 8.3. Meiosis I & Meiosis II (Source: Sakuno T, Hiraoka Y. Rec8 Cohesin: A Structural Platform for Shaping the Meiotic Chromosomes. *Genes*. 2022; 13(2):200. https://doi.org/10.3390/genes13020200, Creative Commons License)

The genetic diversity generated through meiosis plays a crucial role in species' persistence and evolutionary development. Genetic variation plays a significant role in fostering the observed diversity within populations and facilitating the process of adaptation to dynamic environmental circumstances (Srivastava et al., 1992).

The cell cycle is a highly complex and precisely regulated phenomenon that serves as the fundamental basis for the survival and perpetuation of all living organisms. The seamless execution of various processes is paramount for the optimal growth, development, and maintenance of multicellular organisms. The intricate relationship between genetics and the cell cycle constitutes a captivating field of study that has provided valuable insights into the underlying mechanisms that regulate cellular biology and human well-being (Lambing & Heckmann, 2018).

A comprehensive grasp of the cell cycle's fundamental principles and sequential stages serves as a fundamental basis for delving deeper into the intricate mechanisms governing its regulation and comprehending the consequences of its disruption in diverse pathological conditions, with a particular emphasis on cancer. The accurate regulation of the cell cycle is achieved through the implementation of checkpoints and the coordination of genetic and epigenetic processes, which collectively guarantee the precise and faithful occurrence of each division event (Gorelick & Heng, 2011).

As the scientific community continues to progress in understanding the intricacies of the cell cycle, new opportunities arise for developing innovative therapeutic strategies to address aberrations in cell cycle regulation observed in various diseases, notably cancer. These advancements hold the potential to impact human health and overall welfare significantly (Choi & Henderson, 2015). The intricate choreography of the cell cycle continues to captivate researchers and offers the potential to reveal further insights into the fundamental mechanisms of life (Cai & Xu, 2007).

Remember

During the S phase (DNA replication), the DNA in a human cell unwinds and stretches to about two meters long – yet it fits back into the tiny cell nucleus after replication!

8.3 PHASES

Eukaryotic cell cycles include four phases: G1, S (synthesis), G2 (interphase), and M (mitosis and cytokinesis). Mitosis divides the cell's nucleus, and cytokinesis, which divides its cytoplasm, generates two daughter cells during the

M phase. Each step activates when the previous phase is completed. Quiescent cells in the G0 phase have temporarily stopped dividing.

Each daughter cell enters interphase after cell division. Each cell cycle phase contains a distinct metabolic activity set that prepares the cell for cell division. The morphological variations between interphase stages are usually not noticeable (Tyson & Novak, 2001).

8.3.1 G_0 Phase (quiescence)

The G0 phase denotes a quiescent state in which the cell has exited the cell cycle and ceased its division. The initiation of the cell cycle commences with this particular phase. In multicellular eukaryotes, non-proliferative cells typically transition from the G1 phase to the quiescent G0 state. As observed in neurons, these cells can remain in this quiescent state for extended durations, potentially indefinitely. This phenomenon is frequently observed in cells that have undergone complete differentiation. Certain cells undergo a semi-permanent transition into the G0 phase, rendering them post-mitotic. Examples of such cells include those found in the liver, kidney, and stomach. Numerous cellular entities exhibit a perpetual absence of G0 phase entry, thereby sustaining a continuous cycle of division throughout the lifespan of an organism. Epithelial cells serve as a prime example of such cells (Oki et al., 2014).

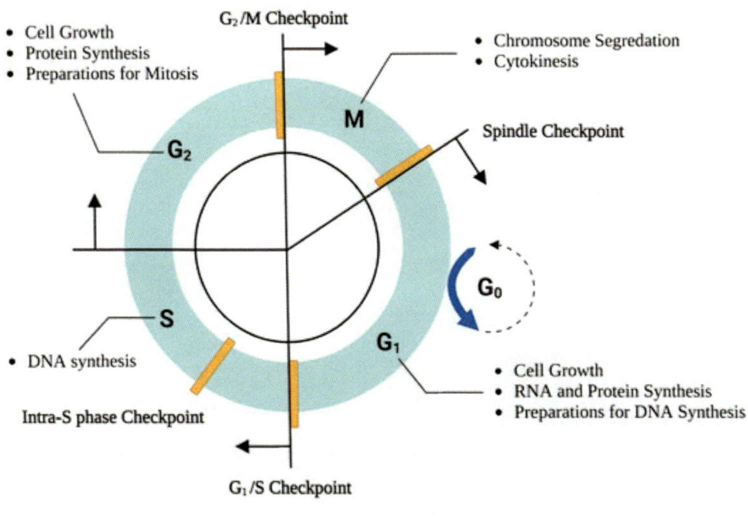

Figure 8.4. Phases of cell cycle (Source: Deng B, Song A, Zhang C. Cell-Cycle Dysregulation in the Pathogenesis of Diabetic Kidney Disease: An Update. *International Journal of Molecular Sciences*. 2023; 24(3):2133. https://doi.org/10.3390/ijms24032133, Creative Commons License)

"Post-mitotic" is occasionally employed to encompass quiescent and senescent cellular states. Cellular senescence is a biological process triggered by DNA damage and external stressors, typically resulting in a halt in the G1 phase of the cell cycle. Cellular senescence can result in the inability of a cell's offspring to survive, serving as a biochemical alternative to the programmed cell death known as apoptosis in cases where the cell is extensively damaged (Yu et al., 2017).

8.3.2 Interphase

The interphase is the cell cycle period that occurs between two consecutive M phases. Interphase refers to a sequence of alterations occurring within a recently formed cell and its nucleus before its ability to undergo division once more. This particular phase is alternatively called the preparatory phase or enter mitosis. In general, interphase accounts for a minimum of 91% of the overall duration of the cell cycle. The process of interphase consists of three distinct stages, namely G1, S, and G2, followed by the cycle of mitosis and cytokinesis. The nuclear DNA contents of the cell undergo duplication during the S phase (Starborg et al., 1996).

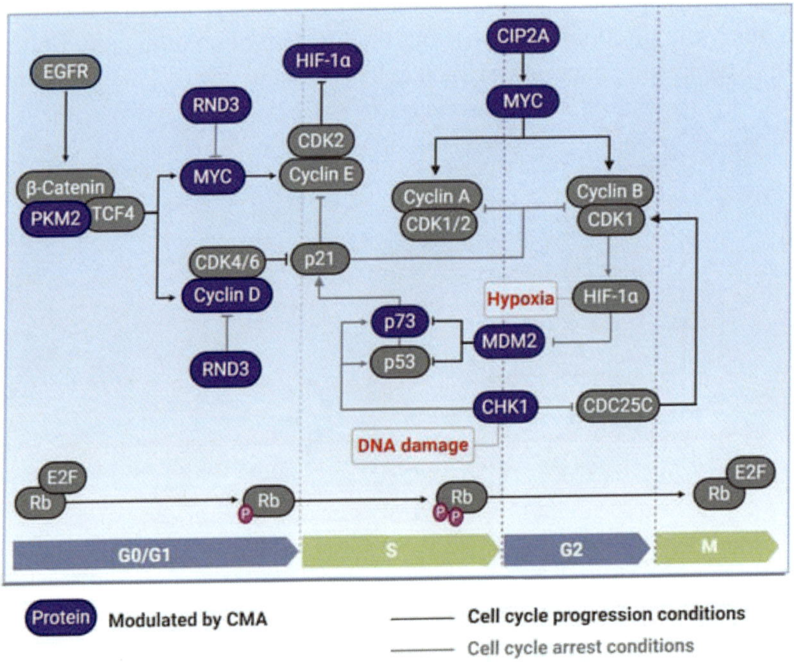

Figure 8.5. The cell cycle involves many proteins (Source: Andrade-Tomaz M, de Souza I, Rocha CRR, Gomes LR. The Role of Chaperone-Mediated Autophagy in Cell Cycle Control and Its Implications in Cancer. *Cells*. 2020; 9(9):2140. https://doi.org/10.3390/cells9092140, Creative Commons License)

8.3.3 G_1 Phase (First growth phase or Post mitotic gap phase)

The initial interphase stage, from the conclusion of the preceding M phase to the commencement of DNA synthesis, is referred to as G1 (where G denotes gap). This phase is alternatively referred to as the growth phase. During the subsequent phase, the cellular biosynthetic activities, which experience a significant deceleration during the M phase, recommence at an elevated pace. The length of the G1 phase exhibits considerable variability, even within cells of the same species. During this stage, the cell undergoes an augmentation in protein synthesis, an increase in the abundance of organelles such as mitochondria and ribosomes, and an overall growth in cellular size. During the G1 phase of the cell cycle, a cell is presented with three distinct options (Marlier et al., 2020).

To progress through the cell cycle and initiate the S phase, it is necessary to undertake specific cellular processes.

Cease progression through the cell cycle and transition into the G0 phase to initiate the process of cellular differentiation.

- The individual may experience arrest during the G1 phase, leading to the possibility of entering the G0 phase or re-entering the cell cycle.

The term used to refer to the determining factor is commonly known as the checkpoint, also referred to as the restriction point. The specific stage in the cell cycle, the restriction point or START, is controlled by G1/S cyclins, which facilitate the transition from the G1 phase to the S phase. The cell's commitment to division is established upon passing through the G1 checkpoint (Owa et al., 2001).

8.3.4 S phase (DNA replication)

The S phase, the synthesis phase, is a pivotal stage within the cell cycle wherein DNA replication occurs. The S phase of the cell cycle succeeds the G1 phase and precedes the G2 phase, thereby occupying the second phase of the cell cycle. During the S phase of the cell cycle, a process known as DNA replication occurs, wherein each chromosome within the cell undergoes duplication. This crucial step ensures the accurate and faithful replication of the genetic material before cell division. DNA replication is characterized by high precision and complexity, as it involves numerous enzymes and checkpoints to guarantee the faithful duplication of the entire genome (Xu et al., 2001).

1. **Initiation of DNA Replication**

   DNA replication occurs at distinct loci present on the chromosomes, referred to as the origins of replication. During the initial stages, a cluster of proteins assembles to create a pre-replication complex, thereby initiating the replication process. The origin recognition complex (ORC) is a pivotal protein that plays a crucial role in identifying and binding replication origins. Additional factors, such as CDC6 and Cdt1, are recruited to the protein complex, thereby facilitating the mini-chromosome maintenance (MCM) helicase loading onto the DNA strands.

2. **Unwinding the DNA Helix**

The MCM helicase assumes a pivotal function in the process of DNA unwinding, effectively separating the two double-helix strands to facilitate the replication machinery's access. The procedure mentioned above initiates the formation of a replication bubble comprising two replication forks that traverse in divergent directions from the point of origin of replication. The sites at which DNA replication occurs are known as replication forks (Fuss & Tainer, 2011).

3. **DNA Polymerization**

DNA replication depends on the activity of DNA polymerases, which are enzymes responsible for facilitating the synthesis of fresh DNA strands. The primary DNA polymerase responsible for replication in eukaryotic cells is DNA polymerase α (pol α), which initiates DNA synthesis. Following the initial synthesis, DNA polymerase δ (pol δ) continuously elongates the leading strand. In contrast, DNA polymerase ε (pol ε) synthesizes the lagging strand in discrete Okazaki fragments.

DNA replication involves adding nucleotides to the elongating DNA strand by DNA polymerases, which accurately pair complementary bases. As an illustration, it can be observed that an adenine (A) situated on the template strand forms a complementary base pair with thymine (T) on the newly synthesized strand. Similarly, a guanine (G) on the template strand pairs with a cytosine (C) on the new strand. This procedure guarantees that the newly synthesized DNA strands are replicas of the original strands.

4. **RNA Primers and DNA Ligase**

Before initiating nucleotide addition to the template strand, the enzymatic activity of primase is employed to synthesize a brief RNA primer. The present primer serves as the primary initiation site for DNA polymerases to facilitate the addition of nucleotides. Subsequently, the RNA primers are enzymatically eliminated and substituted with DNA by additional enzymes.

After the synthesis of the new DNA strands, the RNA primers are eliminated through the action of DNA polymerase δ. The RNA primers create gaps that are subsequently filled with DNA by the action of DNA polymerase δ and pol ε. Ultimately, DNA ligation is responsible for the closure of discontinuities between neighboring DNA fragments, leading to the formation of uninterrupted and intact DNA strands (Sinha et al., 2006).

5. **Replication Fidelity and Checkpoints**

   DNA replication is a highly precise biological process, exhibiting a remarkably low error rate of approximately one error per billion base pairs. The preservation of genetic material's integrity and the prevention of mutations, which may result in detrimental outcomes like cancer, necessitate the presence of high fidelity.

During the S phase, some checkpoint mechanisms are employed to verify the precision and entirety of DNA replication. The checkpoints mentioned above monitor various aspects of DNA integrity, including detecting DNA damage, assessing replication fork stability, and verifying the accurate completion of DNA replication. If errors or abnormalities are identified, the cell cycle can undergo a temporary pause, enabling sufficient time for DNA repair before advancing to the subsequent phase of the cell cycle. The S phase is a critical phase of the cell cycle, wherein the genetic material is accurately replicated. DNA replication is a highly intricate and meticulously controlled mechanism wherein many enzymes and proteins work harmoniously to guarantee DNA strands' precise replication. Cells can maintain the continuity of life and facilitate the development and maintenance of multicellular organisms by accurately duplicating their genome and transmitting their genetic information to their progeny during cell division. Comprehending the complexities of DNA replication is of utmost importance in cellular biology, as it has profound implications for human health and disease, specifically cancer and various genetic disorders (Luciani et al., 2007).

8.3.5 G_2 Phase

The G2 phase, also known as the Gap 2 phase, represents the third stage within the cell cycle, occurring after the S phase, which involves DNA replication and preceding the M phase, encompassing mitosis. During the G2 phase, the cell undergoes continued growth and engages in preparatory activities in anticipation of the subsequent process of cell division. This stage is of utmost importance in guaranteeing the integrity and accuracy of the replicated DNA before entering the mitosis process. The G2 phase of the cell cycle encompasses synthesizing a diverse array of proteins and organelles essential for cell division. Additionally, during this phase, checkpoints are activated to ensure the preservation of DNA integrity (Hwang & Muschel, 1998).

1. **Continuing Cell Growth**

   Following the conclusion of DNA replication during the S phase, the cell proceeds into the G2 phase. During the current phase, cellular expansion persists while proteins and organelles necessary for the forthcoming cell division are synthesized. To guarantee the equitable distribution of a comprehensive and fully operational set of organelles and cellular elements to each daughter cell during the process of cell division, the cell must undergo a doubling of said organelles and components (Conlon & Raff, 2003).

2. **Activation of Cell Cycle Checkpoints**

   Checkpoints are of utmost importance in maintaining the precise and error-free advancement of the cell cycle. The G2-M checkpoint, the DNA damage checkpoint, holds significant importance within the G2 phase. The purpose of this checkpoint is to assess the integrity of the DNA to confirm the successful completion of DNA replication and the absence of any damage or errors in the DNA.

   Suppose any irregularities are identified at the G2-M checkpoint. In that case, the progression of the cell cycle can be transiently halted to facilitate sufficient time for DNA repair mechanisms to rectify any potential damage. Activating cell cycle checkpoints serves as a crucial mechanism in preventing the proliferation of cells that harbor DNA damage, thereby mitigating the potential consequences of deleterious mutations or genomic instability.

3. **DNA Damage Repair**

   In the G2 phase of the cell cycle, the cell possesses inherent mechanisms that enable the detection and subsequent repair of any DNA damage that might have transpired throughout the cell cycle. Numerous protein complexes, including the MRN complex (Mre11, Rad50, and Nbs1), play a crucial role in detecting and transmitting signals related to DNA damage. These protein complexes facilitate the recruitment of additional proteins involved in DNA repair, including ATM and ATR kinases. These kinases then activate signaling cascades that initiate the processes of DNA repair.

   Cells are equipped with intricate mechanisms for repairing DNA, encompassing processes such as homologous recombination and non-homologous end-joining. These pathways rectify various forms of DNA damage, including double-strand breaks and mismatches. The effective restoration of DNA damage during the G2 phase guarantees the preservation of the cell's genomic integrity and inhibits the propagation of erroneous genetic material to subsequent cell generations (Yoshida & Miki, 2004).

4. **Preparing for Mitosis**

   G2 prepares the cell for mitosis, which divides duplicated genetic material into two offspring cells. The cell is preparing for mitosis, and each daughter cell inherits a complete and exact set of chromosomes if the G2 phase ends successfully.

   In addition to DNA replication and repair, the cell synthesizes extra microtubules and other proteins to construct the spindle, segregating chromosomes during mitosis, in G2, spindle fiber-organizing centrosomes double.

5. **G2/M Transition**

   The cell enters the M phase after the G2 phase, which includes mitosis and cytokinesis. CDK activation and p53 degradation control the transition from the G2 to the M phase.

Cyclin-dependent kinases (CDKs) and cyclin proteins phosphorylate key substrates to accelerate the cell cycle. During the G2/M transition, unique CDK-cyclin complexes activate mitosis and the cell cycle.

The cell prepares for cell division in the highly controlled G2 phase of the cell cycle. These preparations need to check the replicated DNA and synthesize key cellular components. The cell uses checkpoints and DNA repair processes to fix DNA mistakes before entering mitosis. Genomic stability, proper cell division, and multicellular organism health depend on G2 phase advancement. Understanding the G2 phase and its role in cell cycle control is essential to understanding basic cellular biology. This understanding also affects cancer and genetic problems (Krek & Nigg, 1991).

Did You Know?

Billion of cells will go through the cell cycle throughout a typical human lifespan, dividing and rebuilding tissues..

8.4 MITOTIC PHASE (CHROMOSOME SEPARATION)

Karyokinesis—nuclear division—occurs during the brief M phase. This cell cycle phase is short. M-phase cell cycle control is complex. Events are divided into stages that end one set of actions and start another. Sequentially, these phases are called:

- Prophase
- Prometaphase
- Metaphase
- Anaphase
- Telophase

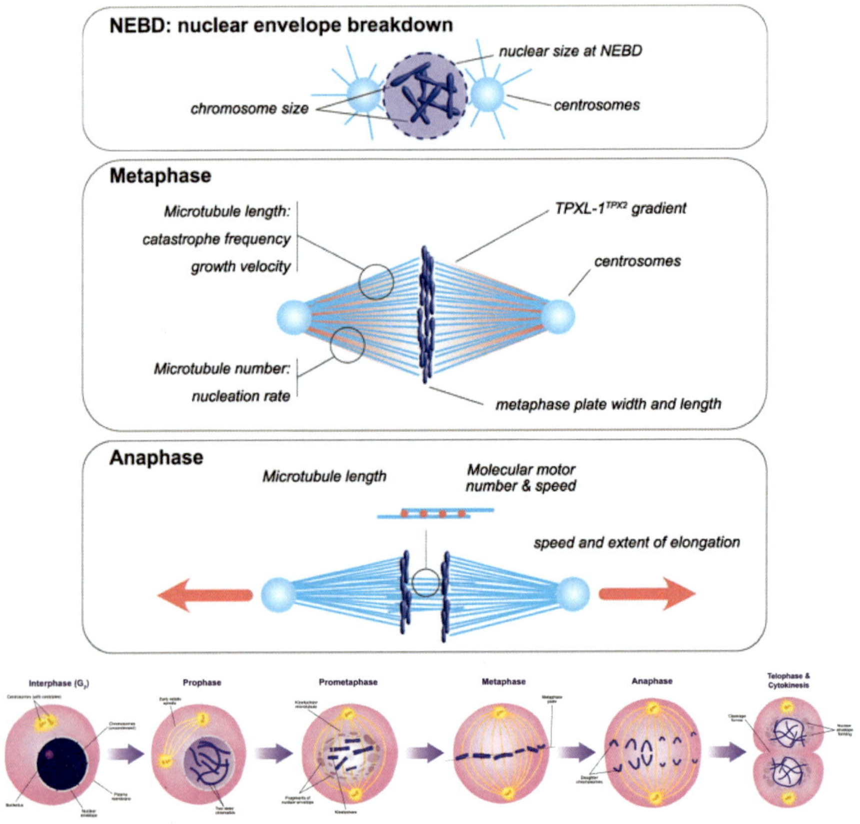

Figure 8.6. Phases of mitosis (Source: Lacroix B, Dumont J. Spatial and Temporal Scaling of Microtubules and Mitotic Spindles. *Cells*. 2022; 11(2):248. https://doi.org/10.3390/cells11020248, Creative Commons License)

Eukaryotes divide their chromosomes into two sets during mitosis, creating two genetically identical nuclei. In mitosis, chromosomal pairs condense and attach to microtubules to separate sister chromatids to opposing cell poles.

Mitosis occurs only in eukaryotic cells, but its processes vary by species. The nuclear envelope disintegrates before chromosome separation in "open" mitosis in animal cells. "Closed" mitosis occurs in fungi like Aspergillus nidulans and Saccharomyces cerevisiae (yeast) (Aist & Berns, 1981).

8.5 CYTOKINESIS PHASE (SEPARATION OF ALL CELL COMPONENTS)

After mitosis, cytokinesis completes cell division. Cytokinesis separates two mitosis-generated offspring cells into two genetically identical cells. The crucial procedure accurately distributes genetic material to daughter cells, allowing them to maintain their roles and let multicellular organisms grow and evolve.

1. **Initiation of Cytokinesis**

   Cytokinesis follows mitosis, where the duplicated genetic material (chromosomes) is separated into two sets. A contractile ring of actin filaments and proteins starts cytokinesis. The contractile ring indicates the cell's physical division position at the equatorial plane (Harris & Hamer, 1995).

2. **Role of the Contractile Ring**

   The contractile ring constricted the cell membrane during cytokinesis, dividing the two offspring cells. The cell membrane constriction is caused by moving actin filaments in the contractile ring. This device works like a drawstring. Cellular division, or "cellular cleavage," involves an inward constriction that separates the daughter cells.

3. **Completion of Cytokinesis in Animal Cells**

   Animal cells form an equatorial contractile ring during cytokinesis. The contractile ring contracts during cytokinesis, creating a furrow that deepens and divides the cell into two daughter cells. The furrow constricts until the two daughter cells are entirely separated with a full complement of chromosomes and organelles.

Remember

The occurrence of cancer can be attributed to a disruption in the cell cycle, whereby uncontrolled cell division arises due to disturbances in the checkpoints of the cell cycle..

4. **Completion of Cytokinesis in Plant Cells**

   Plant cells' stiff cell walls make cytokinesis different from animal cells. Plant cells build a new cell wall between their daughter cells instead of contractile ring development. Initially, vesicles containing cell wall elements like cellulose form a cell plate. The cell plate expands until it fuses with the cell membrane during cytokinesis. This fusion creates a new cell wall, dividing the daughter cells. The cell wall separates the daughter cells, allowing them to develop independently (Kaitna et al., 2000).

5. Coordination with the Cell Cycle

   Cytokinesis is precisely timed with the cell cycle, ensuring effective cell division. Animal cells begin cytokinesis and mitotic spindle breakdown in late anaphase or early telophase. This period ensures correct genetic material segregation before cytokinesis.

   Plant cells begin cytokinesis during late telophase, when the mitotic spindle dissolves. The seamless transition from mitosis to cytokinesis ensures the creation of two daughter cells with identical genetic information.

6. **Implications for Development and Growth**

   Cytokinesis helps multicellular creatures develop and flourish. Cytokinesis helps create tissues and organs by generating new cells. Adult organisms use cytokinesis to regenerate tissues by multiplying cells to replace damaged ones (Zheng & Dean, 2009).

7. **Abnormalities in Cytokinesis**

   Cytokinesis anomalies can harm cells and organisms. Cytokinesis failure can cause multinucleation—cells with numerous nuclei. Multinucleated cells are linked to several disorders, including cancer. Cytokinesis errors can also cause aneuploidy, a pathogenic condition characterized by abnormal chromosomal counts in progeny cells. Cancerous cells have aneuploidy, which disrupts cell genetics. Cytokinesis completes cell division and ensures the formation of two identical daughter cells. Plant cells create cell plates, while animal cells produce contractile rings. These components help the cell divide into two entities. Mitosis and cytokinesis synchronization ensures progeny cells get genetic material and biological components. Cytokinesis is essential for multicellular organisms' development, growth, and tissue repair, and any disruptions can have serious health repercussions. Cytokinesis mechanisms reveal essential biological processes and their effects on human health, development, and illness. The cell cycle's last step of studying cytokinesis may reveal more about biological processes (Kaiser et al., 2002).

8.6 CELL CYCLE DISORDERS

Cell cycle diseases, sometimes called cell cycle-related disorders or cell cycle abnormalities, include cell cycle dysregulation or disruption. These illnesses can cause several diseases and developmental abnormalities. Cell cycle disorders (Michael et al., 2008):

1. **Cancer**: Cancer is a well-known cell cycle disease. Cells divide

uncontrollably, causing neoplastic growth. Genetic abnormalities that disrupt cell cycle progression often cause cancer. Mutations often influence tumor suppressor and oncogene genes, including p53 and Ras. These mutations can evade cell cycle checkpoints, allowing damaged or abnormal cells to proliferate and accumulate genetic changes, promoting cancer.

2. **Cell Cycle Checkpoint Disorders**: Cell cycle checkpoint abnormalities can cause cell cycle diseases. For instance, genetic mutations that impair DNA repair or checkpoint proteins can lead to DNA damage and hereditary illnesses like Fanconi anemia or Ataxia-telangiectasia (Kerzendorfer & O'Driscoll, 2009).

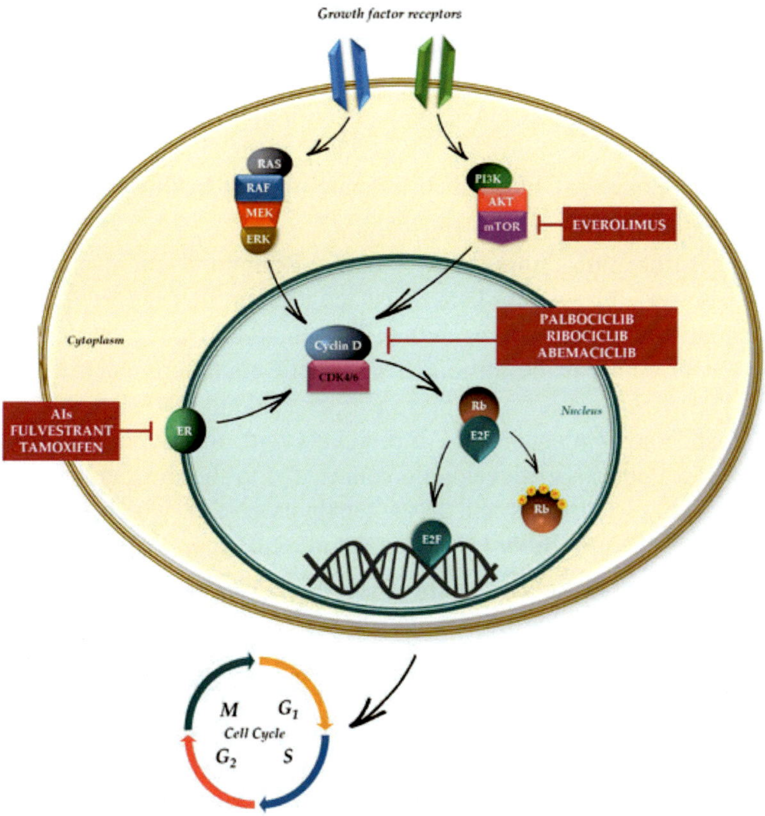

Figure 8.7. Cell cycle checkpoint & DNA damage response defects as anticancer targets (Source: Piezzo M, Cocco S, Caputo R, Cianniello D, Gioia GD, Lauro VD, Fusco G, Martinelli C, Nuzzo F, Pensabene M, et al. Targeting Cell Cycle in Breast Cancer: CDK4/6 Inhibitors. *International Journal of Molecular Sciences*. 2020; 21(18):6479. https://doi.org/10.3390/ijms21186479, Creative Commons License)

3. **Microcephaly**: Microcephaly is characterized by a small head and brain. Cell cycle disruptions during brain development can impede neural progenitor cell growth. Mutations or prenatal exposure to harmful environmental conditions cause these abnormalities.
4. **Progeria:** The rare inherited disorder Hutchinson-Gilford progeria syndrome causes accelerated aging in children. Genetic mutations in the LMNA gene, which regulates nuclear architecture and DNA repair, cause this illness. LMNA gene dysregulation can cause cell cycle disruptions and early senescence (Purhonen et al., 2023).
5. **Bloom Syndrome**: Bloom syndrome is an infrequent hereditary condition distinguished by diminished height, heightened sensitivity to sunlight, and elevated susceptibility to the onset of cancer. The etiology of this condition can be attributed to genetic mutations occurring in the BLM gene, which plays a crucial role in the process of DNA repair. Cells derived from individuals affected by Bloom syndrome demonstrate elevated frequencies of sister chromatid exchanges (SCEs), which indicate impaired cell cycle progression and DNA repair mechanisms.
6. **Rothmund-Thomson Syndrome**: Rothmund-Thomson syndrome is an infrequent hereditary condition that impacts multiple physiological systems within the human body. The etiology of this condition can be attributed to genetic mutations occurring in the RECQL4 gene, which plays a crucial role in DNA helicase activity and repair mechanisms. Individuals with this syndrome are at an elevated risk of developing cancer due to the dysregulation of cell cycle checkpoints and DNA repair processes, which can result in genomic instability (Ajeawung et al., 2019).
7. **Cell Cycle-Related Skin Disorders**: Certain skin disorders are characterized by the dysregulation of the cell cycle. One illustrative instance is psoriasis, a persistent inflammatory dermatological condition distinguished by atypical cellular proliferation and differentiation. The excessive and unregulated multiplication of skin cells results in scaly plaques on the skin's surface.
8. **Cell Cycle-Related Neurological Disorders**: Cell cycle dysregulation causes neurological diseases. Alzheimer's disease and neuronal cell cycle re-entry cause abnormal cell cycle events and neuronal death.

Cell cycle abnormalities are being studied to understand their genetic and molecular causes. Knowledge advances have helped us understand these illnesses. Targeted treatments for some disorders have advanced due to the discovery of cell cycle targets, suggesting improved treatment methods in the future (Joseph et al., 2020).

Remember

The cell cycle is like a traffic light, with checkpoints to ensure everything is in order before moving on.

SUMMARY

The cell cycle chapter covers its phases, key events, and the effects of cell cycle disruption. An introduction to the cell cycle emphasizes its role in cell lifetime regulation. The cell cycle is a biological process with several stages. The cell cycle contains four phases: G1, S, G2, and M (mitosis). These stages coordinate cell development, DNA replication, and division. DNA replication in the S phase of the cell cycle ensures accurate genetic material duplication before cell division. DNA polymerases add nucleotides to the elongating DNA strand via complementary base pairing. Cell cycle checkpoints assess DNA integrity and allow repairs before the cell enters the next phase. Mitosis, the final step of the cell cycle, accurately divides duplicated genetic material into two offspring cells. Prophase, metaphase, anaphase, and telophase make up mitosis. Events in each sub-phase ensure correct chromosomal segregation. Cytokinesis completes cell division after mitosis.

Animal cells divide by contracting the cell membrane using an actin filament contractile ring. Plant cells generate a cell plate, which fuses with the cell membrane to form a new cell wall. Daughter cells divide after this procedure. The chapter then examines cell cycle problems and their effects. Cell cycle dysfunctions can cause cancer and other illnesses. Mutations in tumor suppressor and oncogene genes can disrupt the cell cycle. Genetic changes can cause cell growth and cancers. Cell cycle problems are linked to several hereditary diseases and developmental defects. Microcephaly, a medical disorder with abnormally small head and brain dimensions, can occur from cell cycle regulatory problems during brain development. Targeted therapies must understand cell cycle problems and their molecular underpinnings. Cell cycle research has revealed intriguing targets for cancer therapy and cell cycle abnormalities. The cell cycle chapter covers this vital biological process in detail. This thorough research covers mitosis, cytokinesis, and cell cycle dysregulation. Cell cycle management is crucial for genomic integrity and multicellular organism growth. Understanding cell cycle aberrations is crucial in the context of cancer and hereditary illnesses. It also allows therapeutic research.

REFERENCES

1. Abrieu, A., Brassac, T., Galas, S., Fisher, D., Labbé, J. C., & Dorée, M. (1998). The Polo-like kinase Plx1 is a component of the MPF amplification loop at the G2/M-phase transition of the cell cycle in Xenopus eggs. *Journal of Cell Science*, *111*(12), 1751-1757.
2. Aist, J. R., & Berns, M. W. (1981). Mechanics of chromosome separation during mitosis in Fusarium (Fungi imperfecti): new evidence from ultrastructural and laser microbeam experiments. *The Journal of Cell Biology*, *91*(2), 446-458.
3. Ajeawung, N. F., Nguyen, T. T. M., Lu, L., Kucharski, T. J., Rousseau, J., Molidperee, S., ... & Campeau, P. M. (2019). Mutations in ANAPC1, encoding a scaffold subunit of the anaphase-promoting complex, cause Rothmund-Thomson syndrome type 1. *The American Journal of Human Genetics*, *105*(3), 625-630.
4. Baylin, S. B. (1997). Tying it all together: epigenetics, genetics, cell cycle, and cancer. *Science*, *277*(5334), 1948-1949.
5. Biswas, L., Tyc, K., El Yakoubi, W., Morgan, K., Xing, J., & Schindler, K. (2021). Meiosis interrupted: the genetics of female infertility via meiotic failure. *Reproduction*, *161*(2), R13-R35.
6. Bodwell, J. E., Webster, J. C., Jewell, C. M., Cidlowski, J. A., Hu, J. M., & Munck, A. (1998). Glucocorticoid receptor phosphorylation: overview, function and cell cycle-dependence. *The Journal of Steroid Biochemistry and Molecular Biology*, *65*(1-6), 91-99.
7. Cai, X., & Xu, S. S. (2007). Meiosis-driven genome variation in plants. *Current Genomics*, *8*(3), 151-161.
8. Celis, J. E., & Celis, A. (1985). Cell cycle-dependent variations in the distribution of the nuclear protein cyclin proliferating cell nuclear antigen in cultured cells: subdivision of S phase. *Proceedings of the National Academy of Sciences*, *82*(10), 3262-3266.
9. Choi, K., & Henderson, I. R. (2015). Meiotic recombination hotspots–a comparative view. *The Plant Journal*, *83*(1), 52-61.
10. Chowdhury, R., Bois, P. R., Feingold, E., Sherman, S. L., & Cheung, V. G. (2009). Genetic analysis of variation in human meiotic recombination. *PLoS Genetics*, *5*(9), e1000648.
11. Conlon, I., & Raff, M. (2003). Differences in the way a mammalian cell and yeast cells coordinate cell growth and cell-cycle progression. *Journal of Biology*, *2*, 1-10.
12. De Lichtenberg, U., Jensen, L. J., Fausbøll, A., Jensen, T. S., Bork, P., & Brunak, S. (2005). Comparison of computational methods for the

identification of cell cycle-regulated genes. *Bioinformatics*, *21*(7), 1164-1171.

13. Fuss, J. O., & Tainer, J. A. (2011). XPB and XPD helicases in TFIIH orchestrate DNA duplex opening and damage verification to coordinate repair with transcription and cell cycle via CAK kinase. *DNA repair*, *10*(7), 697-713.
14. Gorelick, R., & Heng, H. H. (2011). Sex reduces genetic variation: a multidisciplinary review. *Evolution*, *65*(4), 1088-1098.
15. Grobelny, J. V., Godwin, A. K., & Broccoli, D. (2000). ALT-associated PML bodies are present in viable cells and are enriched in cells in the G2/M phase of the cell cycle. *Journal of Cell Science*, *113*(24), 4577-4585.
16. Harris, S. D., & Hamer, J. E. (1995). sepB: an *Aspergillus nidulans* gene involved in chromosome segregation and the initiation of cytokinesis. *The EMBO Journal*, *14*(21), 5244-5257.
17. Hartwell, L. H. (1991). Twenty-five years of cell cycle genetics. *Genetics*, *129*(4), 975.
18. Hu, F., Wang, Y., Liu, D., Li, Y., Qin, J., & Elledge, S. J. (2001). Regulation of the Bub2/Bfa1 GAP complex by Cdc5 and cell cycle checkpoints. *Cell*, *107*(5), 655-665.
19. Hwang, A., & Muschel, R. J. (1998). Radiation and the G2 phase of the cell cycle. *Radiation research*, *150*(5s), S52-S59.
20. Johnson, D. G., & Walker, C. L. (1999). Cyclins and cell cycle checkpoints. *Annual review of pharmacology and toxicology*, *39*(1), 295-312.
21. Joseph, C., Mangani, A. S., Gupta, V., Chitranshi, N., Shen, T., Dheer, Y., ... & Gupta, V. (2020). Cell cycle deficits in neurodegenerative disorders: Uncovering molecular mechanisms to drive innovative therapeutic development. *Aging and Disease*, *11*(4), 946.
22. Kaiser, B. K., Zimmerman, Z. A., Charbonneau, H., & Jackson, P. K. (2002). Disruption of centrosome structure, chromosome segregation, and cytokinesis by misexpression of human Cdc14A phosphatase. *Molecular biology of the cell*, *13*(7), 2289-2300.
23. Kaitna, S., Mendoza, M., Jantsch-Plunger, V., & Glotzer, M. (2000). Incenp and an aurora-like kinase form a complex essential for chromosome segregation and efficient completion of cytokinesis. *Current Biology*, *10*(19), 1172-1181.
24. Kamal, K. Y., Herranz, R., Van Loon, J. J., & Medina, F. J. (2018). Simulated microgravity, Mars gravity, and 2g hypergravity affect cell

cycle regulation, ribosome biogenesis, and epigenetics in Arabidopsis cell cultures. *Scientific Reports*, *8*(1), 6424.

25. Kerzendorfer, C., & O'Driscoll, M. (2009). Human DNA damage response and repair deficiency syndromes: linking genomic instability and cell cycle checkpoint proficiency. *DNA repair*, *8*(9), 1139-1152.
26. Krek, W., & Nigg, E. A. (1991). Differential phosphorylation of vertebrate p34cdc2 kinase at the G1/S and G2/M transitions of the cell cycle: identification of major phosphorylation sites. *The EMBO Journal*, *10*(2), 305-316.
27. Lambing, C., & Heckmann, S. (2018). Tackling plant meiosis: from model research to crop improvement. *Frontiers in Plant Science*, *9*, 829.
28. Luciani, M. G., Campregher, C., Fortune, J. M., Kunkel, T. A., & Gasche, C. (2007). 5-ASA affects cell cycle progression in colorectal cells by reversibly activating a replication checkpoint. *Gastroenterology*, *132*(1), 221-235.
29. Lundberg, A. S., & Weinberg, R. A. (1999). Control of the cell cycle and apoptosis1. *European Journal of Cancer*, *35*(14), 1886-1894.
30. MacPherson, D., Sage, J., Crowley, D., Trumpp, A., Bronson, R. T., & Jacks, T. (2003). Conditional mutation of Rb causes cell cycle defects without apoptosis in the central nervous system. *Molecular and cellular biology*, *23*(3), 1044-1053.
31. Marlier, Q., D'aes, T., Verteneuil, S., Vandenbosch, R., & Malgrange, B. (2020). Core cell cycle machinery is crucially involved in both life and death of post-mitotic neurons. *Cellular and Molecular Life Sciences*, *77*(22), 4553-4571.
32. Menges, M., Hennig, L., Gruissem, W., & Murray, J. A. (2002). Cell cycle-regulated gene expression in Arabidopsis. *Journal of Biological Chemistry*, *277*(44), 41987-42002.
33. Michael, S., Trave, G., Ramu, C., Chica, C., & Gibson, T. J. (2008). Discovery of candidate KEN-box motifs using cell cycle keyword enrichment combined with native disorder prediction and motif conservation. *Bioinformatics*, *24*(4), 453-457.
34. Murray, A. W. (1995). The genetics of cell cycle checkpoints. *Current Opinion in Genetics & Development*, *5*(1), 5-11.
35. Nakanishi, M., Shimada, M., & Niida, H. (2006). Genetic instability in cancer cells by impaired cell cycle checkpoints. *Cancer Science*, *97*(10), 984-989.
36. Nasmyth, K. A., & Reed, S. I. (1980). Isolation of genes by complementation

in yeast: molecular cloning of a cell-cycle gene. *Proceedings of the National Academy of Sciences*, *77*(4), 2119-2123.

37. Natarajan, A. T., Obe, G., Van Zeeland, A. A., Palitti, F., Meijers, M., & Verdegaal-Immerzeel, E. A. (1980). Molecular mechanisms involved in the production of chromosomal aberrations II. Utilization of neurospora endonuclease for the study of aberration production by X-rays in G1 and G2 stages of the cell cycle. *Mutation Research/Fundamental and Molecular Mechanisms of Mutagenesis*, *69*(2), 293-305.
38. Oki, T., Nishimura, K., Kitaura, J., Togami, K., Maehara, A., Izawa, K., ... & Kitamura, T. (2014). A novel cell-cycle-indicator, mVenus-p27K−, identifies quiescent cells and visualizes G0–G1 transition. *Scientific Reports*, *4*(1), 4012.
39. Owa, T., Yoshino, H., Yoshimatsu, K., & Nagasu, T. (2001). Cell cycle regulation in the G1 phase: a promising target for the development of new chemotherapeutic anticancer agents. *Current Medicinal Chemistry*, *8*(12), 1487-1503.
40. Papavasiliou, F. N., & Schatz, D. G. (2000). Cell-cycle-regulated DNA double-strand breaks in somatic hypermutation of immunoglobulin genes. *Nature*, *408*(6809), 216-221.
41. Pramila, T., Wu, W., Miles, S., Noble, W. S., & Breeden, L. L. (2006). The Forkhead transcription factor Hcm1 regulates chromosome segregation genes and fills the S-phase gap in the transcriptional circuitry of the cell cycle. *Genes & development*, *20*(16), 2266-2278.
42. Purhonen, J., Banerjee, R., Wanne, V., Sipari, N., Mörgelin, M., Fellman, V., & Kallijärvi, J. (2023). Mitochondrial complex III deficiency drives c-MYC overexpression and illicit cell cycle entry leading to senescence and segmental progeria. *Nature Communications*, *14*(1), 2356.
43. Roncalli, M., Bosari, S., Marchetti, A., Buttitta, F., Bossi, P., Graziani, D., ... & Coggi, G. (1998). Cell cycle-related gene abnormalities and product expression in esophageal carcinoma. *Laboratory investigation; a journal of technical methods and pathology*, *78*(9), 1049-1057.
44. Ruijtenberg, S., & van den Heuvel, S. (2016). Coordinating cell proliferation and differentiation: Antagonism between cell cycle regulators and cell type-specific gene expression. *Cell cycle*, *15*(2), 196-212.
45. Russell, P., & Nurse, P. (1987). Negative regulation of mitosis by wee1+, a gene encoding a protein kinase homolog. *Cell*, *49*(4), 559-567.
46. Sikorski, R. S., Boguski, M. S., Goebl, M., & Hieter, P. (1990). A repeating amino acid motif in CDC23 defines a family of proteins

and a new relationship among genes required for mitosis and RNA synthesis. *Cell*, *60*(2), 307-317.

47. Sinha, K. M., Hines, J. C., & Ray, D. S. (2006). Cell cycle-dependent localization and properties of a second mitochondrial DNA ligase in *Crithidia fasciculata*. *Eukaryotic Cell*, *5*(1), 54-61.
48. Srivastava, S., Lavania, U. C., & Sybenga, J. (1992). Genetic variation in meiotic behaviour and fertility in tetraploid Hyoscyamus muticus: correlation with diploid meiosis. *Heredity*, *68*(3), 231-239.
49. Starborg, M., Gell, K., Brundell, E., & Höög, C. (1996). The murine Ki-67 cell proliferation antigen accumulates in the nucleolar and heterochromatic regions of interphase cells and at the periphery of the mitotic chromosomes in a process essential for cell cycle progression. *Journal of Cell Science*, *109*(1), 143-153.
50. Tyson, J. J., & Novak, B. (2001). Regulation of the eukaryotic cell cycle: molecular antagonism, hysteresis, and irreversible transitions. *Journal of Theoretical Biology*, *210*(2), 249-263.
51. Weinert, T. A., Kiser, G. L., & Hartwell, L. H. (1994). Mitotic checkpoint genes in budding yeast and the dependence of mitosis on DNA replication and repair. *Genes & development*, *8*(6), 652-665.
52. Williams, G. H., & Stoeber, K. (2012). The cell cycle and cancer. *The Journal of Pathology*, *226*(2), 352-364.
53. Xu, X., Hamhouyia, F., Thomas, S. D., Burke, T. J., Girvan, A. C., McGregor, W. G., ... & Bates, P. J. (2001). Inhibition of DNA replication and induction of S phase cell cycle arrest by G-rich oligonucleotides. *Journal of Biological Chemistry*, *276*(46), 43221-43230.
54. Yoshida, K., & Miki, Y. (2004). Role of BRCA1 and BRCA2 as regulators of DNA repair, transcription, and cell cycle in response to DNA damage. *Cancer Science*, *95*(11), 866-871.
55. Yu, L., Ma, X., Sun, J., Tong, J., Shi, L., Sun, L., & Zhang, J. (2017). Fluid shear stress induces osteoblast differentiation and arrests the cell cycle at the G0 phase via the ERK1/2 pathway. *Molecular medicine Reports*, *16*(6), 8699-8708.
56. Zheng, P., & Dean, J. (2009). Role of Filia, a maternal effect gene, in maintaining euploidy during cleavage-stage mouse embryogenesis. *Proceedings of the National Academy of Sciences*, *106*(18), 7473-7478.

INDEX

D

E

F

G

H

I

K

L

M

N

O

P

V

W

X

Y

Z

FSC
www.fsc.org
MIX
Papier | Fördert
gute Waldnutzung
FSC® C083411

Kolibri 360 GmbH
Ferdinand-Jühlke-Straße 7
99095 Erfurt, Deutschland
produktsicherheit@kolibri360.de